Studies in Space Policy

Volume 25

Series Editor

European Space Policy Institute, Vienna, Austria

The use of outer space is of growing strategic and technological relevance. The development of robotic exploration to distant planets and bodies across the solar system, as well as pioneering human space exploration in earth orbit and of the moon, paved the way for ambitious long-term space exploration. Today, space exploration goes far beyond a merely technological endeavour, as its further development will have a tremendous social, cultural and economic impact. Space activities are entering an era in which contributions of the humanities—history, philosophy, anthropology—, the arts, and the social sciences—political science, economics, law—will become crucial for the future of space exploration. Space policy thus will gain in visibility and relevance. The series Studies in Space Policy shall become the European reference compilation edited by the leading institute in the field, the European Space Policy Institute. It will contain both monographs and collections dealing with their subjects in a transdisciplinary way.

More information about this series at http://www.springer.com/series/8167

Annette Froehlich · Diego Alonso Amante Soria · Ewerton De Marchi

Space Supporting Latin America

Latin America's Emerging Space Middle Powers

European Space Policy Institute

Springer

Annette Froehlich ⓘ
European Space Policy Institute
Vienna, Austria

Diego Alonso Amante Soria ⓘ
European Space Policy Institute
Vienna, Austria

Ewerton De Marchi ⓘ
European Space Policy Institute
Vienna, Austria

ISSN 1868-5307   ISSN 1868-5315   (electronic)
Studies in Space Policy
ISBN 978-3-030-38522-4   ISBN 978-3-030-38520-0   (eBook)
https://doi.org/10.1007/978-3-030-38520-0

This Springer imprint is published by the registered company Springer Nature Switzerland AG
The registered company address is: Gewerbestrasse 11, 6330 Cham, Switzerland

Contents

About the Authors

Dr. Annette Froehlich is a scientific expert seconded from the German Aerospace Center (DLR) to the European Space Policy Institute (Vienna) and a honorary adjunct senior lecturer at the University of Cape Town (SA) at SpaceLab. She graduated in European and International Law at the University of Strasbourg (France), followed by business-oriented postgraduate studies and her Ph.D. at the University of Vienna (Austria). Responsible for DLR and German representation to the United Nations and International Organizations, Dr. Froehlich was also a member/alternate head of delegation of the German delegation to UNCOPUOS. Moreover, Dr. Annette Frochlich is an author of a multitude of specialist publications and serves as a lecturer at various universities worldwide in space policy, law and society aspects. Her main areas of scientific interest are European space policy, international and regional space law, emerging space countries, space security and space and culture. She has also launched as editor the new scientific series "Southern Space Studies" (Springer publishing house) dedicated to Latin America and Africa. e-mail: Annette.Froehlich@espi.or.at; Annette.Froehlich@dlr.de

Diego Alonso Amante Soria holds a bachelor's degree in law from the National Autonomous University of Mexico and a master's degree in international law from the University of Grenoble-Alpes (France). His main areas of interest are space law, international security and the system of the United Nations. His master's dissertation was about the legal challenges of the exploitation of natural space resources. He has acquired professional experience at the Mexican Ministries of the Interior and Foreign Affairs and the European Space Policy Institute (Vienna, Austria). Passionate about astronomy, he has been at a Mexican Astronomical society for several years. e-mail: diego.amantesoria@outlook.com

Ewerton De Marchi is an attorney at law, with a bachelor's in law from the Federal University of the State of Rio de Janeiro (UNIRIO), Brazil. Currently, he is pursuing a master's in International and European Law at NOVA University

Lisbon (Portugal). His main interest and main research are in the development of space law, space activities in Latin America and climate change, as well as their intersections.

The authors specially acknowledge the assistance of Ursula Liliam Castillo Guevara for her preliminary research for this report and the elaboration of various maps and graphics.

Abbreviations

ABAE	Agencia Bolivariana para Actividades Espaciales (Bolivarian Agency for Space Activities—Venezuela)
ABE	Agencia Boliviana Espacial (Bolivian Space Agency)
ACAE	Asociación Centroamericana de Aeronáutica y del Espacio (Central American Association of Aeronautics and Space)
ACE	Agencia Chilena del Espacio (Chilean Space Agency)
ACS	Alcantara Cyclone Space
ACTO	Organización del Tratado de Cooperación Amazónica (Amazon Cooperation Treaty Organization)
AEB	Agência Espacial Brasileira (Brazilian Space Agency)
AEC	ASEAN (Association of Southeast Asian Nations) Economic Community
AEM	Agencia Espacial Mexicana (Mexican Space Agency)
AEP	Agencia Espacial del Paraguay (Space Agency of Paraguay)
ALADI	Asociación Latinoamericana de Integración (Latin America Association for Integration)
ALALC	Asociación Latinoamericana de Libre Comercio (Latin American Free Trade Association)
ALAS	Alianza Latinoamericana de Agencias Espaciales (Latin American Space Agencies Alliance)
ALBA	Alianza Bolivariana Para los Pueblos de Nuestra América (Bolivarian Alliance for the Peoples of Our America)
ALMA	Atacama Large Millimetre/submillimetre Array
APSCO	Asia Pacific Space Cooperation Organization
ARCSSTE-E	African Regional Centre for Space Science and Technology Education
ARRA	Rescue Agreement
ARSAT	Empresa Argentina de Soluciones Satelitales (Argentinian Company of Satellite Solutions)
ASI	Agenzia Spaziale Italiana (Italian Space Agency)

BAEMARI	Ground Satellite Control Station of the aerospace base "Capitán Manuel Ríos—Venezuela"
BAGEM	Ground Applications System of the airbase "Generalísimo Francisco de Miranda—Venezuela"
BELSPO	Belgian Federal Office for Science Policy
BIRDS	Joint Global Multi-Nation Birds Satellite Project
BMI	Federal Ministry of the Interior (Germany)
BRICS	Brazil, Russia, India, China and South Africa
BRS	Convention Relating to the Distribution of Programme-Carrying Signals Transmitted by Satellite
CACM	Central American Common Market
CAN	Comunidad Andina (Andean Community of Nations)
CANTV	Compañía Anónima Nacional Teléfonos de Venezuela
CARICOM	Comunidad del Caribe (Caribbean Community)
CAST	Chinese Academy of Space Technology
CATE	Congreso Argentino de Tecnología Espacial (Argentine Congress of Space Technology)
CATHALAC	Centro de Agua del Trópico Húmedo para América Latina y el Caribe (Centre for Humid Tropics of Latin America and the Caribbean)
CBERS	China–Brazil Satellite Resources Program
CCAD	Central American Commission for the Environment and Development
CCE	Comisión Colombiana del Espacio (Colombian Space Commission)
CDB	China Development Bank
CEA	Conferencia Espacial de las Américas (Space Conference of the Americas)
CEATSA	Centro de Ensayos de Alta Tecnología (Centre of High Technology Tests—Argentina)
CEDOES	Centro de Documentación e Investigación Educativa Espacial (Centre of Space Documentation and Educational Research—Venezuela)
CELAC	Comunidad de Estados Latinoamericanos y Caribeños (Community of Latin America and the Caribbean States)
CELP	Colombo-Ecuadorian Lunar Program
CEMB	Centro Espacial Manuel Belgrano (Manuel Belgrano Space Centre—Argentina)
CENDIT	National Centre for Development and Research on Telecommunications—Venezuela
CEOS	Committee on Earth Observation Satellites—Brazil
CEPI	Centro Espacial Punta Indio (Punta Indio Space Centre—Argentina)
CETT	Centro Espacial Téofilo Tabanera (Téofilo Tabanera Space Centre—Argentina)

CGWIC	China Great Wall Industry Corporation
CIDA	Foundation Centre on Astronomy Research "Francisco J. Duarte —Venezuela"
CIDE	Centre of Space Research and Development—Venezuela
CIDTE	Centro de Investigación, Innovación y Desarrollo en Telecomunicaciones (Research, Innovation and Development Telecommunication Centre—Mexico)
CIECT	Congreso Internacional de Electrónica, Control y Telecomunicaciones (International Congress of Electronics, Control and Telecommunications)
CIREN	Centro de Información de Recursos Naturales (Information Centre of Natural Resources—Chile)
CITEL	Comisión Interamericana de Telecomunicaciones (Inter-American Telecommunications Commission)
CLA	Centro de Lançamento de Alcântara (Alcantara Launch Centre—Brazil)
CLBI	Centro de Lançamento da Barreira do Inferno (Barreira do Inferno Launch Centre—Brazil)
CLIRSEN	Centro de Levantamientos Integrados de Recursos Naturales por Sensores Remotos (Centre for Natural Resources Integrated Survey by Remote Sensors—Ecuador)
CMC	Common Market Council—MERCOSUR
CME	Coronal Mass Ejection
CNAE	Comissão Nacional de Atividades Espaciais (National Commission on Space Activities—Brazil)
CNES	Centre National d'Etudes Spatiales (National Centre for Space Studies—France)
CNIE	Comisión Nacional de Investigaciones Espaciales (National Commission of Spatial Research—Argentina)
CNOIS	National Centre of Satellite Image Operations—Peru
CNPq	National Research Council (Brazil)
CNSA	China National Space Administration
COBAE	Brazilian Space Activities Commission
COGNAE	Grupo de Organização da Comissão Nacional de Atividades Espaciais (Organization Group of the National Commission of Space Activities—Brazil)
COMCyT	Comisión Interamericana en Ciencia y Tecnología (Inter-American Committee of Science and Technology)
COMIP	Comisión Mixta Argentino-Paraguaya del Río Paraná (Argentinian-Paraguayan Parana River Joint Commission)
CONACYT	Consejo Nacional de Ciencia y Tecnología (National Council on Science and Technology—Mexico)
CONAE	Comisión Nacional de Actividades Espaciales (National Commission on Space Activities—Argentina)

CONATEL	Comisión Nacional de Telecomunicaciones (National Telecommunications Commission—Venezuela)
CONEE	Comisión Nacional del Espacio Exterior (National Commission of Outer Space—Mexico)
CONICET	Consejo Nacional de Investigaciones Científicas y Técnicas (National Scientific and Technical Research Council—Argentina)
CONICYT	Comisión Nacional de Investigación Científica y Tecnológica (National Commission for Scientific and Technological Research—Chile)
CONIDA	Comisión Nacional de Investigación y Desarrollo Aeroespacial (National Commission for Aerospace Research and Development—Peru)
CPC	Joint Parliamentary Commission—MERCOSUR
CPLP	Comunidade dos Países de Lingua Portuguesa (Community of Portuguese Speaking Countries)
CPTEC	Weather Prevision Centre and Climate Studies—Brazil
CRASTE-LF	African Regional Centre for Space Science and Technology Education
CRECTEALC	Centro Regional de Enseñanza en Ciencia y Tecnología del Espacio para América Latina y el Caribe (Regional Centre for Space Science and Technology Education for Latin America and the Caribbean)
CSA	Canadian Space Agency
CSC	Comisión Colombiana del Espacio (Colombian Space Commission)
CSSTEAP	Centre for Space Science and Technology Education in Asia and the Pacific
CTA	Aerospace Technical Centre—Brazil
CVT-E	Centro Vocacional Tecnológico Espacial (Space Technological Vocational Centre—Brazil)
DCTA	Departamento de Ciência e Tecnologia Aeroespacial (Department of Aerospace Science and Technology of Brazil)
DICAE	Technical Division on Space Sciences and Applications (Peru)
DINCI	Division on Scientific Instruments (Peru)
DIVLA	Division of Launch Vehicles (Peru)
DLR	Deutsches Zentrum für Luft- und Raumfahrt (German Aerospace Centre)
DSRI	Danish Space Research Institute
DTDTE	Technical Division on Space Technology Development (Peru)
DTEE	Division on Space Studies (Peru)
E2T	Plataforma Espaço, Educação e Tecnologia (Space, Education and Technology—Brazil)
ECLAC	Economic Commission for Latin America and the Caribbean
ECSL	European Centre of Space Law

ECSP	Ecuadorian Civilian Space Plan
EEAS	European External Action Service
EGNOS	European Geostationary Navigation Overlay Service
EMAVI	Escuela Militar de Aviación "Fidel Suárez" (Military Aviation School "Fidel Suarez"—Colombia)
END	Estratégia Nacional de Defesa (National Defence Strategy—Brazil)
EO	Earth Observation
EO4SD	Earth Observation for Sustainable Development
ESA	European Space Agency
ESRIN	ESA's Centre for Earth Observation
EU	European Union
EXA	Agencia Espacial Civil Ecuatoriana (Ecuadorian Civil Space Agency)
FAE	Fuerza Aérea Ecuatoriana (Ecuadorian Air Force)
FAO	Food and Agriculture Organization
FCES	Social Economic Advisory Forum—MERCOSUR
G20	Group of 20 (Argentina, Australia, Brazil, Canada, China, France, Germany, India, Indonesia, Italy, Japan, México, Russia, Saudi Arabia, South Africa, Korea, Turkey, UK, USA and the European Union)
GCTC	Yuri Gagarin Cosmonaut Training Centre—Russia
GDP	Gross Domestic Product
GEO	Geostationary Orbit
GETEPE	Executive Group of Work and Study Space Projects—Brazil
GIS	Geographic Information System
GLOBE	Global Learning and Observations for the Benefit of the Environment
GLONASS	Russian-operated satellite navigation system
GMC	Common Market Group—MERCOSUR
GNSS	Global Navigation Satellite System
GOCNAE	Organizing Group of the National Commission of Space Activities—Brazil
GOE	Grupo de Operaciones Espaciales (Space Operations Group—Chile)
GOP	Geostationary Orbital Position
GRULAC	Group of Latin America and the Caribbean
HSA	Hellenic Space Agency—Greece
IAC	International Astronautical Congress
IACHR	Inter-American Commission on Human Rights
IADC	Inter-Agency Space Debris Coordination Committee
IAE	Institute of Aeronautics and Space—Brazil
IAEA	International Atomic Energy Agency
IAF	International Astronautical Federation
IAR	Argentinian Institute of Radio Astronomy

ICAO	International Civil Aviation Organization
ICAS	International Centre for Advanced Studies (Russia)
ICG	International Committee on Global Navigation Satellite Systems
ICSP	Irvine CubeSat STEM Program—United States
IEE	Instituto Espacial Ecuatoriano (Ecuadorian Space Institute)
IFRN	Instituto Federal de Educação, Ciência e Tecnologia do Rio Grande do Norte (Institute of Education, Science and Technology of Rio Grande do Norte—Brazil)
IFT	Instituto Federal de Telecomunicaciones (Federal Telecommunications Institute—Mexico)
IGAC	Instituto Agustín Codazzi (Institute Agustín Codazzi—Colombia)
IIRS	Indian Institute of Remote Sensing
IISL	International Institute of Space Law
IMF	International Monetary Fund
IMSO	International Mobile Satellite Organization
INACH	Instituto Antártico Chileno (Antarctic Institute of Chile)
INICTEL	National Institute for Research and Training on Telecommunications—Peru
INPE	Instituto Nacional de Pesquisas Espaciais (National Institute for Space Research—Brazil)
IO	International Organization
IPSEN-2	National Business Sector Participation Index
ISA	Israel Space Agency
ISF	International Space Forum
ISP	Institutional Strategic Plan—Paraguay
ISPRS	International Society of Photogrammetry, Remote Sensing and Space Information Systems
ISRO	Indian Space Research Organization
ISS	International Space Station
ITA	Aeronautical Technological Institute—Brazil
ITSO	International Telecommunications Satellite Organization
ITU	International Telecommunication Union
JAMSS	Japan Manned Space Corporation
JAXA	Japan Aerospace Exploration Agency
KARI	Korea Aerospace Research Institute
KYUTECH	Kyushu Institute of Technology—Japan
LAFTA	Latin American Free Trade Association
LASA	Latin America Space Agency
LDO	Law of Budgetary Guidelines—Brazil
LEO	Low Earth Orbit
LIAB	Liability Convention
LOA	Lei Orçamentária Anual (Annual Budget Law—Brazil)
LOT	Ley Orgánica de Telecomunicaciones (Organic Law on Telecommunications—Ecuador)

LSC	UNCOPUOS Legal Subcommittee
MAI	Moscow Aviation Institute—Russia
MCCA	Mercado Común Centroamericano (Central American Common Market)
MCT	Secretaría de Comunicaciones y Transportes (Ministry of Communications and Transportation—Mexico)
MDGs	Millennium Development Goals
MECB	Brazilian Complete Space Mission
MERCOSUR	Mercado Común del Sur (Common Market of the South)
MILO	Mission and Innovation and Launch Opportunity—United States
MINCYT	Ministerio de Ciencia, Tecnología e Innovación Productiva (Minister of Science, Technology and Productive Innovation—Argentina)
MOON	Moon Agreement
MoU	Memorandum of Understanding
MPPEUCT	Ministerio del Poder Popular para la Educación Universitaria, Ciencia y Tecnología (Ministry of Popular Power for University Education, Science and Technology—Venezuela)
MTCR	Missile Technology Control Regime
MTSI	Ministerio de Telecomunicaciones y de la Sociedad de la Información (Ministry of Telecommunications and the Information Society—Ecuador)
MXN	Mexican Peso
NAFTA	North American Free Trade Agreement
NASA	National Aeronautics and Space Administration United States
NDP	Plan Nacional de Desarrollo (National Development Plan)
NFP	National Focal Points
NSAP	National Space Activities Program—Mexico
NTB	Treaty Banning Nuclear Weapon Tests in the Atmosphere, in Outer Space and Under Water
OAS	Organization of American States
ODECA	Organización de Estados Americanos (Organization of Central American States)
OECD	Organisation for Economic Cooperation and Development
OEI	Organization of Ibero-American States
OST	Outer Space Treaty
PARLASUR	Parlamento de Mercosur (Mercosur Parliament)
PESE	Programa Estratégico de Sistemas Espaciais (Strategic Space Systems Program—Brazil)
PISA	Programme for International Student Assessment
PNAE	Programa Nacional de Actividades Espaciales (National Programme of Space Activities—Brazil)
PNDAE	Política Nacional de Desenvolvimento das Atividades Espaciais (National Policy for the Development of Space Activities—Brazil)

POLSA	Polish Space Agency
PPDEC	Programa Presidencial para el Desarrollo Espacial Colombiano (Presidential Program for Colombian Space Development)
PROARCO	Programa de Prevenção e Controle de Queimadas e Incêndios Florestais na Amazônica Legal (Project to Control Fires and Forest Fire Prevention—Brazil)
PSLV	Polar Satellite Launch Vehicle
PTBT	Treaty Banning Nuclear Weapon Tests in the Atmosphere, in Outer Space and Under Water
PUCP	Pontificia Universidad Católica del Perú (Pontifical Catholic University of Peru)
R&D	Research and Development
RCCSTEAP	Regional Centre for Space Science and Technology Education in Asia and the Pacific
RCSSTEWA	Regional Centre for Space Science and Technology Education for Western Asia
REG	Registration Convention
ROSCOSMOS	Roscosmos State Corporation for Space Activities—Russia
RSO	Regional Support Offices
SAI	Sistema Andino de Integración (Andean Integration System)
SANSA	South African National Space Agency
SASA	South American Space Agency
SA-SGAC	Space Generation Advisory Council—South America
SA-SGW	South American Space Generation Workshop
SBDA	Associação Brasileira de Direito Aeronáutico e Espacial (Brazilian Association of Aeronautics and Space Law)
SBPC	Sociedade Brasileira para o Progresso da Ciência (Brazilian Society for the Progress of Science)
SCAP	Sociedad Científica de Astrobiología del Perú (Scientific Society of Astrobiology of Peru)
SDG	Sustainable Development Goals
SELA	Latin American and Caribbean Economic System
SELPER	Society of Latin American Specialists in Remote Sensing and Geographical Information
SGAC	Space Generation Advisory Council
SGDC	Geostationary Defense Satellite and Strategic Communications—Brazil
SGW	Space Generation Workshops
SIASGE	Sistema Italo-Argentino de Satélites para la Gestión de Emergencias (Argentinian Satellite System for Emergencies Management)
SICA	Sistema de Integración Centroamericana (Central American Integration System)
SME	Small-to-medium enterprise
SNSB	Swedish National Space Council

SPC	Space Program for Chile
SPFG	Satellite Policy of the Federal Government—Mexico
SPT	Secretaría Pro Tempore (Pro Tempore Secretariat)
SRI	Indian Remote Sensing Satellites
SSAU	Ukraine Space Agency
SSGAT	Sistema Satelital Geoestacionario Argentino de Telecomunicaciones (Argentinian Geostationary Telecommunications Satellite System)
STC	UNCOPUOS Scientific and Technical Subcommittee
SUBTEL	Subsecretaría de Telecomunicaciones (Telecommunications Under Ministry—Chile)
SWF	Secure World Foundation
SWOT	Strengths, Weaknesses, Opportunities and Threats
SWSU	Southwest State University—Russia
TELECOMM	Telecomunicaciones de México
TIC's	Telecommunications, information and communications
TSA	Technological Safeguards Agreement
UAP	University Alas Peruanas—Peru
UKSA	United Kingdom Space Agency
UN	United Nations
UNA	National University of Asuncion—Paraguay
UNAM	Universidad Nacional Autónoma de México (National Autonomus University of Mexico)
UNC	Universidad Nacional de Córdoba (National University of Córdoba—Argentina)
UNCOPUOS	United Nations Committee on the Peaceful Uses of Outer Space
UNECOSOC	United Nations Economic and Social Council
UNEFA	National Experimental University of the Armed Forces—Venezuela
UNESCO	United Nations Educational, Scientific and Cultural Organization
UNGA	United Nations General Assembly
UNISPACE	United Nations Conference on the Exploration and Peaceful Use of Outer Space
UNOOSA	United Nations Office for Outer Space Affairs
UN-SPIDER	United Nations Platform for Space-based Information for Disaster Management and Emergency Response
USA	United States of America
USAID	United States Agency for International Development
USD	United States Dollar
USSR	Union of Soviet Socialist Republics
UTE	Universidad Tecnológica Equinoccial (Technological Equinoctial University—Ecuador)
UTEC	Universidad de Ingeniería y Tecnología (University of Engineering and Technology—Peru)

UTP	Peruvian Technological University
UWI	University of the West Indies—Jamaica
VLM	Microsatellite Launch Vehicle
VLS	Satellite Launch Vehicle

Part I
Space in Latin America

Chapter 1
General Situation in Latin America

Contents

Abstract This chapter overviews the general situation in Latin America, including the concept of Latin America, the ideals of integration, and its divisions, both geographic and ethnic. French Guiana is analysed separately, precisely because it is an overseas French territory and partly fits in with the Latin American concept. From these concepts, the development of the region is shown and the stages of integration, through the cooperation and creation of essentially economic blocs between a few countries, even to the union of the entire South America region into one single block fostering development. Finally, the chapter assesses current prospects for regional integration, given the political challenges in the region.

1.1 Historical Background and Context

The term "Latin America" was first used by Francisco Bilbao at a conference in 1856.[1] The countries of Latin America are distributed over three geographic regions of the American continent (North, Central and South America). Mexico (North America)

---

[1]Clarín, ¿América latina o Sudamérica?, 2005, https://www.clarin.com/ediciones-anteriores/america-latina-sudamerica_0_BkHMQbYJCYl.html (accessed 15 July 2019).

© Springer Nature Switzerland AG 2020
A. Froehlich et al., *Space Supporting Latin America*, Studies in Space Policy 25,
https://doi.org/10.1007/978-3-030-38520-0_1

and all nations of Central and South America have similar historical and cultural backgrounds, both geopolitical and socioeconomic, which is the key element for the division of America between Anglo-Saxon America and Latin America. In this report, the expression Latin America is used as an umbrella term unless otherwise stated.

Latin America consists of 33 countries[2] that cover an area that stretches from the northern border of Mexico to the southern tip of South America, including the Caribbean. It has an area of approximately 19,197,000 km^2 (7,412,000 sq mi),[3] almost 13% of the Earth's land surface area. As of 2016, its population was estimated at more than 639 million[4] and in 2014, Latin America had a combined nominal Gross Domestic Product of USD 6,289.304 million[5] and a Gross Domestic Product Purchasing Power Parity of USD 13,938.675 million.[6]

The process of integration in Latin America was driven by the ideas of the Economic Commission for Latin America (ECLAC) that emerged on 28 February 1948 as a regional branch of the UN Economic and Social Council.[7] To achieve its broader goal of import substitution industrialization, ECLAC considered it sensible to pursue regional integration as opposed to local isolation, particularly by harnessing economies of scale.[8]

1.1.1   Geography and Political Structures

Latin America is entirely located in the western hemisphere, crossed by the Tropic of Cancer through central Mexico; the Equator, through Brazil, Colombia, Ecuador and Peru; and the Tropic of Capricorn, which crosses Brazil, Paraguay, Argentina and Chile. The territory is distributed throughout the northern and southern hemispheres due to the long extension of land south of the Equator.

---

[2]CEPAL, Estados membros, https://www.cepal.org/pt-br/estados-miembros (accessed 15 July 2019).

[3]The World Bank, http://wdi.worldbank.org/table/3.1 (accessed 15 July 2019).

[4]World Population Prospects 2019, UN, Population Division, https://population.un.org/wpp/DataQuery/ (accessed 15 July 2019).

[5]Berube, A., Trujillo, J., Ran, T., Parilla, J., Global Metro Monitor Report, 2015, https://www.brookings.edu/research/global-metro-monitor/ (accessed 15 July 2019).

[6]IMF, Data, Report for Selected Country Groups and Subjects, https://www.imf.org/external/pubs/ft/weo/2010/02/weodata/weorept.aspx?sy=2008&ey=2015&scsm=1&ssd=1&sort=country&ds=.&br=1&pr1.x=101&pr1.y=11&c=205&s=NGDPD%2CPPPPC&grp=1&a=1 (accessed 22 September 2019).

[7]About ECLAC, UN, https://www.cepal.org/en/about-eclac-0 (accessed 15 July 2019).

[8]Prebisch, R., El desarrollo económico de la América Latina y algunos de sus principales problemas, CEPAL, UN, https://repositorio.cepal.org/bitstream/handle/11362/40010/4/prebisch_desarrollo_problemas.pdf (accessed 15 July 2019).

Latin America is still the most democratic developing region on the planet.[9] The only countries in the region classified as authoritarian regimes are Cuba and Venezuela. Bolivia, Nicaragua, Honduras, Guatemala and Haiti are classified as hybrid regimes, where elections have significant irregularities and opposition is often pressured by the government.

1.1.1.1 South America

South America is a subcontinent that comprises the southern portion of America, made up of twelve countries plus French Guiana (French overseas territory). Its area is 17,819,100 km^2, covering 12% of the terrestrial surface and 6% of the world-wide population.

Portuguese and Spanish are the most widely spoken languages in South America, a geographical region that is part of the great cultural region called Latin America. Portuguese is the official language of Brazil, which has almost 50% of the South American population. Spanish is the official language of most countries on the continent. There are also other languages, such as Dutch (official language of Suriname), English (official language of Guyana), French (official language of French Guiana), and several indigenous languages.[10]

South America is considered as the most climatically, ecologically and biologically diverse places on Earth due to its unique geographical features. The Patagonia Region is located at the southernmost tip of the continent, at only 1,000 km from Antarctica. The Amazon jungle with the Amazon river (world's largest river by discharge) is the largest rainforest in the world and occupies the flat plains of eastern South America, mainly in the country of Brazil. On the western edge of the region, an entirely different environment spans the Andes Mountains, the most extensive mountain range in the world, containing sharp peaks and fertile mountain valleys at elevations well over 6,000 m above sea level.[11]

In the south, a massive grassland "Pampas" also creates a unique mid-elevation biome between mountains and rainforests. The continent can be divided into three geophysical regions: mountains and highlands, river basins, and coastal plains. In the north-south direction mountains and coastal plains are found, while highlands and river basins cover the east-west direction, and the coastal plains desert biome rises to the alpine biome of the Andes mountains. The extreme geographic variation that characterises the continent also contributes to the unique number of plants and animal species. The region counts with rich and unique biodiversity among the world's continents, holding extreme characteristics, such as the largest river, the Amazon, as well as the world's driest place, the Atacama Desert, located in Chile.[12]

---

[9]The Economist, Democracy Index 2017: Free speech under attack, https://www.eiu.com/public/topical_report.aspx?campaignid=DemocracyIndex2017 (accessed 15 September 2019).

[10]Geografia, UOL, https://educacao.uol.com.br/disciplinas/geografia/america-do-sul-2-relevo-clima-vegetacao-e-populacao.htm (accessed 15 July 2019).

[11]Spotlight on South America, Nature Ecology and Evolution, volume 1, article number: 0129, 2017, https://rdcu.be/bLFTV (accessed 15 July 2019).

[12]Ibid.

The most industrialized countries in South America are Brazil, Argentina, Chile, Colombia, Venezuela and Uruguay respectively. These countries account for over 75% of the region's economy (Fig. 1.1).[13]

Fig. 1.1   South American countries

[13]O Sistema Econômico/América Do Sul, Atlas Mundial, São Paulo, 1999, pp. 26–27, 88–107.

1.1.1.2 Central America, the Caribbean and Mexico

Central America consists of two parts. One of them, known as Continental Central America, is a narrow strip of land that connects North America to South America and has an area of approximately 522,760 km^2. The other part consists of the islands of the Caribbean Sea, beyond the archipelago of the Bahamas and the islands of Turks and Caicos.

Continental Central America is politically divided into seven independent countries: Belize, Guatemala, El Salvador, Honduras, Nicaragua, Costa Rica and Panama. The Caribbean region is formed by the Greater Antilles and the Lesser Antilles, which have approximately 239,000 km^2 of land area.

The Greater Antilles consists of Cuba, Jamaica, Haiti and the Dominican Republic, plus Puerto Rico, which is a territory controlled by the United States. To the south of Cuba are the small Cayman Islands, controlled by the United Kingdom.

The Lesser Antilles are the smaller islands, which comprise several countries: Antigua and Barbuda, Barbados, Dominica, Grenada, Saint Lucia, Saint Kitts and Nevis, Saint Vincent and the Grenadines, and Trinidad and Tobago. In the Lesser Antilles there are also several constituent countries: Aruba, Bonaire, Curaçao, Saba, St. Eustatius and the southern part of the island of St. Martin, which belong to the Netherlands; Guadeloupe and Martinique, which belong to France; Anguilla, British Virgin Islands and Montserrat, which belong to the United Kingdom; and the US Virgin Islands, which belong to the United States. Outside the Caribbean, at the Atlantic Ocean, there are the Bahamas, an independent country, and Turks and Caicos, islands that belong to the United Kingdom.

Mexico, that is part of North America, also belongs to the Latin America region, because of its similar historical, cultural, geopolitical and socioeconomic background. Mexico is a separate chapter in all these hemispheric processes of trade liberalization. Having operated from the 1980s onwards towards unilateral economic openness and trade integration on a purely free exchange basis, Mexico is comfortable with the existing—and under negotiation—schemes of deepening trade preferences with almost all countries and regional partners as well as the preservation of its exclusive ties with the major trading powers of the hemisphere—and beyond—insofar as this guarantees privileged access to the most dynamic markets without competitors in the region itself.[14]

Mexico has complied with a broad network of trade liberalization agreements that provides it with access to markets in countries that together account for between 2/3 and 4/5 of world GDP (Fig. 1.2).[15]

---

[14]Almeida, P., Integração regional e inserção internacional dos países da América do Sul: evolução histórica, dilemas atuais e perspectivas futuras, 2008, https://fundacaofhc.org.br/files/papers/407.pdf (accessed 16 July 2019).

[15]Ibid.

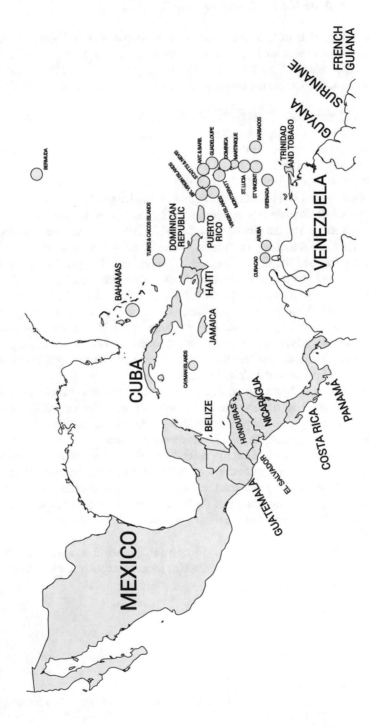

Fig. 1.2 Central America, Caribbean Countries and Mexico

1.1.1.3 Territories of Other States

There are 35 internationally recognized sovereign states with territory located in America. All of them are members of the UN and the Organization of American States. However, there are 19 American entities that are dependent territories constituents of other states, as well as eight territories that are fully integrated into non-American countries and are therefore not considered dependent territories.

An interesting case is that of the French Overseas Department: French Guiana.[16] At just 83,000 km^2 in length, French Guiana has about 296,000 inhabitants.[17] Located in South America, the department has increasingly gained administrative skills. It is the second largest French region, after Metropolitan France: it is rich in natural resources, notably gold, has engaged French military contingents in the fight against illegal mining since 2008 (Operation Harpy) and hosts the ESA's rockets and satellite launch base, the Kourou Space Centre—the result of its geostrategic position for launch into orbit.

French Guiana borders Brazil and Suriname. The border between Brazil and French Guiana is 730 km (the largest French land border), of which 427 km are rivers. Its territory—and therefore part of the European Union (EU)—is wedged in South America, among countries with which it has always had little economic and political ties.

French Guyana is, thus, at the crossroads of various political and economic groups: EU, AEC, MERCOSUR, CARICOM and ACTO. Immigrants were the pioneers of a progressive but limited integration of French Guiana into its Amazonian and therefore South American environment, leading to the beginning of reterritorialization thanks to the reciprocal awareness of the advantages of being a European territory in South America.[18]

---

[16]Maurice, E., Le préfet face aux enseignants autonomistes en Guyane de 1946 au tournant des années 1960, Une inédite rencontre administrative en contexte postcolonial, https://spire.sciencespo.fr/hdl:/2441/4bnoro60588b78jg2csr38n9pu/resources/pox-116-0053.pdf (accessed 16 July 2019).

[17]Ferreira, J., Guiana Francesa: um pedacinho da Europa que fica na América do Sul, https://revistagalileu.globo.com/Sociedade/noticia/2019/07/guiana-francesa-um-pedacinho-da-europa-que-fica-na-america-do-sul.html (accessed 16 July 2019).

[18]Ganger, S., Guiana francesa: um território europeu e caribenho em via de "sul-americanização"?, https://journals.openedition.org/confins/5003 (accessed 16 July 2019).

1.1.2   Linguistic and Ethnic Traits

The historic cultures of Latin American countries developed in connection with distinct regional landscapes and cultures. The region encompasses a diverse number of people, with unifying traits, mostly linguistic and ethnic, due to the long history of Spanish and Portuguese colonialism that is a shared heritage across the region.

The primary language of the continent is Spanish due to Spanish colonization and influence in the region. Portuguese colonization is reflected in Brazil, where the official language is Portuguese. In the region, English, Dutch and French are spoken in some countries. In addition, Latin American nations have formally granted one or more Amerindian languages official status as national languages.

The linguistic diversity found on the continent is also reflected in the ethnic diversity, where European heritage is an essential influence in Latin American cultures, but also, Amerindian ancestry still holds a strong presence on the continent. African populations, imported initially as slaves, also make up a significant ethnic population, as they represented a significant shift in the cultural landscape of Latin America (Fig. 1.3).

In the ethnic formation of the South American population, three ethnic groups predominated: American indigenous, Europeans and Africans. In many countries, Spanish-indigenous mestizos[19] predominate, such as Colombia, Ecuador, Paraguay and Venezuela. In only two countries is the indigenous population in the majority: Peru and Bolivia. Large populations of African descent are found in Brazil and Colombia. Brazil has the largest population of African descent outside Africa in the world, with a population of African descent greater than the sum of all other South American countries combined. Countries of strong European descent are Argentina, Uruguay, Chile, and Brazil (Fig. 1.4).[20]

---

[19]Mestizo is a person of combined European and Indigenous American descent, regardless of where the person was born.

[20]Stavenhagen, R., Derecho Indígena y Derechos Humanos en América Latina, Instituto Interamericano de Derechos Humanos, 1988, https://archive.org/stream/Stavenhagen-DerechoIndigenaYDerechosHumanosEnAmericaLatina/Stavenhagen-Derecho-indigena-y-ddhh-en-AL_djvu.txt (accessed 16 July 2019).

Fig. 1.3 Latin America speaking languages

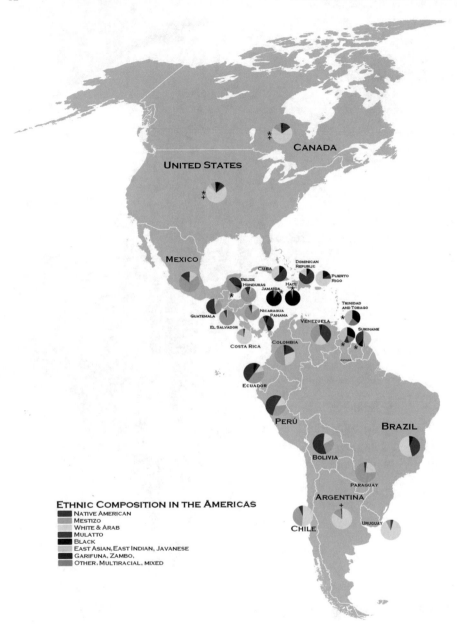

Fig. 1.4 Ethnic composition in Latin America. Ethnic composition of the Americas according to Lizcano and the CIA World Factbook, https://upload.wikimedia.org/wikipedia/commons/3/37/Ethnic_Composition_of_the_Americas.PNG (accessed 20 September 2019)

1.2 Evolution of Regionalism and Regional Cooperation

Latin American integrationist movements began in the 1950s when the process of regional integration was seen as the route to industrialization as all Latin American countries at the time were exporters of primary products and raw materials. These integration movements appear with substantial variations in the degree of reduction of trade barriers, establishing external tariffs and promoting regional industrialization. Integration presupposes changes in economic structures that affect a society's established interests.

Regional integration is usually divided into three phases. The first began in the second half of the 20th century and was based on the developmental environment of the region, strongly influenced by ECLAC, which advocated the industrialization of the region through import substitution. In the 1960s, the region adopted import-substitution models to boost development. This process implied stable governments and intervention in the economy, leading to strong protectionist policies to avoid excessive import rates. This created a defensive system against other regional markets that were more industrialized and followed the idea of creating a broader regional market. The first models include the Central American Common Market (MCCA)[21] established in 1958, the Latin American Free Trade Association (ALALC)[22] created in 1960, and the Andean Pact established in 1969 (until 1996), to promote regional industrialization.

The second phase occurred between the second half of the 1970s and the beginning of the 1980s, characterized by a setback resulting from an international economic crisis and the Oil Shock that negatively impacted the Latin American economy. In the 1980s, the second Treaty of Montevideo was signed, creating the Latin America Association for Integration (ALADI)[23] with less administrative structure allowing a more flexible regional model based on bilateral and multilateral agreements. Later, with the oil crisis in 1973, and the accumulation of public debt, new objectives were pursued with more protectionist measures to make economies more flexible to integrate them into the world economy. In this sense, regional integration was considered as a tool to promote international competitiveness and exports.

The third phase "Open Regionalism" began in the second half of the 1980s, in the context of the crisis of national-developmentalism and the global economic crisis. At that time regional integration was seen as an alternative to a new cycle of development in the region, but in line with the principles of neoliberalism, such as the Andean Community (1996).

---

[21] Mercado Común Centroamericano (MCCA).

[22] Asociación Latinoamericana de Libre Commercio (ALAC).

[23] Asociación Latinoamericana de Integración (ALADI).

It may be possible to speak of a fourth phase, beginning in 2001, with a shift towards a more left-wing political spectrum derived from the political environment in the region,[24] where new models of regional bodies were created, with a vision of open regionalism. Post-liberal regionalism influenced new initiatives that shaped the new models: ALBA[25] was established in 2004, UNASUR[26] in 2008, CELAC[27] in 2010; SICA[28] and MERCOSUR[29] emerged.

Described below are the main blocs in Latin America, historic and current.

1.2.1   Organization of American States

The Organization of American States (OAS) considers itself the oldest regional organization in the world, dating back to the First American International Conference, held in Washington in 1889 and 1890,[30] or, prior to that, to the Panama Congress, convened by Simon Bolivar in 1826. The OAS had its Charter signed in 1948 and entered into force in 1951, with 35 independent American and Caribbean States as current members.[31] The OAS calls itself the main political, legal, and social government forum in the hemisphere.

The OAS is a multilateral, regional organization, an important forum for regional diplomacy focused on human rights, electoral oversight, social and economic development and security in the western hemisphere.[32]

The organization is the premier political forum of the Americas, where countries of North, Central and South America and the Caribbean come together to advance their common goals, following the four main pillars of OAS. Furthermore, it promotes cooperation in the region, supporting member states in building institutional and human capacity to meet new challenges.

---

[24]Vitte, C., Geopolítica e Relações Internacionais: as organizações de integração regional na América Latina, Revista Meridiano, http://www.revistameridiano.org/ (accessed 16 July 2019).

[25]Alianza Bolivariana para los Pueblos de Nuestra América (ALBA).

[26]Unión de Naciones Suramericanas (UNASUR).

[27]Comunidad de Estados Latinoamericanos y Caribeños (CELAC).

[28]Sistema de la Integración Centroamericana (SICA).

[29]Mercado Común del Sur (MERCOSUR).

[30]OAS, "Who we are", http://www.oas.org/en/about/who_we_are.asp (accessed March 2019).

[31]Meyer, P. J., Organization of American States: Background and Issues for Congress, https://fas.org/sgp/crs/row/R42639.pdf (accessed 25 July 2019).

[32]Council on Foreign Relations, "The Organization of American States", https://www.cfr.org/backgrounder/organization-american-states (accessed 18 March 2019).

The organization has also granted permanent observer status to 69 states, as well as to the EU, who are strategic partners to the organization, providing cooperation in the form of political engagement, financial contributions, technical expertise, educational and training opportunities and the exchange of experiences and best practices.[33] OAS headquarters, which comprises three main bodies, the General Assembly, the Permanent Council and the General Secretariat, is located in Washington D.C.

The main organ for decision-making is the General Assembly. The Permanent Council is responsible for managing the day to day affairs. The General Secretariat has the role of implementing policies made by the other two bodies. OAS activities include coordinating security and law enforcement operations by providing technical and financial assistance for disaster management, and the development of projects for monitoring human rights through the inter-American Legal system. Autonomous Institutions such as the Inter-American Commission on Human Rights (IACHR) and the Inter-American Juridical Committee also carry out OAS functions. OAS funding is supported by the financing of the General Assembly that has a regular budget from country quotas, based on members' capacities to pay.

In 2019, the OAS's approved budget is USD 82.7 million. The regular fund supports the General Secretariat (Fig. 1.5).[34]

---

[33] OAS Doc, "Report to permanent observers 2015–2016", http://www.oas.org/fpdb/press/Report-to-Permanent-Observers-2017.pdf (accessed 18 March 2019).

[34] OAS, Program-Budget of the Organization of American States, 2019, http://www.oas.org/budget/2019/Approved%20Budget%202019.pdf (accessed 18 March 2019).

Fig. 1.5 OAS member states

1.2.2 Organization of Ibero-American States

The Organization of Ibero-American States (OEI) was founded in 1949 as an international governmental body for cooperation between Ibero-American Member States to ensure that education systems advance knowledge in the field of education through the planning and development of regional projects.[35] Its efforts are oriented towards fulfilling the gaps in education, science, technology, and culture in the context of development, democracy, and regional integration to guarantee the integral and harmonious formation of new generations. The main objective is to ensure equal educational opportunities, social progress and democratization.[36]

Their member states are Portuguese and Spanish speaking nations of Latin America, Europe, and Equatorial Guinea, Africa. Member States include Andorra, Argentina, Bolivia, Brazil, Colombia, Costa Rica, Cuba, Chile, Dominican Republic, Ecuador, El Salvador, Equatorial Guinea, Guatemala, Honduras, Mexico, Nicaragua, Panama, Paraguay, Peru, Portugal, Spain, Uruguay and Venezuela.

The Headquarters of the General Secretariat is in Madrid, Spain, with regional offices in Argentina, Bolivia, Brazil, Chile, Colombia, Costa Rica, Ecuador, El Salvador, Guatemala, Honduras, Mexico, Nicaragua, Panama, Paraguay, Peru, Portugal, the Dominican Republic and Uruguay.

The OEI is financed by mandatory and voluntary contributions from member states and from contributions by institutions, foundations and other bodies interested in contributing to the fortification of knowledge, mutual understanding, integration, solidarity and peace among the Ibero-American peoples with the objective of encouraging development of education and a culture for peace. Its structure comprises three bodies.

The General Assembly holds the supreme authority of the organization, composed of representatives or official delegations of the highest level from member states. The legislative body establishes OEI's general policies, studies, and evaluates and approves the organization's Plan of Activities, the Global Program, and Budget, sets the annual quotas and elects the Secretary-General for the corresponding period. The delegated body of the General Assembly directs the Council to control the government and the OEI Administration, composed of Ministers of Education of member states, or their representatives. Their primary mission is to approve the Activity Report, the Biennial Program and Budget. The permanent delegate body is the General Assembly, responsible for the execution of OEI programs and projects, and is the General Secretariat that develops the policies, strategies, and Plan of Activities of the organization (Fig. 1.6).

---

[35]Denvex, Organization of Ibero-American States, https://www.devex.com/organizations/organizationof-ibero-american-states-oei-70324 (accessed 26 March 2019).

[36]OEI, Què es la OEI, https://www.oei.es/en/about/que-es-la-oei (accessed 26 March 2019).

Fig. 1.6 OEI member states

1.2.3   Andean Community

The Andean Community (CAN),[37] previously called the Andean Pact, brings together the countries that are cut by the Andes in north-western South America. The formation of the bloc was an initiative of the Chilean president, Eduardo Frei, motivated by the

---

[37] Comunidad Andina (CAN).

difficulties faced by the Latin American Free Trade Association (LAFTA),[38] created by the Treaty of Montevideo in 1960 to advance the integration of Latin American countries.[39] CAN was established in 1969 by the Cartagena Agreement,[40] by Bolivia, Colombia, Ecuador and Venezuela.

The group was also startled by the destabilization processes of the 1980s and the open regionalism of the 1990s. In 1993, however, they achieved the establishment of a free trade area with the elimination of customs duties. In 1997, the designation was changed to the Andean Community of Nations and over time its structural conformation was changed to the present, incorporating a Presidential Council, a Council of Foreign Ministers and a Commission of Delegates. This whole composition is part of the so-called Andean Integration System (SAI) that also has the Andean Parliament, with 20 parliamentarians (five elected in each member country), a development aimed at inducing effective citizen participation.

The current institutional framework of the CAN was created in 1996, through the Trujillo Protocol.[41] In 2003, the Cartagena Agreement was recoded, incorporating the changes it underwent in the Trujillo Protocol and subsequent agreements, merging the most recent decisions into just one text.[42]

The CAN represents a quarter of the population of the South America, but its GDP is only 15% of the subcontinent's total, which is half MERCOSUR's GDP per capita and is also lower than the South American average. Among CAN countries, Colombia has indicators that are closer to the South American average, but still lower.[43]

CAN deals with issues on trade in goods, services, customs union, circulation of persons, common market, foreign policy, border development, sustainability, and social agenda and economic policies. In 2011 CAN members reaffirmed their commitment to deepen Andean integration and launched an Andean Integration System (SAI).[44] For this purpose, the Andean Parliament met in Lima on 29 October 2011 to solve problems faced by migrants abroad. This meeting included the participation of 20 parliamentarians from Ecuador, Colombia, Bolivia, Peru, and Chile, that

---

[38] Asociación Latinoamericana de Libre Comercio (ALALC).

[39] Goldbaum, S., Luccas, V., Comunidade Andina de Nações, Escola de Economia FGV, 2012, https://bibliotecadigital.fgv.br/dspace/bitstream/handle/10438/9650/TD%20309%20-%20Sergio%20Goldbaum.pdf (accessed 26 July 2019).

[40] CEPAL, Acuerdo de Cartagena, https://idatd.cepal.org/Normativas/CAN/Espanol/Acuerdo_de_Cartagena.pdf (accessed 26 July 2019).

[41] OAS, Protocolo Modificatorio del Acuerdo de Integración Subregional Andino (Acuerdo de Cartagena), http://www.sice.oas.org/Trade/Junac/Carta_Ag/Trujillo.asp (accessed 26 July 2019).

[42] A atual Codificação do Acordo de Integração Subregional Andino (Acordo de Cartagena) foi estabelecida por meio da Decisão 563 (2003) da Comissão, Codificações anteriores ocorreram por meio das Decisões 117 (1977), 147 (1979), 236 (1988), 406 (1997) e 472 (1999).

[43] Goldbaum, S., Luccas, V., Comunidade Andina de Nações, Escola de Economia FGV, 2012, https://bibliotecadigital.fgv.br/dspace/bitstream/handle/10438/9650/TD%20309%20-%20Sergio%20Goldbaum.pdf (accessed 26 July 2019).

[44] Sistema Andino de Integración (SAI).

addressed the regulations on the functioning of the Parliament of the Union of South American Nations (UNASUR) (Fig. 1.7).[45]

Fig. 1.7 CAN members and associate members

[45]Parlamento Andino, https://parlamentoandino.org/ (accessed 12 February 2019).

1.2.4 Caribbean Community

The Caribbean Community (CARICOM) was founded in 1973 by the Treaty of Chaguaramas (Trinidad and Tobago)[46] and replaced the Caribbean Free Trade Association that had been created in 1965.[47] It is an organization of 15 Caribbean nations and British entities. The full members are Antigua and Barbuda, Barbados, Belize, Dominica, Grenada, Guyana, Jamaica, Montserrat, Saint Kitts and Nevis, Saint Lucia, Saint Vincent and the Grenadines, Trinidad and Tobago. The British Virgin Islands, The Turks and Caicos Islands are associate members.

The headquarters of CARICOM is in Georgetown, Guyana. The Caribbean community has three main activities: economic cooperation through the Caribbean Common Market, the coordination of foreign policy, and collaboration in fields such as agriculture, industry, transport and telecommunications.[48]

The main objectives of CARICOM are to promote economic integration and cooperation among its members, as well as to ensure that the benefits of integration are distributed equitably, and to coordinate foreign policy. Its main activities include the coordination of economic policies and development planning, and the elaboration and institution of special projects for less developed countries within its jurisdiction. It functions as a single regional market for many of its members (CARICOM single market) and supports the solution of regional trade disputes.[49]

In 2001, the Heads of Government signed a Revised Treaty of Chaguaramas,[50] thus paving the way for the transformation of the idea of a Common Market of the Caribbean, with a single market and economy. Part of the revised treaty between the member states includes the establishment and application of the Caribbean Court of Justice.[51]

The main organ of CARICOM is the Conference of Heads of Governments who meet in regular annual sessions. The Council of Ministers, responsible for strategic planning and coordination in the areas of economic integration, functional cooperation and external relations is composed of ministers responsible for community affairs. Institutions such as the Council for the Development of Trade and Economy, Council for External Relations and Community Affairs, Council for Human and Social Development, Council for Finance and Planning Council, support the main bodies of the community (Fig. 1.8).[52]

---

[46]CARICOM, Treaty Establishing the Caribbean Community, https://caricom.org/documents/4905-original_treaty-text.pdf (accessed 26 July 2019).

[47]R. L. Abbott, The Caribbean Free Trade Association, http://repository.law.miami.edu/umialr/vol1/iss2/2 (accessed 26 July 2019).

[48]CARICOM, "Who we are", https://caricom.org/about-caricom/who-we-are (accessed 26 July 2019).

[49]Ibid.

[50]Ibid.

[51]Ibid.

[52]EcuRed, "CARICOM", https://www.ecured.cu/CARICOM (accessed 19 March 2019).

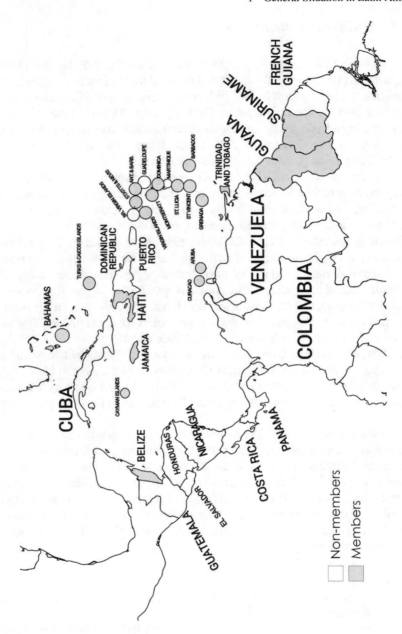

Fig. 1.8 CARICOM member states

1.2.5 Latin American and Caribbean Economic System

The foundation of the Latin American and Caribbean Economic System (SELA)[53] took place in the midst of a difficult protectionist environment in the mid-1970s. Formed as a regional intergovernmental body through the Panama Convention on 17 October 1975, the SELA was intended as a forum for consultation, coordination, cooperation, and economic and social promotion.

The SELA is a regional intergovernmental forum based in Caracas, capital of Venezuela, made up of 26 countries in Latin America and the Caribbean.[54] The objectives of SELA are to promote and coordinate common positions and strategies related to the economy of Latin America and the Caribbean with other countries, groups of nations, forums and international organizations, as well as to stimulate and boost cooperation and integration among Latin American and Caribbean countries.[55] The sectors of activity are economic and social issues, intra-regional relations, extra-regional relations, trade facilitation, integration process and economic growth, information and communication technologies and small and medium-sized enterprises (SMEs). Its main objectives are the acceleration of members' economic and social development and the adoption of common positions and strategies on issues in question *vis-à-vis* other international bodies and fora (Fig. 1.9).

---

[53] Sistema Económico Latinoamericano y del Caribe (SELA).

[54] OECD, SELA Profile, https://www.oecd.org/gov/regulatory-policy/SELA_profile.pdf (accessed 19 July 2019).

[55] Sistema Económico Latinoamericano y del Caribe, Qué es el SELA, https://web.archive.org/web/20081202171415/, http://www.sela.org/sela2008/sela.asp (accessed 19 July 2019).

Fig. 1.9 SELA member states

1.2.6 Southern Common Market

The Southern Common Market (MERCOSUR)[56] has its origin with the signature of the Foz de Iguazú Declaration, in December 1985, by the Brazilian and Argentine Presidents,[57] which was the basis for the economic integration of the so-called Southern Cone. Both countries had just emerged from a dictatorial period and faced the need to reorient their economies to the world, since they had incurred a large foreign debt under the military governments and had no credit abroad. On 6 July 1990, the Presidents of Brazil and Argentina signed the "*Ata de Buenos Aires*", aimed at full customs integration between the two countries.[58] Then, on 26 March 1991, the presidents of Argentina, Brazil, Paraguay and Uruguay signed the Treaty of Asuncion,[59] aimed at building a free trade zone between the four countries called the Southern Common Market.

Based on the Ouro Preto Protocol,[60] signed on 17 December 1994 and entered into force on 15 December 1995, MERCOSUR has a basic institutional structure composed of the Common Market Council (CMC),[61] the supreme body whose function is the political conduct of the integration process. The CMC is composed of the Ministers of Foreign Affairs and Economy of the states parties; and the Common Market Group (GMC)[62] is the executive decision-making body responsible for setting the work programs and negotiating agreements with third parties on behalf of MERCOSUR, by express delegation from the CMC. The GMC acts through resolutions and is composed of representatives of the ministries of foreign affairs, economy and the central banks of states parties. The MERCOSUR Trade Commission (CCM),[63] a technical decision-making body, supports the GMC on the bloc's trade policy.

In addition, MERCOSUR has other advisory bodies, such as the Joint Parliamentary Commission (CPC),[64] a parliamentary representative body composed of up to 64 parliamentarians, 16 from each state party. The CPC is consultative, deliberative, and formulates declarations, provisions, and recommendations. The MERCOSUR

---

[56]Mercado Común del Sur (MERCOSUR).

[57]Barthelmess, E., As relações Brasil-Argentina no aniversário da Declaração do Iguaçu, Cadernos de Política Exterior, v. 3, pp. 27–43, 2016, http://www.funag.gov.br/ipri/images/pdf/3.03_Brasil-Argentina.pdf (accessed 19 July 2019).

[58]Brasil, Ministério da Economia, SISCOMEX, http://www.mdic.gov.br/index.php/comercio-exterior/exportacao/novoex-siscomex-exportacao-modulo-comercial/9-assuntos/categ-comercio-exterior/337-certificado-form-7 (accessed 20 July 2019).

[59]Brasil, Ministério das Relações Exteriores, Mercosur, http://www.mre.gov.py/tratados/public_web/DetallesTratado.aspx?id=0GXnoF+V0qWCz+EoiVAdUg%3d%3d&em=lc4aLYHVB0dF+kNrtEvsmZ96BovjLlz0mcrZruYPcn8%3d (accessed 20 July 2019).

[60]MERCOSUR, Protocolo de Ouro Preto, https://www.MERCOSUR.int/pt-br/documento/protocolo-de-ouro-preto-adicional-ao-tratado-de-assuncao-sobre-a-estrutura-institucional-do-mercosul/ (accessed 20 July 2019).

[61]Consejo del Mercado Común (CMC).

[62]Grupo Mercado Común (GMC).

[63]Comisión de Comercio del Mercosur (CCM).

[64]Comisión Parlamentaria Conjunta del MERCOSUR (CPC).

Fig. 1.10 MERCOSUR members, associate members and observers

Parliament was constituted on 6 December 2006, replacing the CPC. The Social Economic Advisory Forum (FCES)[65] is an advisory body representing the sectors of the economy and society, which acts through recommendations to the GMC.

The main aims of this trade area are the free movement of goods, services and productive factors between countries through the elimination of customs duties and non-tariff restrictions on the movement of goods and any other measure having the same effect, with the establishment of a common foreign tariff and the adoption of a common commercial policy towards third countries or economic blocs. Thus, products originating in the territory of a signatory country will have the same treatment in another signatory country as products of national origin. One practical aim is the coordination of foreign, agricultural, industrial, fiscal, monetary, foreign exchange and capital policies, as agreed upon, in order to ensure adequate conditions of competition among members, with the commitment of these countries to harmonize their laws, especially areas of general importance in order to strengthen the integration process. In relations with non-signatory countries, members of the bloc will ensure a level playing field. In this way, members will apply their national laws to inhibit imports whose prices are influenced by subsidies, dumping or any other unfair practice. At the same time, the bloc's countries are to coordinate their national policies with a view to developing common rules on trade competition.[66]

Currently, MERCOSUR has a GDP of approximately USD 3 trillion,[67] with about 70% of this amount coming from Brazil. Therefore, the asymmetries of current markets in the block are large. This has caused a lot of friction within the bloc and is one of the factors that make it difficult to create a single currency for the economic bloc. MERCOSUR has established a common external tariff and the adoption of common commercial policies concerning third states or groupings of states, by coordinating positions in regional and international economic, commercial forums (Fig. 1.10).

1.2.7 Central American Integration System

The Central American Integration System (SICA)[68] succeeds the Organization of Central American States (ODECA),[69] created by Costa Rica, El Salvador, Guatemala, Honduras and Nicaragua by the 1951 San Salvador Charter. SICA was created on 13 December 1991, by signing the protocol to the Charter of the Organization of

---

[65]Foro Consultivo Económico-Social del Mercosur (FCES).

[66]Ibid.

[67]The World Bank, Data, GDP (constant 2010 USD), https://data.worldbank.org/indicator/NY.GDP.MKTP.KD?end=2018&locations=BR-AR-UY-PY&name_desc=false&start=1960&view=chart (accessed 20 July 2019).

[68]Sistema de la Integración Centroamericana (SICA).

[69]Organización de Estados Centroamericanos (ODECA).

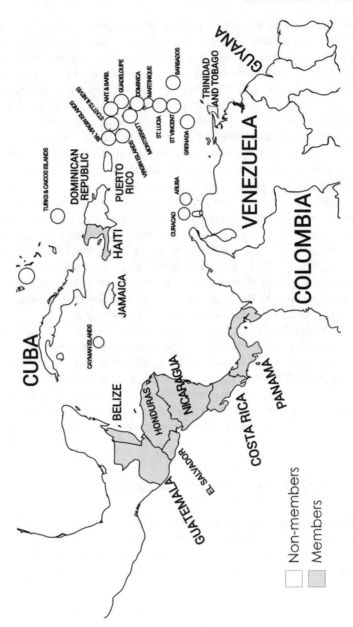

Fig. 1.11 SICA member states

American States, known as "The Tegucigalpa Protocol", being the first legal instrument related to the integration process. The headquarters of the Secretary General of SICA is in El Salvador.

Regional observers include Mexico, Argentina, Chile, Colombia, the United States, Peru, Uruguay, and Brazil; while extra-regional observers include Australia, Spain, China, Germany, South Korea, Japan, Italy, France, the EU and others.

SICA was designed to examine the integration process of the region and consider the lessons learned from the historical events of the region, such as the political crises and armed conflicts. The existence of democratic regimes in Central America is a fundamental objective for Central America to constitute a region of peace, democracy, and development.

Among the objectives of SICA is the vision of consolidating democracy in the region and establishing a regional security model. Furthermore, the organization aims to reach an economic union and strengthen the Central American financial system, in order to successfully insert the region in the international economy. Compared to other Latin American groups, the Central American Common Market (CACM)[70] has been more successful in lowering trade barriers among its members (Fig. 1.11).

1.2.8 Union of South American Nations

The Constitution of the Union of South American Nations (UNASUR) seeks the deepening of South American integration in terms of development strategy, anchored in the practices of foreign policies that emerge from a process of transformation. The Constitutive Treaty (Treaty of Brasilia), signed on 23 May 2008, in Brasilia, brought together the twelve countries of the region: Argentina, Bolivia, Brazil, Chile, Colombia, Ecuador, Guyana, Paraguay, Peru, Suriname, Uruguay and Venezuela. The headquarters is in Quito, Ecuador and the South American Parliament is in Cochabamba, Bolivia. The headquarters of the Bank of the South, established in 2009, is in Caracas, Venezuela.

UNASUR was founded on the following principles: Unrestricted respect for the sovereignty, integrity and territorial inviolability of states; self-determination of peoples; solidarity; cooperation; peace; democracy, citizen participation and pluralism; universal, indivisible and interdependent human rights; reduction of asymmetries; and harmony with nature for sustainable development.

The bloc's path follows the advances already made in the regional process, with reference to MERCOSUR and CAN and, like these, the path states that the implementation of UNASUR should be flexible and gradual, i.e. that each member must act according to their reality in achieving the treaty's objectives.

In this regard, the organization has defined a full range of objectives in the economic sphere. Mechanisms in monetary and financial coordination have been created to meet a broader range of objectives focused on the integration efforts in the region to

---

[70]Mercado Común Centroamericano (MCCA).

address problems in the international economic order.[71] UNASUR represents a consolidation, modernization and deeper integration of the CAN countries, furthering the mechanisms established with CAN, MERCOSUR, CARICOM and SICA.

The general objective of UNASUR reveals the absence of presumptions of prior arrangements in the implementation of an integration space: UNASUR aims to build, in a participatory and consensual way, a space for integration and union in the cultural, social, economic and political sphere among its peoples, prioritizing political dialogue, social policies, education, energy, infrastructure, financing and the environment among others, with a view to eliminating socioeconomic inequality, achieving social inclusion and citizen participation, strengthening democracy and reducing asymmetries within the framework of strengthening each state's sovereignty and independence.[72]

Presidents of each member state have annual meetings, and meetings are implemented according to MERCOSUR's and CAN's mechanisms. The objective of the UNASUR is to strengthen political dialogue among the member states to reinforce South American integration and participation in the international arena in economic and social fields.

With its statute approved on 16 December 2008, by the Summit of Heads of State and Government, held in Salvador, Brazil, the South American Defence Council (CDS)[73] was created as a forum for consultation, cooperation, and coordination in defence matters through meetings between the Ministers of Defence or similar portfolios. Its general objective is to contribute to democratic stability, to consolidate a zone of peace and to generate consensus in this sensitive area.

In November 2011, UNASUR's Defence Ministers agreed on the creation of a South American Space Agency during a meeting of the Defence Council of UNASUR, with the exception of Brazil that argued against the proposal, based on the costs involved in creating new structures. The Brazilian argument is also that disparity in South American space capabilities will reduce Brazil's advantages. Nowadays, there are still factions supporting the creation of a regional space agency, but governments still cannot agree about whether to support and fund the creation of such an entity. The South American Space Agency remains a goal that is being pursued.[74]

---

[71] Union of South American Nations, XLI Regular Meeting of the Latin American Council Caracas, Venezuela 25–27 November 2015, SP/CL/XLI.O/Di, no. 10–15, http://www.sela.org/media/2087751/di-10-unasur-ing.pdf (accessed 15 February 2019).

[72] Carvalho, G., A América do Sul em Processo de Transformação: Desenvolvimento, Autonomia e Integração na UNASUL, 2013, http://www.ie.ufrj.br/images/pos-graducao/ppge/Dissertao_-_Glauber_Carvalho.pdf (accessed 15 July 2019).

[73] Consejo de Defensa Suramericano (CDS).

[74] Sarli, B., Journal of Aeronautical History, 2018, https://www.aerosociety.com/media/9320/review-of-space-activities-in-south-america.pdf (accessed 15 July 2019).

In 2018, as a result of the permanent blockade on Venezuela and, to a lesser extent, of Bolivia, six countries decided to suspend indefinitely their participation in UNASUR. From now on, Argentina, Brazil, Chile, Colombia, Paraguay and Peru will not participate in the different activities of the regional bloc that was created in 2008 (Fig. 1.12).[75]

Fig. 1.12 UNASUR member states

[75]Dapelo, S, La Nacion, 2018, https://www.lanacion.com.ar/politica/la-argentina-y-otros-cinco-paises-abandonan-la-unasur-nid2127623 (accessed 17 July 2019).

1.2.9 Pacific Alliance

The Pacific Alliance[76] is a Latin American trade bloc, established through the Lima Declaration,[77] at the initiative of the President of Peru, who extended the invitation to his colleagues from Chile, Colombia, Mexico and Panama, with the purpose of deepening integration between these economies and defining joint actions for commercial relations with the Asian countries of the Pacific basin based on the existing bilateral trade agreements between the states parties. The proposal for the alliance was announced on 28 April 2011, in Lima, Peru.[78]

The Pacific Alliance has adopted key measures to promote regional integration, including significant advances in the liberalisation of trade in goods; the creation of a joint platform for stock markets; joint diplomatic and trade undertakings in Asia; and numerous social achievements such as the Alliance scholarship programme. A further relevant characteristic of this integration mechanism has been its openness to collaborate with observer countries and to learn from their good practices and lessons learnt.[79]

According to this Declaration, the intention of the alliance is "to encourage regional integration, as well as greater growth, development and competitiveness"[80] of the economies of their countries, while also committing to "progressively progress towards the goal of achieving the free movement of goods, services, capitals and people". This joint effort is considered one of the most successful integration processes in the region and has generated wide international interest, as evidenced by the 55 States that are now observers, as well as the countries that are conducting negotiations to become associated states.[81]

Among the conditions that a country must meet in order to join the Pacific Alliance, are the validity of the rule of law, democracy, constitutional order and the free market. The four founding nations of the Pacific Alliance account for 40% of Latin America's GDP. The only three Latin American countries members of the OECD, Chile, Colombia and Mexico, are part of the Pacific Alliance (Fig. 1.13).

Country members designed this alliance to create a regional gateway to Asian markets, pursuing commercial, economic and political integration. The strategic platform has been created to further growth, development, and competitiveness on the

---

[76]Alianza del Pacífico.

[77]Declaración Presidencial de la Alianza del Pacífico, http://www.sice.oas.org/TPD/Pacific_Alliance/Presidential_Declarations/I_Summit_Lima_Declaration_s.pdf (accessed 17 July 2019).

[78]Mexico, Memorias Documentales, Unidad de Coordinación de Negociaciones Internacionales, Alianza del pacífico, http://www.economia.gob.mx/files/transparencia/informe_APF/memorias/6_md_alianza_pacifico_sce.pdf (accessed 17 July 2019).

[79]OECD, Pacific Alliance and Observer Countries: an agenda for cooperation, Global Policy Perspective Report, https://www.oecd.org/latin-america/home/Global-Policy-Perspective-Report-Pacific-Alliance-and-Observer-Countries.pdf (accessed 17 July 2019).

[80]Ibid.

[81]Costa Rica, Panama and Ecuador are conducting negotiations to become associated States.

Fig. 1.13 Pacific Alliance member states

economies of the members, and is focused on achieving well-being and overcoming socioeconomic inequality. Member states pursue deep integration on services, capital, investments, and movement of people.

In 2019, the Alliance was the eighth economic power and eighth export force worldwide, representing 28% of GDP of CAN countries with a total trade attracting 45% of the Foreign Direct Investment. The alliance is oriented towards modernity, moving progressively towards the free mobility of goods, services resources, and people, as the alliance is open to free trade.[82]

Within this framework, the group accounts for more than one-third of Latin America's GDP, and exports about 92% more than MERCOSUR.[83] In 2013, in Cali, Colombia, the Presidents of the Alliance countries signed an Additional Protocol to the Framework Agreement, which immediately eliminated 92% of tariffs among members.[84] This allowed the Alliance to move towards a broader scope than merely free trade agreements, such as the free movement of people, and measures to integrate the stock markets of member countries (Table 1.1). [85]

Table 1.1 Size, membership and performance of MERCOSUR and the Pacific Alliance[a,b]

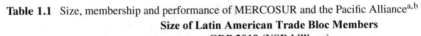

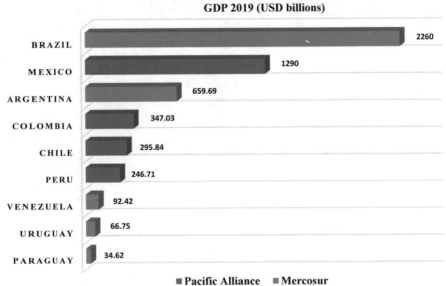

[a]Ferreira, S., What are trade blocs and how do two of Latin America's largest compare?, World Bank Blogs, 2015, https://blogs.worldbank.org/opendata/what-are-trade-blocs-and-how-do-two-latin-america-s-largest-compare (accessed 19 September 2019)
[b]IMF, GDP Ranked by Country 2019, http://worldpopulationreview.com/countries/countries-by-gdp/ (accessed 29 September 2019)

---

[82]The Pacific Alliance, "What is the Pacific Alliance?", https://alianzapacifico.net/en/what-is-the-pacificalliance/ (accessed 15 February 2019).

[83]Americas Society/Council of the Americas, "Explainer: What Is the Pacific Alliance?", https://www.ascoa.org/articles/explainer-what-pacific-alliance (accessed 15 February 2019).

[84]Villarreal, M., The Pacific Alliance: A Trade Integration Initiative in Latin America, https://www.u-tokyo.ac.jp/content/400038391.pdf (accessed 25 July 2019).

[85]Congressional Research Service, The Pacific Alliance: A Trade Integration Initiative in Latin America, https://fas.org/sgp/crs/row/R43748.pdf (accessed 13 February 2019).

1.2.10 Perspectives of Integration and Regionalism Development

The foreign policy of Latin American countries usually acts in favour of the immediate national interests of the time, hardly focusing on long-term policies.

Latin American integrationist movements began in the 1950s when the process of regional integration was seen as the route to industrialization and to change the picture that all Latin American countries were exporters of almost solely primary products. These integration movements appeared with substantial variations in the degree of reduction of trade barriers, establishing external tariffs and promoting regional industrialization. Integration presupposes changes in economic structures that affect society's established interests.

One of the basic explanations for this failure is related to the union of developing countries. The non-competitive industrial output could not compete with other blocks that were formed, such as the case of the EU. Another fact is related to the structural problems of these countries, such as low income per capita, income concentration, and low educational level. Regional integrations have not changed this picture.

In particular, as a result of the second Oil Shock and the rise in international interest rates in 1979, Latin America's external fragility imposed severe restrictions on these integrationist schemes. Indeed, with the worsening foreign debt crisis that erupted in 1982 with the Mexican moratorium, intra-regional trade fell dramatically in the 1980s, putting the future of Latin America's regional blocs in check. Latin American economies were increasingly dependent on the international market rather than the regional.

The models that followed the Latin American regionalism evolution have covered different objectives due to the diverse characteristics of the continent. The cooperation mechanisms already established have allowed Latin American and Caribbean countries to cooperate and solve internal crises. At the same time, there is substantial evidence of the willingness for advancing regional governance in the region. Moreover, the strong inter-governmentalism seen in almost every regional process of integration is because Latin American states reject any sovereignty transfer.[86]

To current date, there are more than 15 regional cooperation organizations with different levels of institutionalization and development. In these significant areas, two groups can be identified. The first group supports the integration processes, with the aim of progressively establishing a Free Trade Area and then, gradually, a Customs Union. Within the first group, SICA, CAN, CARICOM, MERCOSUR and the Pacific Alliance can be mentioned as they were formed from integration processes with formal established institutional frameworks, like the European project. Within this process, one of the elements that could contribute to the integration of the region is the enlargement process of the Pacific Alliance and MERCOSUR; this could build better understanding and stronger trade convergence among the states.

---

[86]Gulf Research Centre, Integration Processes in Latin America, 2014, https://www.files.ethz.ch/isn/184537/Unity_Anna_Ayuso_fin_9127.pdf (accessed 23 March 2019).

The second group covers a diversity of organizations, with sectorial characteristics and with classic intergovernmental cooperation bodies. This group has heterogeneous characteristics and their integration mechanisms have developed from sectoral bodies and classical intergovernmental cooperation organizations. They are characterized by institutional-level cooperation mechanisms in the field of security or social policies. In this regard, UNASUR and ALBA can be mentioned.

Furthermore, Latin America has one of the lowest intraregional trade coefficients of all continents, probably below 20% of the total, compared to significantly higher volumes in other regions.[87] While Europe has already surpassed 60% of intra-regional trade, and Asia has exceeded half of its total trade, Latin America continues to bear the same eccentric character that has marked the entire history of its foreign trade.[88]

Even though the process of regionalism in the region has not advanced to full support of multilateral agreements in the space sector, current regional cooperation in the region can be used to advance the development of space activities in Latin America.

---

[87] Bown, B., Lederman, D., Pienknagura, S., Robertson, R., Better neighbors: toward a renewal of economic integration in Latin America, 2017, World Bank Group, http://documents.worldbank.org/curated/en/402861490788215893/Better-neighbors-toward-a-renewal-of-economic-integration-in-Latin-America (accessed 21 September 2019).

[88] Almeida, P., Integração regional e inserção internacional dos países da América do Sul: evolução histórica, dilemas atuais e perspectivas futuras, 2008, https://fundacaofhc.org.br/files/papers/407.pdf (accessed 16 July 2019).

Chapter 2
Latin American Space Arena

Contents

Abstract This chapter describes the Latin American space arena, from the first activities developed by each country with the creation of its space agency, to the launch and use of satellites. There is detailed analysis of the knowledge in Latin America of communications, Earth observation and science, technology and education. Hence, the regional satellites launch partnership, as well as launch capabilities, concerning projects that are under development by Brazil and Argentina are addressed. The history of Brazil as a launch pioneer in Latin America and how its satellite launch vehicle struggled, is discussed. Once again, in this chapter, French Guiana's launch capacities are separately reported.

2.1   Space Principles and Fundamentals in Latin America

Most countries in Latin America rely heavily on space industries and agencies in countries with a developed space industry, especially regarding technical assistance. Argentina, Brazil and Mexico are among the countries that have signed several bilateral agreements with NASA and ESA in different areas. Argentina, Brazil and Chile are also working together with the French remote sensing programs SPOT and US LANDSAT, as well as launch tracking, satellites and data collection.[1]

---

[1] UN/Brazil Workshop on Space Law, ST/SPACE/28, Disseminating and Developing International and National Space Law: The Latin America and Caribbean perspective, 2005, ISBN 92-1-100977-4, http://www.unoosa.org/pdf/publications/st_space_28E.pdf (accessed 22 September 2019).

© Springer Nature Switzerland AG 2020   37
A. Froehlich et al., *Space Supporting Latin America*, Studies in Space Policy 25,
https://doi.org/10.1007/978-3-030-38520-0_2

Among the Latin American countries with a national space law, the Brazilian Administrative Decree no. 27, issued in 2001, dealing with the most important aspects of private participation in space activities, is to be highlighted.[2] Through this, Brazil became not only the first Latin American, but also the first developing, country with adequate national space legislation. There are several regional initiatives to achieve closer cooperation between Latin American economies, within Mercosur and the Andean Pact, which can serve as a starting point for the development of cooperation policies, especially on law-making on space issues.

Regarding space education, science and technology, training programs have been established as part of the missions of space organizations due to the lack of institutions in these areas. Argentina, Brazil, Chile and Mexico have substantial space programs, but in none of them has the space program achieved the same level of investment in space education as in Brazil in recent decades. From 1968, INPE began to offer postgraduate courses and, nowadays, offers master's and doctorate courses in Astrophysics, Space Engineering and Technology, Space Geophysics, Applied Computing, Meteorology, Remote Sensing and Earth System Science. INPE also receives students from other Latin American countries in its courses. By 2019, INPE had graduated 159 Ph.D.s and 1,072 masters.[3] Recently, Mexico created a master's degree in the field of digital image processing, primarily focused on satellite image processing.[4]

In the 1980s, Argentina developed an interesting general space education program for elementary and high school students, in partnership with CNIE and the Argentine Ministry of Education, with an emphasis on remote sensing. Chile has similar programs for children, in cooperation with NASA, in microgravity experiments. In Bolivia, a centre has been set up in Cochabamba, in partnership with the ITC of the Netherlands, the Centre for Aerospace Surveys and GIS Applications for the Sustainable Development of Natural Resources (CLAS), which provides training and specialization courses in remote sensing and geoprocessing. After hosting the III CEA, Uruguay began to invest in training courses in the areas of remote sensing and geoprocessing, meteorology, GPS and astronomy. Ecuador, Colombia, Mexico, Nicaragua, Honduras, Venezuela, Costa Rica invest primarily in the area of

---

[2] AEB, Administrative Edict no. 27, Regulamento sobre procedimentos e definição de requisitos necessários ao requerimento, avaliação, expedição, controle, acompanhamento e fiscalização de licença para execução de atividades espaciais de lançamento no território brasileiro, 2001, http://www.mctic.gov.br/mctic/opencms/legislacao/portarias/migracao/Portaria_AEB_n_27_de_20062001.html?searchRef=lan%C3%A7amento&tipoBusca=expressaoExata (accessed 22 September 2019).

[3] INPE, Pós-Graduação, http://www.inpe.br/posgraduacao/sobre/sobre.php (accessed 22 September 2019).

[4] Cinvestav, Departamento de Computación, https://www.cs.cinvestav.mx/cursos/ProcesamientodeImagenes.html (accessed 22 September 2019).

Table 2.1 Space agencies in Latin America[a]

Country	Name	Acronym	Founded	Capabilities of the space agency			
				Astronauts	Operates satellites	Sounding rockets	Recoverable biological sounding rockets
Argentina	National Space Activities Commission (Spanish: Comisión Nacional de Actividades Espaciales)	CONAE	28 May 1991	No	Yes	Yes	No
Bolivia	Bolivian Space Agency (Spanish: Agencia Boliviana Espacial)	ABE	10 February 2010	No	Yes	No	No
Brazil	Brazilian Space Agency (Portuguese: Agência Espacial Brasileira)	AEB	10 February 1994	Yes	Yes	Yes	No
Colombia	Colombian Space Commission (Spanish: Comisión Colombiana del Espacio)	CCE	18 July 2006	No	Yes	No	No
Mexico	Mexican Space Agency (Spanish: Agencia Espacial Mexicana)	AEM	31 July 2010	Yes	Yes	Yes	No
Paraguay	Paraguayan Space Agency (Spanish: Agencia Espacial del Paraguay)	AEP	26 March 2014	No	No	No	No

(continued)

Table 2.1 (continued)

Country	Name	Acronym	Founded	Capabilities of the space agency			
				Astronauts	Operates satellites	Sounding rockets	Recoverable biological sounding rockets
Peru	National Commission for Aerospace Research and Development (Spanish: Comisión Nacional de Investigación y Desarrollo Aeroespacial)	CONIDA	11 June 1974	No	Yes	Yes	No
Venezuela	Bolivarian Agency for Space Activities (Spanish: Agencia Bolivariana para Actividades Espaciales)	ABAE	28 November 2005	No	Yes	No	No

[a]Observing Systems Capability Analysis and Review Tool, List of all space agencies, https://www.wmo-sat.info/oscar/spaceagencies (accessed 21 September 2019)

remote sensing, meteorology, geoprocessing and astronomy.[5] Overall, Latin American countries continue to train their professionals in different areas, but without a specific focus on space. Thus, its professionals specialize in excellence centres in the major space powers.

A list of Latin American space agencies and their capabilities (e.g. operates satellites, sounding rockets etc., is presented below (Table 2.1).

2.2 Satellites Capacity in Latin America

Some Latin American countries entered the space sector in its early stages, with a long history of commitment to space activities. In this context, foremost were Argentina and Brazil. Mexico, Peru, Paraguay, and Venezuela have just recently demonstrated interest, and have accomplished milestones in a brief time. Other countries have

---

[5]Sausen, T., A Educação Espacial na América Latina e a posição do Brasil no contexto regional, http://seer.cgee.org.br/index.php/parcerias_estrategicas/article/download/86/79 (accessed 22 September 2019).

little or no capacity to engage in space technology development, for example, the Caribbean countries.

Argentina and Brazil are the two countries with major capacities and capabilities both owning communication, Earth Observation (EO), and science and technology satellites. Brazil had the first satellite in the region and has 16 satellites in communications, eight EO satellites and 20 satellites dedicated to science, technology and education, with specific specialization in the fields of astronomy, the ionosphere, and the magnetosphere.

Argentina followed Brazil's steps and has half Brazil's capacity in communications satellites, with eight satellites. In the sphere of EO, both countries have a similar number of satellites. Argentina has nine EO satellites, one more than Brazil. In the sector of science, technology and education, Brazil has twice Argentina's capacities, with ten general satellites dedicated to scientific exploration.

Chile, Peru, Mexico and Venezuela are middle emerging space nations in the region, with capacities in two sectors, either in communications, EO and science, or technology and education. Peru and Chile have satellite capacities in both EO and science, technology and education. Within the sphere of EO, both countries have one operating satellite. In the military sphere, the Peruvian Military PERÚSAT 1 can be highlighted.[6] In the field of science, technology and education, Peru holds higher capabilities than Chile, with four satellites on technology research, while Chile, has two satellites with multidisciplinary purposes. In the area of satellite capacity in communications and EO, Venezuela has one satellite for communications purposes and two satellites on EO.

Mexico has capacities in communications and science, technology and education, with eleven telecommunications satellites, following the steps of Brazil, that has sixteen satellites with this purpose. Mexico does not have capacity in the area of EO but has two operating satellites in the field of science, technology and education, with multidisciplinary purposes. The countries that have capacity only in one sector are: Bolivia, Costa Rica, Colombia, Guatemala, Ecuador, Nicaragua and Uruguay. None of them have capacity in EO, but only in communication and science, technology and education.

No other Latin American country possesses its own national satellite capabilities. Capacities, services and products from space-based disciplines are acquired from national, international and private commercial providers. Overall, satellites are the basis for development, security and independence. Continental and national capacities of Latin American countries must be developed through national/international policies, legal frameworks and academic education, in order to promote the technological and industrial development of all countries in the region (Fig. 2.1).

---

[6]Portal Directory, PERÚSAT-1 EO Minisatellite, https://directory.eoportal.org/web/eoportal/satellite-missions/p/perusat-1 (accessed 15 August 2019).

Fig. 2.1 Latin America satellite capacity map. Gunter's Space Page, https://space.skyrocket.de (accessed 12 March 2019)

2.2.1 Latin American Communications Satellites

In the area of communications, the countries with own capacities in this sector are Brazil, Mexico and Argentina. Brazil has the highest number of satellites, with sixteen communications satellites while Mexico has eleven satellites and Argentina has seven. Emerging countries in the communications field having one satellite with this purpose are Bolivia, Nicaragua and Venezuela (Table 2.2).

Recently, the Andean Community launched an orbiting satellite, a project that had been expected since the 1970s. This satellite launching was the result of the signing of

Country	Communication satellites
Argentina	**AMSAT-LU**: 1. LUSAT (LO 19, OSCAR 19) **AR-SAT**: 2. ARSAT 1 (ARSAT-3K) 3. ARSAT 2 (ARSAT-3K) 4. ARSAT 3 (ARSAT-3K) **LatinSat**: 5. LATINSAT A, B, C, D **Nahuelsat**: 6. NAHUEL 1A (Spacebus-2000) 7. NAHUEL I2 (HS-376)
Bolivia	**Agencia Bolivariana Espacial (ABE)**: 1. TÚPAC KATARI 1 (TKSat 1) (DFH-4 bus)
Brazil	**BRAMSAT**: 1. DOVE (DO 17, OSCAR 17) **Embratel/Star One**: 2. BRASILSAT A1, A2 (HS-376) 3. BRASILSAT B1, B2, B3, B4 (HS-376 W) 4. STAR ONE C1, C2 (Spacebus-3000B3) 5. STAR ONE C3 (Star-2) 6. STAR ONE C4 (SLL-1300) 7. STAR ONE C12 (Spacebus-4000) 8. STAR ONE D1 (SSL-1300) 9. STAR ONE D2 (SSL-1300) **INPE**: 10. SCD 1 11. SCD 2, 2A 12. SCD 3 **Loral Skynet do Brazil**: 13. BRASIL 1T (HS-376) 14. ESTRELA DO SUL 1 (Telstar 14) (SSL-1300) 15. ESTRELA DO SUL 2 (Telstar 13R) (SSL-1300) **Telebras**: 16. SGDC 1 (Spacebus-4000C4)

Table 2.2 Overview Latin American communication satellites

(continued)

Table 2.2 (continued)

Country	Communication satellites
Mexico	**DirectTV Latin America**: 1. SKY-MEXICO 1 (SKYM 1) (Star-2) **Federal Government of Mexico**: 2. MEXSAT 1, 2 (Centenario, Morelos 3) (BSS-702HP) 3. MEXSAT 3 (Bincentenario) **SATMEX—Eutelsat Americas**: 4. MORELOS 1,2 (HS-376) 5. SOLIDARIDAD 1, 2 (HS-601) 6. SATMEX 5 (HS-601HP) 7. SATMEX 6 (SSL-1300X) 8. SATMEX 7 (BSS-702SP) 9. SATMEX 8 (SSL-1300) 10. SATMEX 9 (BSS-702SP) **QuetzSat**: 11. QUETZSAT (SSL-1300)
Nicaragua	1. NICASAT (DFH-4 bus)
Venezuela	**Agencia Bolivariana para Actividades Espaciales (ABAE)**: 1. VENESAT 1 (Simon Bolivar 1) (DFH-4 bus)

a contract with the Dutch company New Skies Satellites BV, known commercially as SES (in 2010) and in accordance with the reactivation of the Andean satellite network. The SES-10 Satellite will provide commercial services that will benefit telecommunications, transmission and service providers in the Andean countries, which will benefit Latin American operators in general, as the SES-10's transmission capacity will be extended throughout Latin America. The satellite capacity will be used autonomously by each member country based on their national development interests and on education, medicine and cultural outreach activities (Fig. 2.2).[7]

Most Latin American countries are dependent on communication satellite services from third providers. The overall technical requirements are provided through cooperation agreements mainly with the biggest worldwide space agencies.

---

[7]AETecno, Los beneficios para cuatro países andinos del satélite multipropósito, https://tecno.americaeconomia.com/articulos/los-beneficios-para-cuatro-paises-andinos-del-satelite-multiproposito (accessed 22 August 2019).

Fig. 2.2   Latin American communication satellites presence map

2.2.2   Latin American Earth Observation Satellites

In the field of EO, only three countries have own national capabilities: Brazil, Argentina and Venezuela. Brazil has eight satellites, followed by Argentina with seven satellites and Venezuela with two. All other countries buy or lease products either on a permanent basis or, depending on needs and circumstances such as local disasters or projects, acquire products from the commercial markets.[8] Overall the Latin American countries are dependent on provided data and resources from third parties (Table 2.3; Fig. 2.3).

Table 2.3 Overview of Latin American EO satellites

Country	Earth observation satellites
Argentina	**CONAE:** 1. SABIA-Mar 1 (SAC E) 2. SABIA-MAR 2 3. SAC D (Acquarius, ESSP 6) 4. SAOCOM 1A, 1B 5. SAOCOM-CS **Satellogic:** 6. BUGSAT 1 (Tita) 7. ÑUSAT 1, …, 98 (ALEPH-1 1, …, 98)
Brazil	1. AMAZÔNIA 1, 1B (SSR 1, 1 B) 2. AMAZÔNIA 2 3. CBERS 1, 2, 2B 4. CBERS 3, 4, 4° 5. CBERS 5, 6 6. MAPSAR 7. SABIA-MAR 1 (SAC E) 8. SABIA-MAR 2)
Chile	1. SSOT (FASat-Charlie)
Peru	1. PERÚSAT-1
Venezuela	**Agencia Bolivariana para Actividades Espaciales (ABAE):** 1. VRSS 1 (Francisco Miranda) 2. VRSS 2 (Antonio José de Sucre)

---

[8]Gunter's Space Page, https://space.skyrocket.de (accessed 15 March 2019).

Fig. 2.3 Latin American EO capacities map

2.2.3 *Latin American Science, Technology and Education Satellites*

Latin American countries are significantly involved in the science, technology and education sector. The leading nations investing in this field are Brazil, Argentina and Peru—Brazil with 19 satellites, Argentina with ten, and Peru with four satellites. Argentina's satellites have a general focus on the science and technology sector, while Peru is mainly concentrated in technology. Moreover, Brazil as the main actor in this field, has three general satellites, two satellites in astronomy research, one in the field of the study of the magnetosphere and nine satellites on technology research.

Emerging countries investing in this field, with two and three satellite capabilities are Chile, Colombia, Ecuador and Mexico. Colombia has three satellites with multidisciplinary purposes, followed by Chile and Mexico, with two satellites each in the same sector. Ecuador has three satellites for technology research. Some countries that have just started to work in this field are Costa Rica, Guatemala and Uruguay. All three countries have one satellite, Costa Rica with a general multidisciplinary satellite and Guatemala and Uruguay with technology-oriented satellites (Table 2.4).[9]

Country	Science, technology and education satellites
Argentina	**General**: 1. BUGSAT1 2. CUBEBUG 1, 2 (El Capitán Beto, Manolito) 3. MUSAT1 (VICTOR 1) 4. PEHUENSAT 1 5. SAC A 6. SAC B 7. SAC C 8. SAC D (Acquarius, ESSP 6) 9. SAC E 10. SARE-1B 1, 2, 3, 4

Table 2.4 Overview Latin American science, technology and education satellites

(continued)

---

[9]Ibid.

Table 2.4   (continued)

Country	Science, technology and education satellites
Brazil	**Astronomy, X-ray**: 1. LATTES 2. MIRAX **General/multi discipline**: 3. AST 1 4. AST 2 5. SACI 1 6. SACI 2 **Ionosphere**: 7. EQUARS 8. LATTES 9. NANOSATC-BR 2 10. SPORT **Magnetosphere**: 11. NANOSATC-BR 1 **Technology**: 12. 14BISAT (QB50 B501) 13. AESP-14 14. DOVE 15. FBM 16. ITASAT 1 17. SATEC 18. SERPENS 19. TANCREDO 1 20. UNOSAT 1
Chile	**General/multi discipline**: 1. FASAT ALFA, BRAVO 2. SUCHAI
Costa Rica	**General/multi discipline**: 1. IRAZÚ
Colombia	**General/multi discipline**: 1. FACSAT 1 2. FACSAT 2 3. LIBERTAD 1
Guatemala	**Technology**: 1. QUETZAL 1 (GUATESAT 1)
Ecuador	**Technology**: 1. NEE-01 PEGASO 2. NEE-02 KRYSAOR 3. UTE-EUSOR (Ecuador-Ute-Yuzgu)
Mexico	**General/multi discipline**: 1. AZTECHSAT 1 2. UNAMSAT A, B
Peru	**Technology**: 1. CHASQUI 1 2. POCKET-PUCP 3. PUCP-SAT 1 4. UAPSAT 1
Uruguay	**Technology**: 1. ANTELSAT

Latin America's abilities in science, technology and education are developing. Established nations and emerging countries in this sector appreciate the need for academic and other research assisted by and with satellite assets. This tendency reflects their investments in science, technology and education, supported by national or cross-regional efforts, promoting development, prosperity and stability (Fig. 2.4).

Fig. 2.4 Latin American science, technology and education capacities map

2.3 Launch Capacities in Latin America

Latin American space technology has been growing rapidly in recent years, but will likely continue to depend in the near future on major foreign powers in technology, expertise and launch capability. The region will move forward through support from the USA, Europe, China and Russia, among others.

As seen below, due to the proximity of some countries to the Equator, launch capacities in Latin American have favorable conditions for outer space launches especially into geostationary orbit. The advantage of geographical location on the Equator best utilizes the momentum of Earth's rotation, thus avoiding the need to lift considerable amounts of fuel used to change an inclined orbit to an equatorial one.[10]

The main actors that have developed launch capacities in the region are Brazil and Argentina, with orbital and suborbital launch sites. In this context, the orbital launch site of French Guiana will be discussed in the next chapter.

Some emerging countries that have been active in the sectors of satellite development on EO, communications, science and technology also have suborbital launch sites. They include Mexico, Panama and Peru, with one suborbital launch site each (Fig. 2.5).

---

[10]Logsdon, D, U.S., Brazil should act now to forge a partnership in space, https://spacenews.com/op-ed-u-s-brazil-should-act-now-to-forge-a-partnership-in-space/ (accessed 15 August 2019).

Fig. 2.5 Latin American launch capability map

2.3.1 Launch Vehicles

The most significant steps have been initiated by Brazil and Argentina. Brazil currently is continuing its efforts to develop its own launch capability, by reason of the disaster with the VLS-1.[11] Since 2007, on behalf of the Argentine Commission for Space Activities, Argentina embarked on an enterprise known as VENG to develop space vehicles, particularly satellite launch vehicles and launch services. The initial version of this rocket family (TRONADOR I)[12] was a 3.4 m high single-stage rocket successfully launched in 2007, followed by the next version, TRONADOR Ib, 6 m in height, launched in 2008. The rocket TRONADOR II is being designed to be a standalone navigation vehicle—once programmed, it will be able to reach its orbit—a major step forward for the program's technological capabilities. The rocket will have two stages. The first will work in the first two minutes of flight, until it can overcome the force of gravity, dropping after achieving the height of 100 km and falling into the ocean—in the process consuming 90% of all fuel. The remaining 10% will be used in the second stage, which will place the satellite in the predetermined planned orbit. TRONADOR II will be able to launch satellites up to 250 kg in low orbits, up to 700 km from the Earth's surface. Instead of testing with a complete vehicle, Argentina's National Space Activities Commission (CONAE) decided to have several launches to separately test the different elements of the launcher system. The TRONADOR II is designed to inject high payloads with precision into Earth's orbits for EO. All of its engines are locally developed and run on two stage liquid fuels and oxidants, also locally developed. Completed, the TRONADOR will weigh about 70,000 lb, of which 63,000 lb will be fuel. The vehicle itself, which measures just over 30 m in height by 2.5 m in diameter, weighs only 7,000 lb.

In 2008, in partnership with DLR, Brazil started developing a microsatellite launch vehicle, which resulted in the successful joint development of two drill rockets: the VS-30/Orion[13] and the VSB-30.[14] The Institute of Aeronautics and Space (IAE)[15]

---

[11] On 22 August 2003, 13h26, at the CLA, the VLS was triggered ahead of time and was ready for launch. With the premature ignition of the VLS—which was 21 m high and was to orbit two EO satellites—the tower eventually exploded resulting in the death of 21 professionals that were working there at the time. According to the final investigation report, completed by the Brazilian Air Force in February 2004, there was a "sudden start-up" of one of the four engines of the VLS, caused by a small part that started the engine, but it is not known why this detonator went off, although two hypotheses have been raised: electric current or electrostatic discharge (contact energy transfer between two bodies). The investigation commission ruled out the possibility of sabotage, gross human error or meteorological interference, but pointed to "latent failures" and "degradation of working and safety conditions".

[12] CONAE, TRONADOR, https://www.argentina.gob.ar/ciencia/conae/acceso-al-espacio/tronador (accessed 15 August 2019).

[13] AEB, PEB, VS-30, http://www.aeb.gov.br/programa-espacial-brasileiro/transporte-espacial/vs-30orion/ (accessed 15 August 2019).

[14] IAE, VSB-30, http://www.iae.cta.br/index.php/espaco/vsb-30 (accessed 15 August 2019).

[15] Instituto de Aeronáutica e Espaço (IAE).

and the DLR were looking for a transport solution for the SHEFEX-II Experiment,[16,17] the first negotiations began between the two institutions for the VLM-1 project. The VLM-1 has three solid propellant stages in its configuration: two stages with the 12 ton S50 propellant engine, innovative composite envelope motor development; and an orbiting stage with the S44 engine, spin-off of the VLS-1 project.[18] The objective is to develop a rocket intended for launching special payloads or microsatellites (up to 150 kg) in equatorial and polar or reentry orbits, with a three-stage solid propellant in its basic configuration.

Brazil is responsible for the development of the S50 and S44 propulsive systems, the backup navigation system, the launch and flight safety infrastructure and the management of project documentation. DLR is responsible for the development and qualification of the other sensitive systems of both the VLM-1 and a new VS-50 sounding rocket. The optimistic forecast of its first qualifying flight (which will take place from the Alcântara Launch Centre) is for the first half of 2020, possibly having on board a payload that could be a Technological Satellite (SATEC) developed in Brazil, or even a payload of German origin.

2.3.2 Brazil Launch Pioneer in Latin America

In the 1950s, Brazil began developing space rockets. However, only in the mid-1960s, with the beginning of the development of SONDA rockets by the Aeronautical Technical Centre, did this type of activity reach larger dimensions.[19] The SONDA rockets were a series of small sounding rockets (SONDA I, II and III), used mainly for meteorological research (Fig. 2.6).

The SONDA rockets were the first space vehicles manufactured in Brazil, launched from the Barreira do Inferno Launch Centre in the state of Rio Grande do Norte. The knowledge gained from these rockets was the basis for the creation of the engines that would compose the stages of the future Satellite Launch Vehicle (VLS-1).

From 1971 to 1994 the Brazilian Space Program was linked to the armed forces. With the foundation of AEB in 1994, the program started having a civil purpose. Even so, part of the research, such as the Aerospace Technology Centre, one of the main aerospace laboratories in the country, is still linked to military Aeronautics. Most

---

[16]MORABA,   Shefex,   https://www.moraba.de/index.php/missions/articles/shefex-79.html (accessed 15 August 2019).

[17]Brazilian Space, Operação SHEFEX II, https://brazilianspace.blogspot.com/2012/06/operacao-shefex-ii.html (accessed 15 August 2019).

[18]At a public hearing held at the Federal Senate, the Commission for Science, Technology, Innovation, Communication and Informatics, it was pointed out by the AEB President that the VLS project was cancelled.

[19]Ramos, I, Brasil e os nossos Foguetes, https://rocketsciencebr.com/2018/01/17/brasil-e-os-nossos-foguetes/ (accessed 20 August 2019).

Fig. 2.6 Brazilian space rocket. Sonda I is a two stage rocket ($1 \times$ S10-1 $+ 1 \times$ S-10-2 rocket stages) with a maximum flight altitude of 65 km, a liftoff thrust of 27 kN a total mass of 100 kg, a diameter of 11 cm and a length of 4.5 m. It was launched nine times between 1965 and 1966

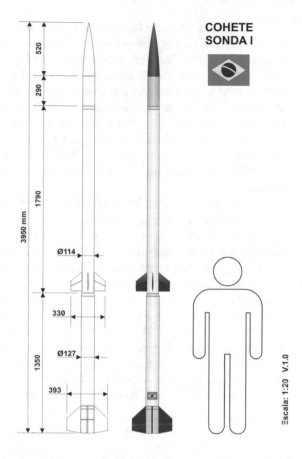

space technicians are trained at the Technological Institute of Aeronautics, another military body.

In 1979, the Brazilian Complete Space Mission (MECB)[20] was presented, with three basic aspects:

(a) the development of the Satellite Launch Vehicle (VLS-1), under the responsibility of the CTA;
(b) the development and production by the National Institute for Space Research of four satellites: SCD-1 and SCD-2 (Data Collection Satellites), and SSR-1 and SSR-2 (Remote Sensing Satellites); and
(c) the construction of all necessary ground infrastructure, including a new launch centre.[21] MECB's schedule was quite optimistic as it waited for the first VLS-1 to be released in the late 1980s.

---

[20]Missão Espacial Completa Brasileira (MECB).

[21]The mission was called "complete" because it followed a vision of total self-sufficiency, intended to cover the three basic elements of space activities: a launch pad, a launching rocket and four satellites.

In 1984, in compliance with the MECB, the Alcântara Launch Centre (CLA) was established in the state of Maranhão, under the responsibility of the military Ministry of Aeronautics. The CLA's main task was to launch the VLS-1 and the previously launched SONDA rockets. In the early 1990s, the United States invoked the MTCR treaty to stop Brazil from the construction of the VLS-1,[22,23] claiming that it was actually a missile for the launch of nuclear weapons.[24] This prevented Russia from providing technology. Since then, tensions have lessened, and Brazil has cooperated in several US-led initiatives such as LANDSAT. Through this program, in which both Brazil and Argentina participate, the United States provides a direct downlink from LANDSAT land remote-sensing satellites to ground receiving stations in cooperating countries.[25]

In 1993, the first Brazilian data collection satellite, SCD-1, fully created at National Institute for Space Research (INPE),[26] was launched by the US rocket Pegasus.[27,28] Blocked internationally by the MTCR, the VLS-1 was only able to make its first launch attempt in 1997.[29] Due to failures, the flight was halted with rocket destruction 66 min after launch. During the second launch attempt in 1999, a new failure occurred, this time in the second stage. The flight lasted about 3 min and 20 s, but had to be stopped, again with the destruction of the rocket.[30] In 2003, on the eve of the third launch attempt, the biggest disaster in the history of the Brazilian Space Program occurred. The rocket, still attached to the Mobile Integration Tower, was being inspected when one of the four first-stage engines caught fire resulting in a large explosion. The third VLS-1 prototype caught fire, taking the lives of 21 CTA technicians and engineers—one-fifth of the project team. The accident happened at 1:26 pm, when many people were still working on the rocket. The fire consumed all

---

[22]Forman, J., Sabathier, V., Faith, G., Bander, A., Toward the Heavens Latin America's Emerging Space Programs, A Report of the CSIS Americas Program and Space Initiatives, 2009, https://csis-prod.s3.amazonaws.com/s3fs-public/legacy_files/files/publication/090730_Mendelson_TowardHeavens_Web.pdf (accessed 12 August 2019).

[23]In 2010, diplomatic files leaked by WikiLeaks showed that the US Embassy in Brasilia tried to convince the Ukrainian government to not join the Cyclone 4 rocket program (Brazil and Ukraine had already spent US$1 billion on Alcântara Cyclone Space, a binational company to launch satellites from other countries) "because of our longstanding policy of not encouraging Brazil's space rocket program".

[24]Valente, R, EUA boicotaram o programa espacial do Brasil nos anos 90, https://www1.folha.uol.com.br/fsp/ciencia/fe0910201101.htm (accessed 12 August 2019).

[25]Ibid.

[26]Instituto Nacional de Pesquisas Espaciais (INPE).

[27]The launch was made from the Wallops Control Centre in Virginia, east coast of the United States. On 9 February 1993, 1 h and 15 min after takeoff, 83 km off the Florida coast and 13 km in altitude, the Pegasus rocket was released from the wing of a Nasa B52 aircraft. As predicted, the rocket freefalls for five seconds before cranking its engines into space. A few minutes later, at 11:41 am (GMT), the SCD-1 is placed in orbit of the Earth at an altitude of 750 km.

[28]INPE, Primeiro satélite brasileiro completa 25 anos em órbita e mantém operação de coleta de dados, http://www.inpe.br/noticias/noticia.php?Cod_Noticia = 4699 (accessed 12 August 2019).

[29]AEB, VLS-1, http://www.aeb.gov.br/programa-espacial-brasileiro/transporte-espacial/vls-1/ (accessed 12 August 2019).

[30]Ibid.

the fuel from the four engines, causing a huge explosion. The tower was reduced to a heap of twisted irons.[31]

After the VLS-1 disaster, no further attempts to launch it were made. At a public hearing held at the Federal Senate by the Commission for Science, Technology, Innovation, Communication and Informatics, it was pointed out by the AEB President that while the VLS was a project to validate critical technologies and train human resources, the largest launcher priority nowadays is the VLM—a Microsatellite Launch Vehicle intended for launching smaller loads than previously scheduled for the VLS.[32]

Over the years, Brazil has increased its efforts to develop launch facilities at Alcântara Launch Centre (CLA). Currently, Brazil has a basic space program that aims to produce EO satellites (to control deforestation in the Amazon, monitor climate, facilitate communications and defend and monitor borders) and reach space with the VLM.

But above all, there is a direct need for the urgent regulation of the sector to make the local industry functional and competitive. The industry needs rules that allow the authorization of launches with agility compatible with market demands, which is beyond the current legal framework of Brazil. The Brazilian Space Code is currently under discussion within the government and the Brazilian Association of Aeronautical and Space Law (SBDA) is drafting a preliminary proposal on the fundamentals of the Brazilian legal system applicable to the country's space activities, complemented by the relevant national legislation and the instruments of international law to which Brazil is committed.[33] The Space Code has chapters for the organization of space activities, space infrastructure, safety and accident investigation, the private sector, space regulation, remote sensing, small and mini satellites and coercive actions (Fig. 2.7; Table 2.5).[34]

---

[31]Bugge, A., Folha de S. Paulo, Desastre em Alcântara: Foguete nacional explode no chão e deixa 16 mortos, https://www1.folha.uol.com.br/fsp/ciencia/fe2308200301.htm (accessed 12 August 2019).

[32]Tribunal de Contas da União, Relatório TC 016.582/2016-0, https://portal.tcu.gov.br/data/files/5F/81/63/41/6F109510EE89EF851A2818A8/016.582-2016-0%20-%20Programa%20Espacial%20Brasileiro.pdf (accessed 16 August 2019).

[33]Brasil, Código Brasileiro do Espaço, https://sbda.org.br/wp-content/uploads/2019/05/Proposta-CBE.pdf (accessed 31 August 2019).

[34]Ibid.

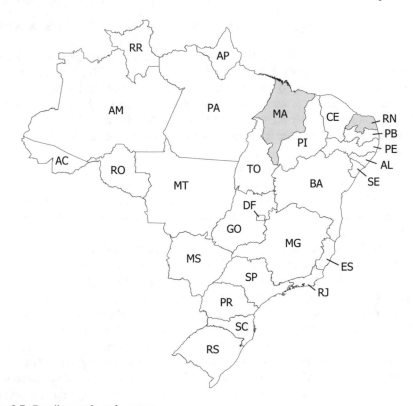

Fig. 2.7 Brazil space launch centres

Table 2.5 Main legal instruments of Brazil

Policy and legal framework	Theme
1994 Brazilian National Policy for the Development of Space Activities	Strategic goal for the development of national space technology capabilities
National Programme of Space Activities 2012–2021 (PNAE 2012–2021)	Priority actions for investment for the needs of the country. It refers to the international cooperation areas that could be developed with international partners. Defines the space missions calendar
2004 Technological Innovation Law	Instrument to foster partnerships between universities, technological institutes and industry

2.3.3 French Guiana

French Guiana is an overseas department and region of France on the North Atlantic coast of South America. Since 1981, when Belize became independent from the United Kingdom, French Guiana has been the only continental territory in the Americas that is still under the sovereignty of a European country.

In 1961, Michel Debré, Head of Government, and Pierre Guillaumat, Minister in Charge of Scientific Research, adopted the law that created the National Centre for Space Studies (CNES). At the time, 14 sites were studied for the installation of a Launch Centre by France. In 1964, the French Prime Minister Georges Pompidou chose French Guiana for the construction of the Launch Centre, which has the advantage of being located near the equator and enjoys a very wide opening in the Atlantic Ocean that considerably reduces the risks to the population.[35]

In 1968, the Guiana Space Centre in Kourou opened with the launch of the VÉRONIQUE[36] sounding rocket. In 1975, the ESA was founded and the first ARIANE-1 rocket launch took place in Kourou, on 24 December 1979.[37] The ESA set up there a decade later. Currently, the launch centre is shared by three major partners: ESA, the Arianespace Society, in which ten European countries participate, and the CNES, the French Space Agency.[38] ESA pays two thirds of the spaceport's annual budget and has also financed the upgrades made during the development of the ARIANE launchers (Fig. 2.8).

---

[35]France, Inauguration du centre spatial guyanais de Kourou avec le lancement de la fusée-sonde "VERONIQUE", https://www.gouvernement.fr/partage/8638-9-avril-1968-inauguration-du-centre-spatial-guyanais-de-kourou-avec-le-lancement-de-la-fusee-sonde (accessed 16 August 2019).

[36]Ibid.

[37]CNES, Historique des lancements de 1979 à (…), http://www.cnes-csg.fr/automne_modules_files/standard/public/p9776_631ebab2570943304156c15414e8d6b3Historique_des_lancements_au_CSG_-_2013.pdf (accessed 16 August 2019).

[38]Gazeta do Povo, EFE, Guiana abriga principal centro espacial europeu, https://www.gazetadopovo.com.br/mundo/guiana-abriga-principal-centro-espacial-europeu-8opmp7oavuoxu2w2v0lwumury/ (accessed 16 August 2019).

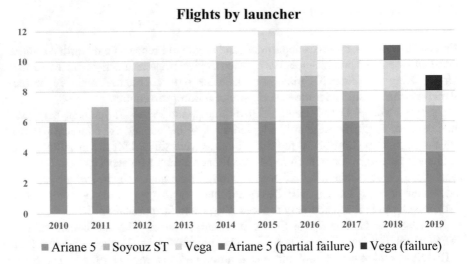

Fig. 2.8 Guiana Space Centre—space launches in Kourou

Chapter 3
Regional Space Cooperation: Latin American Experience

Contents

Abstract This chapter illustrates Latin American experience in regional cooperation in space activities, mainly by summarizing the evolution of the Space Conferences of the Americas and the South American Space Generation Advisory Council. The Conferences addressed topics such as the prospects of cooperation for development and regional integration, security and development, and promotion of technology and education for the use of space science in the Americas. Also presented is the work of the South American Space Generation Advisory Council that aims to provide opportunities to share regional perspectives on space activities, bringing together students and young professionals, and organizing workshops in different countries. Finally, other bodies that contribute to successful cooperation are noted.

© Springer Nature Switzerland AG 2020   61
A. Froehlich et al., *Space Supporting Latin America*, Studies in Space Policy 25,
https://doi.org/10.1007/978-3-030-38520-0_3

3.1   Space Conference of the Americas

Against the background of the 1982 UNISPACE II conference, in 1988 the 43rd
UN General Assembly supported the "establishment and strengthening of regional
mechanisms of co-operation and their promotion and creation through the UN
system".[1]

Within this UN framework, a hemispheric cooperation mechanism was created,
known as the Space Conference of the Americas (CEA),[2] to have an "institutional
cooperation forum that brings together the countries of the region to advance the
development of space activities and promote the application and the peaceful use of
technologies derived from them".[3]

The main objective of CEA is to create a continental forum to foster regional and
international cooperation on space matters. The forum is attended by Latin American
and international space experts, space agencies and industries, non-governmental
organizations, UNOOSA representatives, academia and observers.[4]

The Convention's aim is to achieve a convergence of positions on common interest
issues in the field of outer space, by designing regional strategies that promote the
use of space applications in support of actions and programs of social content for the
region. Regional projects are designed to support space matters in order to accelerate
social and economic development in the nations of Latin America and the Caribbean.

The government and experts of the space sector of member states work on the
design of regional and multilateral projects to strengthen education programs in
space science and technology to advance science programs and space applications
in the Latin American region. Member states participating at CEA also collaborate
to promote the application and development of space legislation by disseminating
national and regional programs for the benefit of the population, the protection of
the environment and the prevention and mitigation of natural disasters.[5]

3.1.1   Space Conference of the Americas: Evolution

The Space Conference of the Americas is one of the main southern hemispheric
forums for discussions on the space sector. Throughout the conferences held since
1990, CEA has acted as a catalyst to bring cohesion in space activities and has

---

[1] See UN General Assembly Resolution 43/56 of 6 December 1988, https://undocs.org/en/A/RES/
43/56 (accessed 4 September 2019).

[2] Conferencia Espacial de las Américas (CEA).

[3] Gobierno de México, Agencia Espacial Mexicana, Acciones y Programas, Conferencia Espacial de las Américas CEA, https://www.gob.mx/aem/acciones-y-programas/conferencia-espacial-
de-las-americas-cea (accessed 25 August 2019).

[4] Ibid.

[5] Conferencia Espacial de las Américas, Objetivos, http://www.astroscu.unam.mx/congresos/ALF/
cea/acerca.html (accessed 25 August 2019).

addressed numerous possibilities for regional and international cooperation on space issues.

Since the beginning, the forum has promoted the knowledge and the application of space science and technology in support of UNOOSA and has been supported by space agencies from around the world, such as ESA as well as UNESCO. In this sense, it is important to highlight that advancing the definition of common goals and issues of interest on space matters in the region can assist in individualizing development needs (Table 3.1).[6]

Table 3.1 CEAs

Year	Date	Theme	Outcome
1st CEA 1990	12–16 March 1990, San José, Costa Rica	Perspectives of Collaboration for Development	The activities of the first Conference were focused on the initiation of a research process for regional integration and collaboration, through space technology, with the aim of improving the quality and life of Latin American countries
2nd CEA 1993	26–39 January 1993, Santiago de Chile, Chile	Cooperation, Development and Regional Integration	The Santiago Declaration established a Pro-Tempore Secretariat to monitor and continue the results of the last CEA
3rd CEA 1996	4–8 November 1996, Punta del Este, Uruguay	Science and Technology, Sustainable Development, Environment, Education and Communications	The Declaration of Punta del Este decided to adopt a Regional Cooperation Action Plan empowering the Pro-Tempore Secretariat to follow up, with the assistance of an Intersessional Group

(continued)

---

[6]Martínez, J., Presente y futuro de la Conferencia Espacial de las Américas, https://afese.com/img/revistas/revista50/presfutesp.pdf (accessed 25 August 2019).

Table 3.1 (continued)

Year	Date	Theme	Outcome
4th CEA 2002	14–17 May 2002, Cartagena de Indias, Colombia	The Application of Space Science and Technology in the Americas and their Benefits for Civil Society	The Declaration of Cartagena de Indias and its Plan of Action recommends seeking financial resources to make viable the commitments made in this CEA. The advances of Mexico and Brazil, supported by UNOOSA, stood out for the establishment of the Regional Centre for Space Science and Technology Education for Latin America and the Caribbean (CRECTEALC)
5th CEA 2006	24–28 July 2006, Quito, Ecuador	A Regional Spatial Collaboration for Security and Human Development	The Declaration of San Francisco de Quito and its Plan of Action established the Troika as a mechanism for coordination and cooperation between the past, current and future Secretary Pro-Tempore and the need to strengthen the CEA as a process to deepen knowledge of space science and technology, boosting cooperation among members
6th CEA, 2010	15–19 November 2010, Pachuca, Mexico	Space and Development: Space Applications to serve Humanity and for the Development of the Americas	The Pachuca Declaration was an important step towards the consolidation of the CEA, with the purpose of strengthening it in each of its sessions and projecting its development as an intergovernmental and inter-agency coordination process among member countries

(continued)

Table 3.1 (continued)

Year	Date	Theme	Outcome
7th CEA, 2015	17–19 November 2015, Managua, Nicaragua	Science and Technology for Human Development in an Environment of Cooperation, Culture of Peace and Respect for International Space Law	The Conference highlighted the interest of the people in promoting the exploration and use of outer space for peaceful purposes. The Declaration of Managua and its Plan of Action highlighted the need to reaffirm the region's commitment to advance in the development of space activities

3.1.2 First CEA Conference 1990 in Costa Rica

The objective of the 1st Space Conference of the Americas in San Jose, Costa Rica in March 1990 on Prospects of Cooperation for Development was to build bases to promote cooperation in the areas of science and technology for peaceful uses of outer space among Latin American countries. 20 official delegations, four extra-continentals, and eleven space agencies participated as well as the UN.[7]

Among the outcomes, 66 projects were identified in the areas of remote sensing, tele-detection, education, applications in marine resources, planning of human settlements, meteorology, observation of climate changes in the global scales, use of high-resolution technology for agriculture, telecommunications, and a transmission network with large data banks to serve scientific research. In this regard, the hemispheric body was created to deal with space issues in the region, and the CEA conference began to be considered as part of international forums on space-related issues.[8]

3.1.3 Second CEA Conference 1993 in Chile

The 2nd Space Conference of the Americas "Cooperation, Development and Regional Integration", in Santiago, Chile, comprised decisions, commitments and

---

[7]VII Conferencia Espacial de las Américas, Nicaragua 2015, http://viiceanicaragua2015.gob.ni/?page_id=160 (accessed 19 February 2019).

[8]Martínez, J., Presente y futuro de la Conferencia Espacial de las Américas, https://afese.com/img/revistas/revista50/presfutesp.pdf (accessed 25 August 2019).

the Declaration of Santiago,[9] where the importance of continuing progress in the elaboration of norms that contribute to the development of international space law was reiterated. The Pro Tempore Secretariat (SPT)[10] was created to follow up and continue the results of the conference in order to promote cooperation in space matters between the countries of Latin America and the Caribbean region.[11]

3.1.4   Third CEA Conference 1996 in Uruguay

The 3rd CEA, in Punta del Este, Uruguay, addressed issues related to science and technology, sustainable development and the environment, education and communications. The conference led to the adoption of the "Declaration of Punta del Este"[12] and the "Action Plan of Regional Cooperation"[13] which proposed that the members of the International Support Group become the core of the Intersessional Group in support of the SPT for the follow-up of approved projects, the preparation of reports on its progress status, the promotion of work meetings, and the dissemination of information.

UNCOPUOS recommendations for further efforts for UNISPACE III to support sustainable development were also supported by the 3rd CEA. The Conference recalled the attention of the SPT to disseminate the Declaration and the Conclusions reached in the forum. It was recalled that the SPT should monitor the cooperation projects endorsed by the Conference and report on progress.[14]

---

[9]UNOOSA Doc, "Declaración de Santiago", Anexo pp. 24, http://www.unoosa.org/pdf/gadocs/A_48_20S.pdf (accessed 12 March 2019).

[10]Secretaría Pro Témpore (SPT).

[11]Agencia Espacial Mexicana, Acciones y Programas, Conferencia Espacial de las Américas, 2019, https://www.gob.mx/aem/acciones-y-programas/conferencia-espacial-de-las-americas-cea (accessed 12 February 2019).

[12]UNOOSA Doc, "Informe de la Comisión sobre la Utilización del Espacio Ultraterrestre con Fines Pacíficos", Asamblea General, Documentos Oficiales, Quincuagésimo segundo período de sesiones Suplemento No. 20 (A/52/20), "Declaración de Punta del Este", Anexos II, pp. 36, http://www.unoosa.org/pdf/gadocs/A_52_20S.pdf (accessed 15 March 2019).

[13]UNOOSA Doc, "Informe de la Comisión sobre la Utilización del Espacio Ultraterrestre con Fines Pacíficos", Asamblea General, Documentos Oficiales, Quincuagésimo segundo período de sesiones Suplemento No. 20 (A/52/20), "Plan de Acción de Cooperación Regional", Anexos II, pp. 37, http://www.unoosa.org/pdf/gadocs/A_52_20S.pdf (accessed 15 March 2019).

[14]Ibid.

3.1.5 Fourth CEA Conference 2002 in Colombia

The 4th Conference in May 2002 in Cartagena focused on furthering regional cooperation mechanisms and coordination in respect of space science development and technology for space applications for disaster management, tele-education, tele-medicine, public health, environmental protection and space law.[15]

The participating members adopted the "Declaration of Cartagena de Indias"[16] and an Action Plan[17] to further the objectives of the Declaration with the result that the SPT should promote the cooperation and coordination of the programs and projects[18] recommended for the region in the field of the environment, in line with sustainable development.[19]

3.1.6 Fifth CEA Conference 2005 in Ecuador

The 5th CEA, in Quito, Ecuador, aimed to individualize perspectives on the contribution of space to the future of security and human development. It analyzed projects in the fields of information technology and communication in education, in telesurgery, transfer of information, protection of the environment and prevention of disasters caused by natural phenomena.

Regional workshops were held on the role of the Global Navigation Satellite System (GNSS) in the environment, agriculture, telecommunications, and natural disasters and climate change management. A seminar named the "Andean Road" project was also held to establish a virtual library on international spatial legislation and the development of seminars on applications of space science for the protection of historical and archaeological cultural heritage.[20]

---

[15]Memorandum, Grupo de apoyo internacional y coordinadores de la IV CEA Cartagena de Indias, 15 to 17 de Mayo de 2002, http://www.astroscu.unam.mx/congresos/ALF/cea/Descargas/Documentos_historicos/CEAIV_Memorandum.pdf (accessed 15 March 2019).

[16]Declaración de Cartagena de Indias.

[17]UNOOSA Doc, Res. A/AC.105/L.261, "Informe de la secretaría Pro Tempore de la Cuarta Conferencia Espacial de las Américas", 2005, http://www.unoosa.org/pdf/limited/l/AC105_L261S.pdf (accessed 18 February 2019).

[18]Projects included prevention, early warning, rescue operations and mitigation of the effects of natural and anthropogenic disasters, education, research and development in science, technology, space applications and space law.

[19]Universidad de Chile, "Derecho y Política Espacial Latinoamericana, Una vía a la integración", pp. 68, http://repositorio.uchile.cl/bitstream/handle/2250/130070/Derecho-y-Politica-Espacial-Latinoamericana-Unavia.pdf?sequence=1 (accessed 15 March 2019).

[20]UNOOSA Doc, Secretaria Pro-Tempore Conferencia Espacial de las Américas, http://www.unoosa.org/documents/pdf/psa/activities/2008/colombia/presentations/9-6.pdf (accessed 10 September 2019).

The results of this conference were reflected in the "Declaration of San Francisco de Quito"[21] and the Action Plan, which highlighted the support for CRECTEALC. The proposal from Colombia to join the process with the creation of the Regional Telecentre of Geospatial Technology of the Geographical Institute Agustín Codazzi (IGAC) was welcomed to strengthen cooperation projects of the Andean Region and integrate educational actions on the continent. The SPT was recognized as a fundamental mechanism for executing the decisions of the Conference. The benefits of establishing a regional space agency were also discussed, in order to advance the use of space technology in the four axes identified for the conference.[22]

3.1.7 Sixth CEA Conference 2010 in Mexico

The 6th CEA, in Pachuca, Mexico, on "Space and Development: Space Applications at the Service of Humanity and for the Development of the Americas" aimed at innovating the traditional format structure into three distinct sectors: Intergovernmental; Industry and Services; Academia and Research.[23] The Conference discussed the challenges and prospects of space cooperation in Latin America under four themes: Space Policy, Space Law and Youth Vision; Environment and Natural Disasters and Heritage Protection; Education and Health; Technological Development, Industry and Scientific Research.[24]

During this Conference, the path towards a regional Space Agency was considered by establishing a common regional space policy for the promotion of government, industry and academic synergies in research, aerospace science and technology. The development of space law was also discussed and special attention was given to the regulation of digital resources and telecommunication in the region. The Declaration of Pachuca was adopted. In paragraph 9 it recommends that the region use remote sensing applications for water management.[25]

The Conference also developed strategies for the harmonization of the activities of CEA, with CRECTEALC.[26] The Latin America Space Agenda was developed

[21] Declaración de San Francisco de Quito.

[22] CEA, VI Conferencia Espacial de las Americas, http://www.astroscu.unam.mx/congresos/ALF/cea/cea.html (accessed 10 September 2019).

[23] Space Conference of the Americas, "VI Space Conference of the Americas Schedule", http://www.astroscu.unam.mx/congresos/ALF/cea/Descargas/PDF/VI_CEA_VI_Programa_General_Preliminar_V3_ENG.pdf (accessed 10 September 2019).

[24] CRECTEALC Doc, VI Conferencia Espacial de las Américas, Panel Informativo "Participación Institucional y de iES en materia Aeroespacial y de Telecomunicaciones", http://www.cudi.edu.mx/boletin/2010/ipn/04_S_Camacho_VI_CEA.pdf (accessed 10 September 2019).

[25] CRECTEALC Doc, VI Conferencia Espacial de las Américas, Panel Informativo "Participación Institucional y de iES en materia Aeroespacial y de Telecomunicaciones", http://www.cudi.edu.mx/boletin/2010/ipn/04_S_Camacho_VI_CEA.pdf (accessed 10 September 2019).

[26] Centro Regional de Enseñanza de Ciencia y Tecnología del Espacio para América Latina y el Caribe.

to use the CEA process to encourage the adherence of most Latin American and Caribbean countries to CRECTEALC.[27]

Within the Conference, three important events took place: "The International Aerospace Industry and Telecommunications Fair (Feria Internacional de la Industria Aeroespacial y de Telecomunicaciones)",[28] "The Latin America Youth Forum on Space Policy" and the presentation of the Mexican Space Agency.

3.1.8 Seventh CEA Conference 2015 in Nicaragua

Nicaragua hosted the 7th CEA on "Science and Technology of Space for Human Development in an Environment of Cooperation, Culture of Peace and Respect for International Space Law", with the overall objective of promoting knowledge, exchange of experience, and capacity building in space science and technology to contribute to the integral development of the American continent.

Nicaragua entered the space age with a GLONASS Ground Station, the Russian space navigation system, installed in 2017, based on an agreement with ROSCOS-MOS,[29] as well as the Chinese-produced LSTSAT-1 telecommunications satellite, which was expected to be launched in 2019.[30]

Venezuela has applied to host the 8th Space Conference of the Americas within three or four years.

3.2 Central American Association for Aeronautics and Space

On 16 June 1988, the Costa Rican Association for Space Research and Dissemination (Asociación Costarricense de Investigación y Difusión Espacial—ACIDE) was created, an organization that forms the basis of what is today the Central American Association for Aeronautics and Space (ACAE).[31] The formation of ACIDE was due to the first astronaut in Costa Rica, Franklin Chang Diaz, being particularly

---

[27]UN-SPIDER, "Conferencia especial de las américas", 2010, http://www.unspider.org/sites/default/files/Brochure_gen%20inf%20WORD%20ENG.pdf (accessed 18 February 2019).

[28]Feria Internacional de la Industria Aeroespacial y Telecomunicaciones (FIAT).

[29]Ghitis, F., A Russian Satellite-Tracking Facility in Nicaragua Raises Echoes of the Cold War, 2017, https://www.worldpoliticsreview.com/articles/22385/a-russian-satellite-tracking-facility-in-nicaragua-raises-echoes-of-the-cold-war (accessed 10 September 2019).

[30]Nicaragua has announced in 2013, that CAST has been selected to manufacture and launch the NicaSat-1 communications satellite. It will be based on the DFH-4 Bus. Launch is planned for 2019 on a CZ-3B/G2 rocket. The satellite project has apparently been renamed LSTSAT-1, https://space.skyrocket.de/doc_sdat/nicasat-1.htm (accessed 10 September 2019).

[31]Asociación Centroamericana de Aeronáutica y del Espacio (ACAE).

committed to the idea of space development. ACIDE's initial objective was to promote and develop the subject of space technology study and research in the country through courses, seminars, international exchanges, visits by NASA experts and the organization of congresses. Despite the work done, ACIDE was disbanded in the mid-1990s.[32]

In 2010, with the intention of reviving the work of this association, a group of professionals interested in the project and concerned about the abandonment of the aerospace industry in Costa Rica and Central America, decided to create the ACAE, expanding its range of actions to involve the participation of different strategic actors in society.[33]

Nowadays, the association is formed by a network of professionals who work voluntarily to forge the foundations of an aerospace industry in the region. This is done through the development of the areas of law and engineering, as well as the promotion of research, the generation of innovative projects and the promotion and dissemination of knowledge in the aerospace field. ACAE advises the Government of Costa Rica on issues related to the aerospace development strategy and national and international legislation governing this sector.

In addition, ACAE is a founding member of the National Aerospace Research and Development Council (Consejo Nacional de Investigación y Desarrollo Aeroespacial), which is responsible for deciding the strategy of the country for aerospace development. Regarding the industrial sector, the association seeks to lay the groundwork for gradually generating and developing an aerospace industry in the country and in the region, already owning about 36 national affiliated companies.[34]

ACAE's main project was the development of the first Central American satellite made in Costa Rica, IRAZU, which enabled the creation of a platform to apply space technology to climate change monitoring in the country's tropical forests. The Costa Rican Technological Institute (Instituto Tecnológico de Costa Rica) was the main development partner and was responsible for the scientific mission, as well as experts from institutions such as NASA, Ad Astra Rocket Company, TU Delft University (Netherlands), Kyushu Technological Institute (Japan) and from the University of Surrey (United Kingdom).[35]

The satellite was designed under the CubeSat standard—small 10 cm cube-shaped satellites on the side. The aluminum housing of this satellite was built by Atemisa Precision SA, a Costa Rican SME specializing in metallurgy. Also, one of the two onboard computers was built by Costa Rican engineers under the direction of TEC. The objective was to develop a space engineering project in all its stages as proof of the concept of a communication platform capable of transmitting measurements

---

[32] ACAE, Proyecto Irazú: Primer satélite de Costa Rica, https://www.acae-ca.org/proyectos/9-proyecto-irazu-primer-satelite-de-costa-rica (accessed 10 September 2019).

[33] Ibid.

[34] Ibid.

[35] Ibid.

of environmental variables from remote protected areas of Costa Rica to a data visualization centre for climate change monitoring.[36]

IRAZU aims to enable the development of skills among Costa Rican professionals and students for space technology projects, as well as addressing the country's goal of achieving carbon neutrality by serving as a platform for obtaining data on environmental variables, forests growth and carbon sequestration in the country's forests.[37]

The cubesat was launched on 2 April 2018 with the FALCON-9/DRAGON vehicle of SpaceX CRS-14, from Cape Canaveral Air Force Station, Florida, USA.[38] The satellite communicates with two earth stations: the experimental station at San Carlos, and the monitoring station on the central Campus of Costa Rica Institute of Technology in Cartago.[39]

3.3 Space Generation Advisory Council (Latin America)

Conceived at UNISPACE-III, the Space Generation Advisory Council (SGAC) has permanent observer status at the UN Committee on the Peaceful Uses of Outer Space (UNCOPUOS) and consultative status at the UN Economic and Social Council (UNECOSOC). The organization seeks to develop a pragmatic set of advice to decision makers in the space policy field, considering students and young professionals, with a commitment to the exploration of outer space.[40,41] The organization headquarters is in Vienna, Austria; registered as a non-profit organization in the United States. The council has a volunteer network of over 15,000 members in 150 countries, in six different regions with participants aged between 18 and 35 years.[42]

---

[36]Ibid.

[37]Ibid.

[38]EO Sharing EO Resources, EO Portal Directory, IRAZU CubeSat Mission, https://directory.eoportal.org/web/eoportal/satellite-missions/i/irazu (accessed 12 September 2019).

[39]Ibid.

[40]"The States participating in the 3rd UN Conference on the Exploration and Peaceful Uses of Outer Space (UNISPACE III), held in Vienna from 19 to 30 July 1999: 1. Declare the following as the nucleus of a strategy to address global challenges in the future (…) d. Enhancing education and training opportunities and ensuring public awareness of the importance of space activities: action should be taken (…) vi. To create, within the framework of the Committee on the Peaceful Uses of Outer Space, a consultative mechanism to facilitate the continued participation of young people from all over the world, especially young people from developing countries and young women, in cooperative space-related activities…", Report of the 3rd UN Conference on the Exploration and Peaceful Uses of Outer Space, 1999, http://www.unoosa.org/pdf/reports/unispace/ACONF184_6E.pdf (accessed 9 September 2019).

[41]SGAC, SGAC Creation at UNISPACE III, https://spacegeneration.org/cool_timeline/sgac-creation-at-unispace-iii (accessed 9 September 2019).

[42]SGAC, About, https://spacegeneration.org/about (accessed 9 September 2019).

Fig. 3.1 SGAC presence in Latin American countries

The Space Generation Workshops (SGW) were created with the aim to provide opportunities to share regional perspectives on space activities, bringing together students and young professionals. The aim is to promote the voice of the next generation of space leaders in each of the six regions.[43]

---

[43] Ibid.

SGAC is actively present in Latin America, but it is divided in two different regions: SGAC in South America that comprises Argentina, Bolivia, Brazil, Chile, Colombia, Ecuador, Paraguay, Peru, Uruguay, Venezuela, and SGAC in North, Central America and the Caribbean (NCAC) that comprises Barbados, Costa Rica, El Salvador, Guatemala, Jamaica, Mexico, Nicaragua, Panama and St. Lucia. (Fig. 3.1). In recent years, SGAC has promoted aerospace technology in the region and has organized several events related to space technology and applications.

Pursuant to the SGAC vision on expanding knowledge on international space policy issues, four SA-SGW have been organised in South America. The workshops are held for students, young professionals and experts in the space sector that join to share ideas and projects about space programs. Outcomes and results of the workshops are then presented to UNCOPUOS sessions annually.

3.3.1 SA-SGW Workshop 2015 in Argentina

The 1st South American Space Generation Workshop (SA-SGW) was held in Buenos Aires, Argentina on 4 and 5 May 2015. The event took place in conjunction with the 8th Argentine Congress of Space Technology (CATE)[44] on 6 to 8 May 2015. The participants collaborated with SMEs and guests to discuss the following topics: education and space outreach programs for South America; creation of a South American Space Agency; technology and research in South America; potential for Mars missions' simulations in South America.[45]

3.3.2 SA-SGW Workshop 2016 in Peru

The 2nd SA-SGW took place in Lima, Peru on 1 and 2 August 2016. The event was part of the first Peruvian Space Week and the first Latin American Astrobiology Congress (I-CLA).[46] The event was organized by SGAC, the University Pontificia Católica del Perú (PUCP)[47] and the Scientific Society of Astrobiology of Peru (SCAP).[48,49]

---

[44]Congreso Argentino de Tecnología Espacial (CATE).

[45]SGAC, South American Regional Space Generation Workshop 2015, https://archives.spacegeneration.org/index.php?option=com_content&view=article&id=1280 (accessed 11 September 2019).

[46]Congreso Latino Americano de Astrobiología (I-CLA).

[47]Pontificia Universidad Católica del Perú (PUCP).

[48]Sociedad Científica de Astrobiología del Perú (SCAP).

[49]Sociedad Científica de Astrobiología del Perú, 2019, http://astrobiologyperu.wixsite.com/grupoastrobioperu (accessed 18 February 2019).

The workshop gathered students, young professionals and new leaders in space research to promote the latest advances in space research in Latin American astrobiology.[50] The working groups focused on key issues that the space community faces in South America particularly related to science and technology. The SA-SGW discussed astrobiology studies, space research database, emerging spacefaring nations, nanosatellites, CubeSats and research.

The main objectives of this event were to strengthen the regional network of students and young professionals in the South American region, to examine and consider key questions that the regional space community is facing and to provide inputs from the next generation of space professionals as well as to enable tomorrow's space sector leaders in the South American region to have the opportunity to interact with today's space leaders and professionals in the region through cooperation with regional and international institutions.[51]

3.3.3 SA-SGW Workshop 2017 in Brazil

The 3rd SA-SGW was held in São José dos Campos, São Paulo, Brazil on 9 and 10 November 2017. The workshop gathered numerous experts from the space sector to discuss the regional perspective on current and upcoming space sector opportunities and challenges. In the same year, SGAC organized "SpaceUp Colombia" on 4 October, at the University Foundation Los Libertadores (Fundación Universitaria Los Libertadores) in Bogota, Colombia, to further connect students and young professionals from the space sector. The next meeting of SpaceUp took place in January 2018 (SpaceUp Peru) at the University of Engineering and Technology (UTEC)[52] in Lima.[53]

During the 3rd SA-SGW, working groups discussed current space topics on education, sustainability of a Mars analog research station in South America and communications protocols for a South American Space Agency.[54]

---

[50]La Pontificia Universidad Católica del Perú, "First Peruvian Space Week 2016", https://agenda. pucp.edu.pe/evento/first-peruvian-space-week/ (accessed 18 February 2019).

[51]SA-SGW Homepage, 2016, https://archives.spacegeneration.org/index.php?option=com_ content&view=article&id=1649 (accessed 18 February 2019).

[52]Universidad de Ingeniería y Tecnología (UTEC).

[53]Space Generation Advisory Council, "South American Region", 2019, https://spacegeneration. org/regions/south-america (accessed 18 February 2019).

[54]SGAC, SA-SGW 2017, São José dos Campos, São Paulo State, Brazil, https://archives. spacegeneration.org/index.php?option=com_content&view=article&id=2095&catid=931& Itemid=2938 (accessed 9 September 2019).

3.3.4   SA-SGW Workshop 2018 in Colombia

The 4th SA-SGW, "South America: A new world of opportunities for space" was held on 9 and 10 November 2018, in Bogota, Colombia. The event took place at Plaza de Artesanos, in partnership and within the framework of the XIII International Congress of Electronics, Control and Telecommunications (CIECT),[55] as well with the support of the UN Programme on Space Applications.[56] The event gathered around 1,000 delegates, consisting of bachelor's students, young professionals, experts, academia, space institutions, and industry to create a network of exchange know-how in diverse fields such as engineering, business, science, law and their connections with the future of space-based problem solving in the region.[57]

3.4   Latin American Society of Remote Sensing and Spatial Information Systems

The Latin American Remote Sensing Society (SELPER)[58] is an international, non-profit scientific society that was founded in 1980 in Ecuador. Its main objectives are: to support all activities related to remote sensing and spatial information systems in the broadest sense; encourage the improvement of its members so that they contribute positively to the benefit of the Latin American community and its institutions; and bring together all professionals and institutions that are dedicated or interested in research, development, application and operations in remote sensing and spatial information systems, always seeking cooperation between the various sectors of this activity and the exchange of information, thus making knowledge more effective.[59]

SELPER is a regional member of the International Society of Photogrammetry, Remote Sensing and Space Information Systems (ISPRS). One of SELPER's statutory activities is to promote and stimulate development, study and research in the field of remote sensing, geotechnology, geoprocessing and related areas, in a broad sense, covering the phases of data acquisition, treatment, analysis and interpretation, as well as disclosure of the information obtained.[60]

---

[55]Congreso International de Electrónica, Control y Telecomunicaciones (CIECT).

[56]Universidad Distrital Francisco José de Caldas, "Welcome to the XIII International Congress of Electronics, Control and Telecommunications (CIECT)" 2018, https://comunidad.udistrital.edu.co/ciect13eng/2018/05/10/hola-mundo-2/ (accessed 19 February 2019).

[57]Space Generation Advisory Council, "The 4th South American Space Generation Workshop", 2019, https://spacegeneration.org/sa-sgw-2018 (accessed 19 February 2019).

[58]Sociedad Latinoamericana en Percepción Remota y Sistemas de Información Espacial (SELPER).

[59]Sociedad Latinoamericana en Percepción Remota y Sistemas de Información Espacial, "¿Qúe es Selper?", https://selper.info/que-es-selper/ (accessed 27 February 2019).

[60]MundoGeo, SELPER populariza imagens de satélite, https://mundogeo.com/2006/03/01/selper-populariza-imagens-de-satelite/ (accessed 12 September 2019).

Latin American countries (Argentina, Bolivia, Brazil, Chile, Colombia, Cuba, Ecuador, Guatemala, Mexico, Paraguay, Peru, Uruguay and Venezuela) that have SELPER units are classified as "National Chapters". Countries that already have advanced development in the area of remote sensing and support initiatives of this society are called "Special Chapters" (Germany, Canada, Spain, United States, France, Netherlands and Italy).[61] Every two years SELPER International holds its Latin American Remote Sensing Symposium. The National Chapters regularly organize national meetings or symposiums on the theme in their respective countries.

SELPER International has three committees, the International Relations Committee, which is responsible for the international activities carried out by the society, the Editorial Committee, which is responsible for publishing the SELPER Magazine, and the Education Committee, which is responsible for all SELPER educational activities.[62]

The purpose of SELPER is to promote and stimulate development, study and research in the field of remote sensing, geotechnology, cartography, geodesy, photogrammetry and related areas in a broad sense, encompassing the phases of data acquisition, treatment, analysis and interpretation, as well as the respective disclosure of the information obtained; bring together all people and entities that are dedicated to or interested in the problems, development and applications of remote sensing, geotechnology, cartography, geodesy, photogrammetry and related areas, always aiming for closer cooperation between the various sectors of activity and the exchange of information, more effective data and knowledge; and interact with other technical-scientific societies, directly or indirectly involved in the field of remote sensing, geotechnology, cartography, geodesy, photogrammetry and related areas, in order to increase their performance and also promote these other societies so that there is always wide and efficient integration.[63]

3.5   Inter-American Committee of Science and Technology

The Inter-American Committee of Science and Technology (COMCyT) is under the Inter-American Council for Integral Development (CIDI) which is a body of the Organization of American States.[64] The mission of COMCyT is to contribute to the definition and execution of OAS policy on scientific, technological and innovative partnership for development. Its main goal is to coordinate, provide follow-up and evaluate the activities on partnership for development of the Organization in the sector

---

[61] SELPER Brasil, Capítulos Especiais, http://www.selperbrasil.org.br/capitulos_especiais.php (accessed 1 September 2019).

[62] SELPER Brasil, http://www.selperbrasil.org.br/ (accessed 12 September 2019).

[63] SELPER Brasil, Sobre a SELPER, http://www.selperbrasil.org.br/sobre_a_selper.php (accessed 12 September 2019).

[64] OAS, "Member States", 2019, http://www.oas.org/en/member_states/default.asp (accessed 27 February 2019).

of science and technology. The Technical Secretariat of COMCyT is the Office of Science and Technology.[65] Since it was established, several regular meetings have been held, starting from 1998, in Bariloche, Argentina, until 2013 in Washington D.C., USA.[66]

---

[65]OAS, "Who We Are", 2019, http://www.oas.org/en/about/who_we_are.asp (accessed 27 February 2019).
[66]Ibid.

Chapter 4
Latin America's Global Engagement

Contents

Abstract This chapter analyzes the current Latin American global engagement in
the space sector, by reporting on the major initiatives and cooperation that European
countries and the European Space Agency (ESA) maintain in the region. Highlighted
are the cooperation agreements between ESA and Argentina, ESA and Brazil, ESA
supporting Chile's capabilities in the space sector, and ESA and Peru cooperation on
the development of EO capacities. In addition, the quest to create a regional South
American Space Agency to support a space regional program continues to be under
discussion, with the commitment of several countries. Lastly, Latin American membership expansion in UNCOPUOS and the creation of the regional group GRULAC,
as well as the UNISPACE Conferences resulting in the creation of Regional Centres
and the support of UN-SPIDER mechanism in Latin America, are described.

4.1 Europe's Engagement in Latin America

Major initiatives, relationships and cooperation activities between the EU/ESA
and Latin America are based on two main European programs, namely the
COPERNICUS program and the GALILEO program.

In this respect, in 2018, the European Commission and three Latin American
countries (Brazil, Chile and Colombia) signed cooperation agreements to strengthen
and stimulate cooperation in EO. Under these agreements, the European Commission

© Springer Nature Switzerland AG 2020   79
A. Froehlich et al., *Space Supporting Latin America*, Studies in Space Policy 25,
https://doi.org/10.1007/978-3-030-38520-0_4

intends to grant access in these three Latin American countries to data provided by
SENTINEL satellites, using bandwidth connections between data centres. The three
cooperation agreements include reciprocity clauses that benefit both parties. In the
case of Brazil, this reciprocity clause allows COPERNICUS to obtain full, free and
open access to EO data in Brazil (CBERS satellites).[1]

Brazil, Chile and Colombia will provide technical support to the COPERNICUS
program for the calibration and evaluation of data generated by SENTINEL satellites
for Latin America and for the joint development of new applications. The parties
provide free, complete and open access to SENTINEL satellite data and satellite
information to end users.[2]

Cooperation between the EU, ESA and Latin America have made significant
progress in shaping a range of future-oriented programmes. Regarding the development of GALILEO, Europe's satellite navigation system, it offers a wide range of
independent navigation services for commercial and private users and promises to
generate new commercial services in areas such as road vehicle navigation and air
traffic control.[3]

The GALILEO Information Centre for Latin America on satellite navigation is
hosted by the "Regional Centre for Space Science and Technology Education for
Latin America and the Caribbean—CRECTEALC" which uses INPE facilities in
São José dos Campos, São Paulo, Brazil.[4] The Centre is a focal point in the region
and aims to disseminate information about the European Geostationary Navigation
Overlay Service (EGNOS) and GALILEO to increase awareness of GNSS among
main actors in the region such as decision makers, investors, institutions, regulators,
industries, service and data providers and final users; establish links between key
players and stakeholders from Latin America and Europe at different levels (e.g.
research institutes, public institutions, private sector); identify the social, economic
and operational benefits that GALILEO will bring to the users of the region; facilitate
the introduction of GALILEO services and standards in Latin America through the
development of current and emerging GNSS markets; facilitate long-term uptake of
European GNSS in the region, identify new applications that could be developed in
the region based on EGNOS/GALILEO and its services; support the regional development of GALILEO (e.g. local elements, applications, regional components, etc.)

---

[1] Agencia Brasil, Brazil inks cooperation deal on use of EU satellites, 2018, http://agenciabrasil.
ebc.com.br/en/pesquisa-e-inovacao/noticia/2018-03/brazil-inks-cooperation-deal-use-eu-satellites
(accessed 22 August 2019).

[2] Delegación de la Unión Europea en Chile, Espacio: La Comisión Europea firma importantes
Acuerdos de Cooperación con tres socios Latinoamericanos, https://eeas.europa.eu/delegations/
chile/40792/espacio-la-comisi%C3%B3n-europea-firma-importantes-acuerdos-de-cooperaci%
C3%B3n-con-tres-socios_es (accessed 22 August 2019).

[3] ESA, Space serving European Citizens, https://www.esa.int/About_Us/Welcome_to_ESA/Space_
serving_European_citizens (accessed 22 August 2019).

[4] Camacho, S., GNSS Activities of the Regional Centre for Space Science and Technology Education
for Latin America and the Caribbean, National Space-Based Positioning, Navigation, and Timing
15th PNT Advisory Board Annapolis, MD June 11–12, 2015, https://www.gps.gov/governance/
advisory/meetings/2015-06/camacho.pdf (accessed 22 August 2019).

and industrial cooperation between Latin America and European partners; support the integration of GALILEO in the regional plan for GNSS; and facilitate the establishment of an interface between the regional service provider/s and the GALILEO Operating Company.[5]

In March 2006, the Celeste Consortium was created to support GALILEO applications in Latin America. Among the Consortium's missions, there is a broad study on the use of GALILEO services in Latin America and a contribution to the Regional Plan being developed by the UN International Civil Aviation Organization (ICAO).[6]

While the Latin Consortium in which CRECTEALC participates aims to disseminate the technical and scientific scope of GALILEO through an Information Centre, the Celeste Consortium will bring together companies and institutions to provide commercial services, using GALILEO technology, which is the positioning system of the EU.[7] In 2018, the European Commission decided[8] to set up and operate two GALILEO Information Centres.[9]

In the area of cooperation agreements on the advancement of training and education for the development of human resources in Latin American countries, ESA has also played an important role. Programmes related to EO are here described. In 2002 a joint programme was established by the UN and ESA to teach advanced remote-sensing technologies to EO specialists in Chile and other Latin American countries. The UN/ESA course established in 1998 has been coordinated by UNOOSA, and ESA's ESRIN[10] facility in Frascati, Italy. The course aimed to provide national institutions a follow-up to support remote sensing applications for sustainable development activities. The programme started in 1999 and brought together government representatives and university researchers from Bolivia, Argentina, and Chile, who got together with ESA's specialists for a series of regional seminars and pilot projects for training in various areas of EO technologies.[11]

The focus of the training was interferometric techniques for synthetic aperture radar, by combining two radar images of the same spot of the Earth taken at different

---

[5]European Global Navigation Satellite Systems Agency, GALILEO Information Centre for Latin America, https://www.gsa.europa.eu/galileo/international-co-operation/galileo-information-centre-latin-america (accessed 22 August 2019).

[6]INPE, Consórcio dará suporte industrial ao Galileo na América Latina, 2006, http://www.inpe.br/noticias/noticia.php?Cod_Noticia=586 (accessed 22 August 2019).

[7]Ibid.

[8]Commission Implementing Decision adopting the 2018 work program and financing the European satellite navigation programs—C (2018) 3354 final, annex 1, 4.6.2018, pursuant to Regulation 1285/2013 of the European Parliament and of the Council on the implementation and operation of European satellite navigation systems.

[9]European Commission, Internal Market, Industry, Entrepreneurship and SMEs, Call for Proposals, Two GALILEO Information Centres in Latin America, 297/G/GRO/SAT/18/10600, GRANT PROGRAMME 2018, https://agora.mfa.gr/infofiles/Call%20for%20proposals%20br.pdf (accessed 22 August 2019).

[10]ESA Centre for EO (ESRIN).

[11]ESA Portal, "Latin America reaps development dividends with ESA EO Training", https://www.esa.int/Our_Activities/Observing_the_Earth/Latin_America_reaps_development_dividends_with_ESA_Earth_observation_training/(print) (accessed 22 March 2019).

times from different angles. The program was a success, as it was an opportunity for the region to have a more profound impact on space-related technologies on EO benefits. The outcomes of the training enabled Bolivia, Argentina, and Chile to expand their knowledge in these fields, with new advancements and discoveries to further their national projects. At the same time, this included training Latin American professionals that could give lectures on these technologies and broaden the impact of this course to other states of the region.[12]

Another important initiative is the cooperation on space law between Argentina, Brazil and ESA. The meeting on "Science, Technology and Society", in Buenos Aires from 1 to 4 November 2014, had the purpose of developing this field. The scientific communities joined to develop binational programmes (Argentina and Brazil) for scientific and technological research to lead economic growth to improve living conditions. The purpose of joining together to strengthen cooperation in the field of space law between both countries was fundamental in order to develop a framework that could have favourable conditions for space programmes development. In this context, both countries agreed on co-operating on advancing teaching and research initiatives, with universities and research centres. The Declaration of Buenos Aires on Space Law was adopted, for joint research on topics of common interest with a focus on space policy and law in Argentina, Brazil and other Latin American countries.[13]

4.1.1 Cooperation Agreements Between ESA and Argentina

Argentina remains an important regional partner for ESA in South America, since the first agreement on reception and distribution of ERS-1 and ERS-2 signed in 1997, which allowed Argentina to use the data from the ESA's EO missions ERS-1 and ERS-2.[14]

In 2002, one of the most general cooperation agreements concerning space cooperation for peaceful purposes was signed between ESA and Argentina, through which joint training courses have been set up with CONAE and ESA for regular traineeships of Argentinean students. Various workshops organized by CONAE have also been financially supported by ESA.[15]

In July 2003, Argentina also signed the International Charter on Space and Major Disasters. This charter has just been activated for Chile to support monitoring the Chaiten Volcano eruption. In 2007, the UN/Argentina/ESA Workshop on Sustainable

---

[12]Ibid.

[13]ECSL, Declaration of Buenos Aires on Space Law Cooperation, http://www.esa.int/About_Us/ Space_Law_virtual_network_with_Latin_American_countries/declaration_of_buenos_aires_on_ space_law_cooperation_English_Version (accessed 23 March 2019).

[14]ESA Portal, "ESA extiende sus lazos globales", 2016, http://www.esa.int/esl/ESA_in_your_ country/Spain/ESA_extiende_sus_lazos_globales (accessed 23 March 2019).

[15]Space Daily, ESA And Argentina Sign Extension Of Cooperation Agreement, http://www. spacedaily.com/reports/ESA_And_Argentina_Sign_Extension_Of_Cooperation_Agreement_999. html (accessed 22 March 2019).

Development in Mountain Areas of Andean Countries was held in Mendoza, on 26–30 November 2007.[16]

In 2008, the Cooperation Agreement between the Argentine Republic and ESA was renewed for five years. At the time, an ESA delegation was also in Buenos Aires to discuss the possible installation of a Deep Space Ground Station (35 m antenna) that would provide support to ESA's space exploration programme for future scientific missions, in particular ExoMars and Mars Rover.[17]

Under the auspices of the agreement, in 2012 the agency's third deep-space tracking station was inaugurated in Malargüe, near Buenos Aires. ESA support to the local space industry of Argentina continued to advance, and ESA requested a station of Argentina, due to the need to have a connection with a station with the longitudes in the southern hemisphere. The station has enabled ESA to complete global coverage together with the first and second Deep Space Stations of ESA located in Australia and Spain.[18]

In 2012, the Government of Argentina, CONAE, and ESA, organized a Workshop on Space Law on the theme "Contribution of space law to economic and social development", held in Buenos Aires from 5–8 November 2012, which provided an overview of the legal regime governing the peaceful uses of outer space, examined and compared various aspects of existing national space legislation, considered the contribution of space law to economic and social development, discussed global governance of space activities and the role of the UNCOPUOS and its subsidiary bodies in fostering international cooperation in space activities and considered the current state of university-level studies and programmes in space law and ways of enhancing the availability and development of those studies and programmes.[19]

The Space Day of 2016 initiative, organized by ESA and Argentina, reviewed past and current progress developed together and examined possible future projects. The emphasis was put on the potential of the ESA mission SAOCOM CS and the Argentine mission of two satellites SAOCOM, Satellites of Observation with Microwaves that had the role of studying soil moisture and disaster monitoring.[20] The ESA satellite would have the role of flying in formation with SAOCOM-1 to study boreal forests. The development of this project was particularly interesting for Argentina, because the data derived from these satellites would have an important role for the

---

[16]Ibid.

[17]ESA, ESA And Argentina Sign Extension Of Cooperation Agreement, 2008, https://www.esa.int/About_Us/Welcome_to_ESA/ESA_and_Argentina_sign_extension_of_Cooperation_Agreement/(print) (accessed 2 September 2019).

[18]Ibid.

[19]UN, Committee on the Peaceful Uses of Outer Space, Report on the UN/Argentina Workshop on Space Law on the theme "Contribution of space law to economic and social development" (Buenos Aires, 5–8 November 2012), http://www.unoosa.org/pdf/reports/ac105/AC105_1037E.pdf (accessed 2 September 2019).

[20]ESA Portal, ESA boosting its Argentine link with deep space, 2017, https://www.esa.int/Our_Activities/Operations/Estrack/ESA_boosting_its_Argentine_link_with_deep_space (accessed 2 September 2019).

province of Mendoza, where the Malargüe monitoring station is located. Data availability would allow them to have detailed knowledge on the use and conservation of water, that at the same time has a direct effect on the local economy, especially, in wine production in the zone.[21]

4.1.2   ESA and Brazil Cooperation Agreements

Brazil, together with Argentina, can be considered as one of the oldest Latin American partners of ESA. Collaboration has been mostly in the field of launching requirements. The first agreement between ESA and Brazil dates back to 1977.[22]

In 2002, AEB and ESA signed a Cooperation Agreement, which extended it until 2025.[23] This agreement included the National Institute for Space Research (INPE)[24] and allowed ESA and Brazil to extend their cooperation in the fields of space science, EO, telecommunications, experiments in microgravity, space systems and life science. In addition, the agreement also extends to the exchange of experts to participate in studies; the holding of joint conferences and symposiums; and the award of fellowships to enable nominees of either party to undertake training or other scientific or technical activities at the institutions proposed by the awarding party.[25]

The significant years of collaboration between both agencies are concentrated in the area of EO, also considering that Brazil's main priorities are concentrated in this domain, due to the vast size of the country and difficult access to some areas that are affected continuously by climate change episodes.

The Technical Operating Arrangement between ESA/INPE and the COPERNICUS Space Component has advanced negotiations between the European Commission and the Brazilian Ministry for Science, and Technology to access SENTINEL data from the COPERNICUS program. Brazil actively participates in fostering monitoring and CEOS Data Cubes activities for the Committee on EO Satellites (CEOS). INPE is advancing Data Cube projects with CBERS and other datasets to be able to support the Brazil Biome monitoring programme. This programme is supported with funds from the Amazonia Project, that focus on projects tackling deforestation in the country, under the framework of the reforestation programme. There are also

---

[21]ESA Portal, "ESA extiende sus lazos globales", 2016, http://www.esa.int/esl/ESA_in_your_country/Spain/ESA_extiende_sus_lazos_globales (accessed 24 March 2019).

[22]ESA Portal, "ESA and Brazil sign implementing an arrangement for Natal Tracking Station", https://www.esa.int/About_Us/Welcome_to_ESA/ESA_and_Brazil_sign_Implementing_Arrangement_for_Natal_tracking_station (accessed 26 March 2019).

[23]ESA Portal, "ESA and Brazil sign implementing an arrangement for Natal Tracking Station", https://www.esa.int/About_Us/Welcome_to_ESA/ESA_and_Brazil_sign_Implementing_Arrangement_for_Natal_tracking_station (accessed 26 March 2019).

[24]Instituto Nacional de Pesquisas Espaciais (INPE).

[25]ESA Portal, "ESA on the world stage—International agreements with Brazil, Poland and India", 2002, https://www.esa.int/About_Us/Welcome_to_ESA/ESA_on_the_world_stage__international_agreements_with_Brazil_Poland_and_India (accessed 2 September 2019).

discussions on installing an ESA GNSS sensor station at INPE facilities in Cachoeira Paulista, São Paulo.[26]

Most recently, under the Implementation Arrangement of 19 December 2018 AEB will further collaborate with ESA in setting up the use of telemetry and tracking facilities on Brazilian territory. The agreement was signed at CLBI[27] at Natal in Rio Grande do Norte, Brazil.

Cooperation agreements that have been developed between Brazil and ESA represent more than 41 years of cooperation successes. The space transportation domain and EO have been the main areas of collaboration that support Brazil in the objectives outlined in its National Space programme.[28]

4.1.3 ESA Supporting Chile's Capabilities in the Space Sector

Cooperation agreements and initiatives carried out by ESA in Chile have been developed in different fields. In 1998, a joint programme, was established by the UN and ESA, to teach advanced remote-sensing technologies to EO specialists in Chile and other Latin American countries. Its aim was to provide national institutions with follow-on support for remote sensing applications in ongoing sustainable development activities.[29]

Especially in Chile, this training allowed the Antarctic Institute of Chile (Instituto Antártico Chileno) to start a facility for interferometry processing open to scientists in the region who are interested in ice and glacier studies. Hence, the agreement signed between ESA and DLR for operations at Chile's Bernardo O'Higgins Antarctic Base played a role. The base is a receiving station for data from ESA's ERS satellites and supports German Antarctic Research.

The priority of the German Antarctic Receiving Station O'Higgins is an unrivalled Antarctic station for acquiring EO data, satellite operations, geodetic measurements, researching change processes on the Antarctic Peninsula and enabling maritime services in this part of the world.[30] At the moment, the data is recorded on magnetic tapes, which are then sent by ship. The Chilean government is currently exploring

---

[26]ESA Portal, "ESA and Brazil sign implementing an arrangement for Natal Tracking Station", https://www.esa.int/About_Us/Welcome_to_ESA/ESA_and_Brazil_sign_Implementing_Arrangement_for_Natal_tracking_station (accessed 26 March 2019).

[27]Barreira do Inferno Launch Centre (Centro de Lançamento da Barreira do Inferno).

[28]ESA Portal, "ESA and Brazil sign implementing an arrangement for Natal Tracking Station", https://www.esa.int/About_Us/Welcome_to_ESA/ESA_and_Brazil_sign_Implementing_Arrangement_for_Natal_tracking_station (accessed 26 March 2019).

[29]ESA, Latin America reaps development dividends with ESA EO training, https://www.esa.int/Our_Activities/Observing_the_Earth/Latin_America_reaps_development_dividends_with_ESA_Earth_observation_training/(print) (accessed 2 September 2019).

[30]DLR, GARS O'Higgins priority topics, https://www.dlr.de/eoc/en/desktopdefault.aspx/tabid-9476/16314_read-40008/ (accessed 2 September 2019).

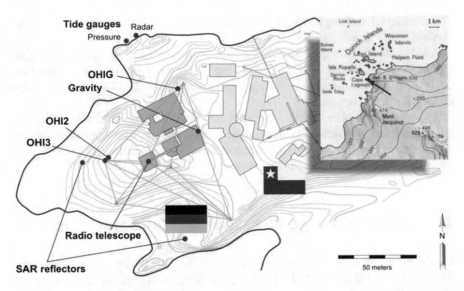

Fig. 4.1 Map showing the Chilean base Bernardo O'Higgins and the German Antarctic Receiving Station with locations of the geodetic instruments (OHIG, OHI2, OHI3). Local survey network is shown in green. Inset shows surrounding of the station. Klügel, T., Höppner, K., Falk, R., Kühmstedt, E., Plötz, C., Reinhold, A., Rülke, A., Wojdziak, R., Balss, U., Diedrich, E., Eineder, M., Henniger, H., Metzig, R., Steigenberger, P., Gisinger, C., Schuh, H., Böhm, J., Ojha, R., Kadler, M., Humbert, A., Braun, M., Sun, J., Earth and space observation at the German Antarctic Receiving Station O'Higgins, 2015, p. 8, http://gfzpublic.gfz-potsdam.de/pubman/item/escidoc:852893: 6/component/escidoc:1423050/852893.pdf (accessed 25 September 2019)

whether the Antarctic research stations could be connected with the mainland by a sea cable (Fig. 4.1).[31]

In 1999, the program began a series of regional seminars and pilot projects for training in various EO technologies, sponsored by the government and universities. In Chile, the training allowed INACH to start an interferometry processing facility open to scientists in the region who are interested in ice and glacier studies.

The university program became commercial as it also undertook a series of technology transfer projects to help private companies and government institutions to implement the technologies taught in the classroom. These projects were funded by interested organizations and brought the university closer to ESA, the French Aerospace Remote Sensing Development Group and CNES, the French space agency.[32]

---

[31] DLR, Working on the eternal ice: DLR's GARS O'Higgins Antarctic station turns 25, https://www.dlr.de/dlr/en/desktopdefault.aspx/tabid-10261/371_read-20483/#/gallery/25353 (accessed 2 September 2019).

[32] ESA, Latin America reaps development dividends with ESA EO training, https://www.esa.int/Our_Activities/Observing_the_Earth/Latin_America_reaps_development_dividends_with_ESA_Earth_observation_training/(print) (accessed 2 September 2019).

Another project that must be highlighted is the three-year project on the study of dunes over the coastal zone of central Chile. In this project, satellite images from LANDSAT and SPOT were used to assess vegetation types and the protection factor of soils in potential erosion problems. This project was financed by Chile's National Science Fund with the cooperation of France's National Centre for Scientific Research and a research project with the University of Nantes. CPR and SIG is currently working on a follow-up program to use imagery from ESA's ERS and ENVISAT satellites to analyze the physical characteristics of the dune's environments.[33]

4.1.4 ESA and Peru Cooperation on the Development of Earth Observation Capacities

From 14–18 September 2009, the Nacional Commission of Aerospace for Research and Development (CONIDA)[34] hosted a workshop on integrated applications for sustainable development in Lima. The event was organized by the UN, Peru, Switzerland and the ESA.[35] ESA's capabilities in the specific field of EO make it a key partner for Peru in applying innovative solutions to address the global challenges of sustainable development.

This supports developing countries with EO systems that can provide scientific data for the management of urban growth, water resources, and coastal areas. At the same time, data is useful to protect the forest, especially in regions such as the Amazon area. The EO Development initiative plays a crucial role in Peru by helping to improve technical assistance capabilities, capacities that are necessary to develop the skills required for reading data obtained by remote sensing.[36]

In addition, the EO4SD (EO for Sustainable Development) project on water resource management aims to provide EO demonstrations on a large-scale in Latin America (Bolivia and Peru), Africa (Sahel, Africa Horn and Zambezi), and Asia (Myanmar and Laos).[37] EO4SD is an initiative by ESA aimed at meeting longer-term strategic geospatial information needs in individual developing countries as well as international and regional development organisations. A key objective for

---

[33] Ibid.

[34] Comisión Nacional de Investigación y Desarrollo Aeroespacial (CONIDA).

[35] UNOOSA, Information Note, "UN/Peru/Switzerland/ESA/Workshop on the integrated Space Technologies Applications in the Mountain Regions of the Andean Countries", 14–18 September 2009, http://www.unoosa.org/documents/pdf/psa/activities/2009/peru/Peru_2009_information_note.pdf (accessed 27 March 2019).

[36] UNOOSA, A/AC.105/968, "Report on the UN/Peru/Switzerland/European Space Agency Workshop on Integrated Space Technology Applications for Sustainable Development in the Mountain Regions of Andean Countries", 8–18 June 2010, http://www.unoosa.org/pdf/reports/ac105/AC105_968E.pdf (accessed 28 March 2019).

[37] ESA, Water resource and management, Seminar and workshop in Peru, 2019, http://eo4sd-water.net/news/seminar-and-workshop-peru (accessed 12 September 2019).

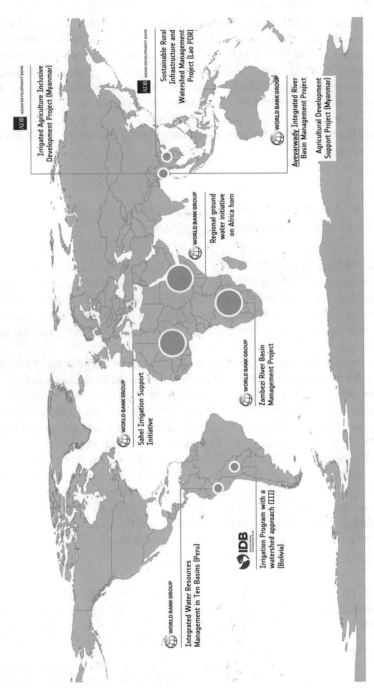

Fig. 4.2 EO4SD project worldwide by ESA. Water resource and management, http://eo4sd-water. net/ (accessed 12 September 2019)

the EO4SD on water resources management is to support the development of the required human, technical and institutional capacity to empower stakeholders with the ability to utilize the EO data and services in an independent and sustainable manner.[38]

From 27–29 March 2019, a technical training workshop at Autoridad Nacional del Agua (ANA) was held to inform participants in greater detail about the European Copernicus Satellite Monitoring Program and to provide insight to and hands-on experience with the production process for surface water monitoring and water quality monitoring—the two services demonstrated for ANA and the World Bank by EO4SD Water (Fig. 4.2).[39]

4.2 South American Space Agency Proposal

With the adoption of the UNASUR treaty of 2008,[40] a new legal instrument with international legal status integrating all the twelve south American countries, built stronger regional cooperation for development was established. UNASUR paths the way to a more integrated region, that furthers the progress achieved by MERCOSUR and CAN by enhancing current legislation. Regional integration also encourages regional cooperation in more general areas, such as politics, social, economic and human development, sustainability and includes technical areas such as aeronautics, and establishing a suitable environment for a regional space agency.

The 6th CEA, approved the Pachuca Declaration, in which the creation of a Space Technical Consulting Group brought together representatives of the national agencies or government bodies in charge of space affairs. In November 2011, the Defence Ministers of Argentina, Brazil, Bolivia, Chile, Colombia, Ecuador, Paraguay, Peru, Surinam, Venezuela and the deputy-minister of Uruguay agreed on the creation of a South American Space Agency (SASA) during a meeting of the Defense Council of UNASUR.[41]

In the same year, Brazilian Defence Minister Celso Amorim received the proposal to create a SASA from his Argentine colleague.[42] From the beginning, Brazil was against the proposal arguing that it would be a body that would yield a lot of

---

[38] Ibid.

[39] Ibid.

[40] UNASUR, South American Union of Nations Constitutive Treaty, http://www.gsdrc.org/docs/open/regional-organisations/unasur,%202008,%20establishing%20treaty.pdf (accessed 23 February 2019).

[41] Bruno Sarli, Marco Cabero, Alejandro Lopez, Josue Cardoso, Diego Jimenez, et al. South American Space Era. 66th International Astronautical Congress—IAC 2015, Oct 2015, Jerusalem, Israel. pp. 11, https://www.researchgate.net/publication/327970898_Review_of_Space_Activities_in_South_America (accessed 5 September 2019).

[42] G1, Argentina propõe ao Brasil criação de agência espacial sul-americana, 2011, http://g1.globo.com/mundo/noticia/2011/08/argentina-propoe-ao-brasil-criacao-de-agencia-espacial-sul-americana.html (accessed 5 September 2019).

bureaucracy and few results.[43] According to the Director of AEB, Brazil has also proposed the establishment of a body that congregates the presidents of the Latin American space agencies. This collegium would discuss proposed actions such as buying remote sensing images, which would be more affordable for everyone and would not require the creation of another body to spend money.[44]

At the 67th International Congress of Astronautics (IAC), held Guadalajara, Mexico, in 2016, the project of a regional agency, this time led by Costa Rica, was again discussed.[45] The SASA could enable countries to exchange more information on natural disasters and climate change, as well as strengthen actions related to the digital divide or security cooperation. The creation of SASA would reinforce the region and would work as a proactive structure to advance new regional alliances. At the same time, a regional agency that can support a space regional program would guarantee stability to the space agenda when the region faces political and economic fluctuations.[46]

In 2014, the Seminar on the Challenges of Latin America in the Space Sector[47] was held in Bariloche, Argentina to join together efforts for integration and cooperation.[48] The event was organized by CONAE[49] and all government agencies of the sector in the countries of the region were present. The Argentine statement furthered their commitment to build together a Latin American Program in the space area, establishing a defined agenda with each of the country's current developments and needs.[50] Argentina also emphasized the importance of building a SASA like ESA. Consensus on this matter was achieved by participants stating that regional cooperation is not only a possibility but a necessity.[51]

---

[43] Defesanet, AEB recua e extingue criação da Agência Espacial Latino Americana, 2015, http://www.defesanet.com.br/space/noticia/19200/AEB-recua-e-extingue-criacao-da-Agencia-Espacial-Latino-Americana/ (accessed 12 September 2019).

[44] Ibid.

[45] Xinhua, Exploran creación de agencia espacial en América Latina, 2016, http://spanish.xinhuanet.com/2016-09/30/c_135723797.htm (accessed 12 September 2019).

[46] Silva-Martinez, J., et al., Study on the development of a South American Space Agency, Conference Paper, 67th International Astronautical Congress (IAC), Guadalajara, Mexico, 23–30 September 2016, Study on the development of a South American Space Agency, https://www.researchgate.net/publication/308691747_STUDY_ON_THE_DEVELOPMENT_OF_A_SOUTH_AMERICAN_SPACE_AGENCY (accessed 24 January 2019).

[47] Desafíos del sector espacial en Latino Americano.

[48] Fibra Tecnologías de la comunicación, "Conae Arsat e Invap organizan seminario orientado al sector espacial latinoamericano", 2014, http://revistafibra.info/conae-arsat-e-invap-organizan-seminario-orientado-al-sector-espacial-latinoamericano/ (accessed 24 February 2019).

[49] Comisión Nacional de Actividades Espaciales (CONAE).

[50] INVAP, "Agencias espaciales de Latinamérica analizan integración y desarrollos conjuntos en Bariloche", http://www.invap.com.ar/es/la-empresa/acerca-de-invap/prefil-de-la-empresa.html (accessed 24 February 2019).

[51] Télam, Agencia Nacional de Noticias, "Agencias espaciales de Latinoamérica analizan integración y desarrollos conjuntos en Bariloche", 2014, http://www.telam.com.ar/notas/201412/87760-bariloche-agencias-espaciales-latinoamerica.html (accessed 22 January 2019).

Even though there is no current harmonization in space technology in the region, the creation of SASA would represent an opportunity for emerging space countries. At the same time, developing countries that currently do not have space technology capability would benefit from space technology and space transfer technologies. Nowadays, space activities have increased considerably in the region and several countries have established or are in the process of establishing space agencies or similar bodies to coordinate national space activities.

At the 3rd International Space Forum (ISF2018),[52] held in Buenos Aires, Argentina, panellists from various space agencies raised the discussion for Southern Cone countries to create the Latin American Space Agency.[53] The discussion still goes on and the main issue to be faced is that a space project in the region would require a large budget. UNASUR's ability to cooperate is affected by the disproportionate size of Brazil's economy, which accounts for about 60% of UNASUR's total economic output.[54]

4.3 Latin America and UN Space

Lately, Latin American voices at UN Space *fora* have focused on sustainability in space exploration and how the advancement of activities such as space mining and exploration would leave room for developing nations to benefit from in-space resources in an equitable way. They have also called for geostationary orbits to be classified as limited resources, a plea they have been pursuing for decades. Specifically, Latin America calls for regulations—legally-binding rules to govern outer-space activities.[55]

---

[52] Argentina, Tercer Foro Internacional del Espacio, 2018, https://www.argentina.gob.ar/noticias/tercer-foro-internacional-del-espacio (accessed 12 September 2019).

[53] Infobae, Avanza el proyecto de formar una agencia espacial Latinoamericana, 2018, https://www.infobae.com/tendencias/innovacion/2018/11/20/avanza-el-proyecto-de-formar-una-agencia-espacial-latinoamericana/ (accessed 12 September 2019).

[54] Bruno Sarli, Marco Cabero, Alejandro Lopez, Josue Cardoso, Diego Jimenez, et al. SOUTH AMERICAN SPACE ERA, 66th International Astronautical Congress—IAC 2015, October 2015, Jerusalem, Israel, p. 11, https://www.researchgate.net/publication/327970898_Review_of_Space_Activities_in_South_America (accessed 5 September 2019).

[55] Sarang, M., Op'Ed: Thoughts on UN COPUOS, https://spacegeneration.org/oped-thoughts-on-un-copuos (accessed 15 September 2019).

4.3.1 UNCOPUOS and Latin American Countries

In 1959, the UN General Assembly established UNCOPUOS as a permanent body, which had 24 members at the time, and reaffirmed its mandate in UN Resolution 1472 (XIV).[56] Since then, COPUOS has been serving as a focal point for international cooperation in the peaceful exploration and use of outer space, maintaining close contacts with governmental and non-governmental organizations concerned with outer space activities, providing for exchange of information relating to outer space activities and assisting in the study of measures for the promotion of international cooperation in those activities.[57]

Membership of UNCOPUOS has grown continuously since 1959, making COPUOS one of the largest committees in the UN. In addition to states, several international organizations, including both intergovernmental and non-governmental organizations, have observer status with COPUOS and its subcommittees.[58]

Concerning Latin American countries' membership, Argentina, Brazil and Mexico have been members since the ad hoc Committee on the Peaceful Uses of Outer Space, in 1958, shortly after the launching of the first artificial satellite. Membership grown continuously from the Group of Latin America and Caribbean Countries (GRULAC) (Table 4.1; Fig. 4.3).

---

[56]UN RES 1472 (XIV), International Co-operation in the Peaceful Uses of Outer Space, 1959, General Assembly 14th session, http://www.unoosa.org/pdf/gares/ARES_14_1472E.pdf (accessed 5 September 2019).

[57]UNCOPUOS, COPUOS History, http://www.unoosa.org/oosa/en/ourwork/copuos/history.html (accessed 5 September 2019).

[58]Ibid.

Table 4.1 GRULAC 2019 countries UNCOPUOS membership evolution

Country	Year of membership adherence	GA Resolution and GA Decision
Argentina	1958	GA Resolution 1345 (XIII)
Brazil	1958	GA Resolution 1345 (XIII)
Mexico	1958	GA Resolution 1345 (XIII)
Chile	1973	GA Resolution 3182 (XXVIII) GA Decision A/9492
Venezuela	1973	GA Resolution 3182 (XXVIII) GA Decision A/9492
Ecuador	1977	GA Resolution 32/196B GA Decision A/32/499
Colombia	1977	GA Resolution 32/196B GA Decision A/32/499
Uruguay	1980	GA Resolution 35/16 GA Decision A/35/791
Peru	1994	GA Resolution 49/33 GA Decision 49/319
Bolivia	2007	GA Resolution 62/217
Costa Rica	2012	GA Resolution 67/113 GA Decision 67/528
El Salvador	2015	GA Decision 70/518
Paraguay	2018	GA Decision 73/517
Dominican Republic	2019	GA Decision 74/408

Fig. 4.3 GRULAC countries members of UNCOPUOS in 2019

4.3.2 UNISPACE Conferences and Regional Centres

The UN, in order to disseminate information and knowledge on the existing legal framework governing activities in outer space and promote the need to ratify the UN treaties, has organized a series of conferences dedicated to promoting a greater international collaboration in outer space law and policies. Within these, the UN has facilitated the recognition of space technology benefits and the potential for socio-economic development from space activities, with three unique global conferences on the Exploration of Peaceful Uses of Outer Space—UNISPACE Conferences.[59]

The first UN Conference on outer space, UNISPACE I, was held in 1968, to raise awareness of the vast potential of space benefits for all humankind. The Conference called for increased international cooperation in space applications and created the UNOOSA Program on Space Applications, established in 1971.[60]

UNISPACE II was held from 9–21 August 1982 and addressed the concerns of how to maintain outer space for peaceful purposes and prevent an arms race in outer space as essential conditions for its peaceful exploration and use. The Conference led to strengthening of the UNOOSA Programme on Space Applications, which increased opportunities for developing countries to participate in educational and training activities in space science and technology and to develop their indigenous capabilities in the use of space technology applications, as well as leading to the establishment of regional centres for space science and technology education, which are affiliated to the UN and focus on building human and institutional capacities for exploiting the immense potential of space technology for socio-economic development.[61]

Within the objectives and the Action Plan of the Programme of Space Application, a series of missions was held to host a centre in the respective regions between 1992 until 1998. The result of these evaluation missions led to the establishment of six Regional Centres for Space and Technology Education, affiliated to the UN. They were established in India in 1995, Morocco and Nigeria in 1998, Mexico and Brazil in 2003, Jordan in 2012 and China in 2014 (Fig. 4.4; Table 4.2).[62]

The quick progress in space exploration and technology, as well as the efforts made by UNISPACE I and II, led to the UNISPACE III Conference in 1999. The committee outlined a variety of lines of action as the Conference addressed themes such as increasing developing countries access to space science, and increasing the use of space applications for human security, welfare, the environment and the protection of natural resources for the benefit of humankind. UNISPACE III created a blueprint for UNISPACE+50 and the basis for peaceful uses of outer space in the 21st century.[63]

---

[59]UNOOSA, UNISPACE Conferences, http://www.unoosa.org/oosa/en/aboutus/history/unispace. html (accessed 5 September 2019).

[60]UN, A/7285, Report of the Committee on the Peaceful Uses of Outer Space, 1968, http://www. unoosa.org/pdf/gadocs/A_7285E.pdf (accessed 20 January 2019).

[61]A/CONF.101/10, Report of the 2nd UN Conference on the Exploration and Peaceful Uses of Outer Space, 1982, UNISPACE 82, http://www.unoosa.org/res/oosadoc/data/documents/1982/aconf/aconf_10110_0_html/A_CONF101_10E.pdf (accessed 5 September 2019).

[62]UNOOSA, "Regional Centres for Space Science and Technology Education affiliated to the UN, Background, mandate and objectives", http://www.unoosa.org/oosa/en/ourwork/psa/regional-centres/index.html (accessed 22 March 2019).

[63]A/CONF.184/6, Report of the 3rd UN Conference on the Exploration and Peaceful Uses of Outer Space, 1999, UNISPACE III, http://www.unoosa.org/oosa/en/oosadoc/data/documents/1999/a/aconf.1846_0.html (accessed 5 September 2019).

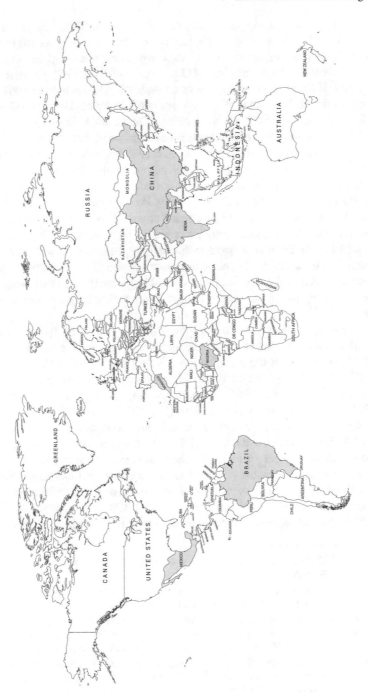

Fig. 4.4 Regional centres for space science and technology and education affiliated to the UN

Table 4.2 Regional centres for space science and technology and education affiliated to the UN

Regional centres for space science and technology education affiliated to the UN	Location	Date of establishment
CSSTEAP[a]	India	1995
CRASTE-LF[b]	Morocco	1998
ARCSSTE-E[c]	Nigeria	1998
CRECTEALC[d]	Mexico	2003
CRECTEALC	Brazil	2003
RCSSTEWA/RJGC[e]	Jordan	2012
RCCSTEAP[f]	China	2014

[a] Indian Institute of Remote Sensing (IIRS)
[b] African Regional Centre for Space Science and Technology Education (CRASTE-LF)
[c] African Regional Centre for Space Science and Technology Education (ARCSSTE-E)
[d] Regional Centre for Space Science and Technology Education for Latin America and the Caribbean (CRECTEALC)
[e] Regional Centre for Space Science and Technology Education for Western Asia (RCSSTEWA)
[f] Regional Centre for Space Science and Technology Education in Asia and the Pacific (RCCSTEAP)

Recommendations highlighted the importance of an integrated global system, primarily through international cooperation, in order to manage natural disaster mitigation through EO and it was recommended to make maximum use of existing capabilities, especially of an international nature to fill the gaps in worldwide satellite coverage. The outcomes of the Conference recognized the achievements of space science and technology to date and accordingly looked forward with confidence to achieving more significant progress in the future, through international cooperation efforts, by furthering on the achievements of the goals defined during UNISPACE III.[64]

4.3.3 Latin American and International Disaster Management

In 2006, the UN Platform for Space-based Information for Disaster Management and Emergency Response (UN-SPIDER) was established under the UN Office for Outer Space Affairs (UNOOSA) by resolution 61/110.[65] UN-SPIDER's objective is to facilitate cooperation between satellite data and information providers allowing a better information flow on disaster risk or disaster impacts between stakeholders and affected populations. Hence, UN-SPIDER builds capacities in developing countries

---

[64] Ibid.

[65] UN, Res A/RES/61/110, Resolution adopted by the General Assembly, on the report of the Special Political and Decolonization Committee (Fourth Committee) (A/61/406), http://www.un-spider.org/sites/default/files/General%20Assembly%20Resolution%2061-110.pdf (accessed 25 March 2019).

to access and use space technologies through a vast network of international partners, bringing together stakeholders from both the space and the disaster communities in order to foster an exchange of innovation and experiences.

In line with this, the UN-SPIDER Knowledge Portal works as a hub for the dissemination of information and resources. The regional network provides technical advisory support to member states and follows the guidelines established by the senior programme officer, who is responsible for planning and coordinating. UN-SPIDER has three offices, one in Vienna (Austria), in Bonn (Germany) and in Beijing (China).

UN-SPIDER has created a global network in order to foster and strengthen strategic alliances and partnerships on a global and regional scale, by Regional Support Offices (RSOs) and National Focal Points (NFP). The RSOs are regional or national centres of expertise, that are set up with an existing entity of a member state or a group of member states, that join up efforts to set up and fund an RSO. RSO offices communicate and coordinate with UN-SPIDER technical advisory support. RSO's are usually hosted by space agencies, research centres, universities or disaster management institutions that cover outreach activities and capacity building.

Within the regional office's presence in Latin America and the Caribbean, UN-SPIDER is present in Argentina, Colombia, Mexico, Panama and Trinidad (Fig. 4.5).[66]

In 2010, the University of the West Indies (UWI) and UNOOSA signed a cooperation agreement establishing the 10th UN-SPIDER RSO, being the first university to host an RSO. The office aims to advance initiatives in disaster risk reduction and disaster management within the University Disaster Management Programme. The UWI is active in developing partnerships to maximize the effectiveness of their interventions in disaster management to advance sustainable development in the Caribbean region.[67]

In 2011, the Colombian RSO was created through a cooperation agreement between the Augustin Codazzi Institute (IGAC) and UNOOSA. The office was created to promote the use of space-based information with Colombia and Latin America and the Caribbean. In line with this, IGAC has offered several training activities, three master's degrees and one doctoral program on geography. The Institute also counts with a virtual training centre that includes courses on remote sensing techniques, basic cartography and GIS.[68] IGAC develops thematic products in risk and disaster management, providing specializations related to cartography, agrology, cadastral, geography and geospatial technologies, supporting the integral development of the country.

---

[66]UN-SPIDER Portal, Regional Support Offices, http://www.un-spider.org/network/regional-support-offices (accessed 27 February 2019).

[67]UN-SPIDER Knowledge Portal, "University of the West Indies—Disaster Risk Reduction Centre", http://www.un-spider.org/network/regional-support-offices/university-west-indies-disaster-risk-reduction-centre (accessed 23 January 2019).

[68]UN-SPIDER Knowledge Portal, "Colombia Regional Support Office", http://www.un-spider.org/network/regional-support-offices/colombia-regional-support-office (accessed 23 February 2019).

Fig. 4.5 UN-SPIDER RSO presence in Latin America and the Caribbean

In 2012, a Memorandum of Understanding (MoU) between UNOOSA and CONAE established the UN-SPIDER's 13th RSO. The office furthers the advancement of knowledge and capacity building by providing cooperation and technical assistance to the organization involved in disaster prevention and mitigation in the region. In line with this, the Mario Gulich Institute for Advanced Space Studies offers a two years Master course on Space Applications for Early Warning and Response to Emergencies.[69]

CONAE also works as a platform for dissemination of know-how on space applications and organizes local and regional training activities and workshops regularly. Initiatives are specifically related to remote sensing applications, such as expertise in the areas of emergency management, human health, socio-economic activities monitoring, environmental and natural resources monitoring, climate, hydrology and oceanography, cartography geology and mining production, territorial, urban and regional planning. CONAE has outreach programs, such as the long-term educational project on space information for young students, the so-called 2Mp,[70] created under the framework of the National Space Plan 2004–2015.[71] CONAE also counts on the Córdoba Ground Station infrastructure operating since 1997, to maintain records for historical optical and SAR satellite data from the region.[72]

In 2016, UNOOSA and the Mexican Space Agency (AEM) signed a cooperation agreement to incorporate AEM as an RSO of the UN-SPIDER programme under the UNOOSA framework. Within the activities carried out by AEM, the office has carried out technical advisory missions with experts to disseminate information in Central American countries.

Within the water management centre, UN-SPIDER counts the Water for Humid Tropics of Latin America and the Caribbean (CATHALAC),[73] in Panama, providing mechanisms of cooperation and technical assistance to countries and organizations involved in disaster prevention and mitigation. CATHALAC has expertise in implementing the regional visualization and monitoring system (SERVIR). SERVIR is a Regional Visualization and Monitoring System that integrates EO and forecast models together with in situ data and knowledge for timely decision. The first SERVIR regional operational facility for the Latin America and the Caribbean region was established in 2005 through the efforts of NASA, the United States Agency for International Development (USAID), the CATHALAC, the Central American Commission for the Environment and Development (CCAD) and other partners.[74]

---

[69] UN-SPIDER Knowledge Portal, "Argentina Regional Support Office", http://www.un-spider.org/network/regional-support-offices/argentina-regional-support-office (accessed 25 January 2019).

[70] The 2Mp Program is created within the framework of the National Space Plan 2004–2015, with the aim of reaching two million children and teenagers, hence the acronym 2Mp. It is managed by the Education Unit of the National Space Activities Commission (CONAE).

[71] Programa 2MP, "Acerca de 2Mp", https://2mp.conae.gov.ar/index.php/2mp/programa-2mp (accessed 22 January 2019).

[72] CONAE Portal, Comisión Nacional de Actividades Espaciales, "Catálogo de Imágenes y Productos", https://catalogos.conae.gov.ar/catalogo/catalogo.html (accessed 19 January 2019).

[73] Centro del Agua del Tropico Humedo para America Latina y El Caribe (CATHALAC).

[74] UN-SPIDER Knowledge Portal, Water Centre for the Humid Tropics of Latin America and the Caribbean (CATHALAC), http://www.un-spider.org/network/regional-support-offices/water-center-humid-tropics-latin-america-and-caribbean-cathalac (accessed 24 January 2019).

Chapter 5
Latin America's Space Legal Framework

Contents

Abstract As in almost every area of human activity, law is a necessary tool to govern human relations in an orderly manner. The same conclusion applies at the international level, where international law regulates relations between states and other subjects. With the advent of space activities in the second half of the last century, the creation of specific legal rules to govern these activities became essential to avoid international conflicts and to guarantee legal security to the international community in the exploration and use of outer space. Latin American countries, as members of the international community of states, have contributed to some extent to the development of this legal discipline. Moreover, some Latin American states have adopted national legislation on space related matters or are in the process of elaborating such legislation to develop a national space sector or become strong space actors at the international level, or both. In this regard, the present chapter is divided in two parts: Part one refers to the participation of Latin American countries in the development and implementation of international space law. The analysis focuses on the views of Latin America on space legal issues within the UN's current discussions, Latin America's implementation of specific conventional obligations, and space law education programmes in the region. Part Two refers to national space law in Latin America. A description of the main national space and telecommunications regulations of selected Latin American countries is briefly presented.

© Springer Nature Switzerland AG 2020
A. Froehlich et al., *Space Supporting Latin America*, Studies in Space Policy 25,
https://doi.org/10.1007/978-3-030-38520-0_5

5.1   International Space Law

As one of the branches of international law, international space law has been a field of high interest by the majority of Latin American states. Negotiations and subsequent ratifications by a significant number of Latin American states of international treaties and others instruments governing activities in outer space, mostly adopted under the auspices of the United Nations (UN), have contributed to a large degree to the development of this legal field. Moreover, many Latin American countries have generally implemented the space treaties as part of their international obligations and have voluntarily engaged in respect of non-binding space instruments.

Furthermore, the necessity to understand and participate in the development and implementation of international space law has led some Latin American states, or specifically some institutions and universities of these states, to adopt education programs in space law in order to train human capital specialized in this field.

5.1.1   Latin America and the Progressive Development of International Space Law

Legal problems concerning space activities have existed since the beginning of the space race. The potential risks of certain space activities prompted the UN to create an institutional framework to discuss legal space issues in order to codify and develop international space rules. The creation in 1958 of the UN Committee on the Peaceful Uses of Outer Space (UNCOPUOS) with the aim of elaborating this new legal framework led the United Nations General Assembly (UNGA) to adopt five space treaties and five declarations on principles on different space issues that are the foundations of international space law. Some Latin American countries have contributed to the progressive development of this legal discipline.[1]

5.1.1.1   The UN and Space Law

The space race between the USSR and the USA with their first successful launch of satellites at the end of the 1950s encouraged the UNGA to adopt in 1963 the Declaration of Legal Principles Governing the Activities of states in the Exploration and Use of Outer Space,[2] a first effort to establish principles to carry out space activities in conformity with international law and the UN Charter.

In those years, a number of significant achievements in human activities and international interaction in outer space became a reality, prompting the UN to function

---

[1]The purpose of this part is to present a general overview of Latin America's participation in the development of international space law. For this reason, will not be analyzed the content of the relevant international space instruments.

[2]UNGA Resolution 1962 (XVIII) of 13 December 1963.

as a focal point for international cooperation on space issues. Within this panorama, the UN has played a fundamental role in the formulation of international legal rules to facilitate peaceful cooperation in the exploration and use of outer space.

The international law applicable to operations in space is mostly the product of UNCOPUOS. In 1958, UNGA adopted Resolution 1348 (XIII) establishing UNCOPUOS, which was originally composed of 18 Member states, including three Latin America countries: Argentina, Brazil and Mexico. Resolution 1348 (XIII) requested the Committee to report to UNGA on "the nature of legal problems which may arise in the carrying out of programmes to explore outer space".[3]

UNCOPUOS has two subsidiary bodies: the Scientific and Technical Subcommittee (STC), and the Legal Subcommittee (LSC), both established in 1961. The LSC discusses legal questions related to the exploration and use of outer space. It meets every year for two weeks and presents a final report to UNCOPUOS, which at the same time reports to UNGA's Fourth Committee (the Special Political and Decolonization Committee), which adopts an annual resolution on international cooperation in the peaceful uses of outer space. The United Nations Office for Outer Space Affairs (UNOOSA) functions as the secretariat of UNCOPUOS and keeps, among others, the UN Register of Objects Launched into Outer Space (see Sect. 5.1.2.1).

Over the years, Latin American countries have become integrated in the LSC and have participated in its work. Nowadays, there are 16 Latin American countries that are members of UNCOPUOS and its subcommittees (see Table 5.1 and Fig. 5.1).[4]

Table 5.1 Latin American countries who are members of UNCOPUOS[a]

Year of accession	UNCOPUOS Latin America's Members
1958	Argentina, Brazil and Mexico
1973	Chile and Venezuela
1977	Colombia and Ecuador
1980	Uruguay
1994	Cuba, Peru and Nicaragua
2007	Bolivia
2012	Costa Rica
2015	El Salvador
2018	Paraguay
2019	Dominican Republic

[a]Cf. Committee on the Peaceful Uses of Outer Space: Membership Evolution, UNOOSA, http://www.unoosa.org/oosa/en/ourwork/copuos/members/evolution.html (accessed 1 May 2019)

---

[3]UNGA Resolution 1348 (XIII), Question of the peaceful use of outer space, 13 December 1958, available at http://www.unoosa.org/pdf/gares/ARES_13_1348E.pdf.

[4]In June 2019, UNCOPUOS welcomed the application of the Dominican Republic for membership in the Committee and decided to recommend to UNGA that the Dominican Republic should become a member of the Committee. See Report of the Committee on the Peaceful Uses of Outer Space, Sixty-second session (12–21 June 2019), A/75/20, p. 46, paragraph 367. In December 2019,

Fig. 5.1 UNCOPUOS Latin America's Members

Moreover, many Latin American countries, even those that are not part of the LSC, have indirectly expressed their views through one or more members of UNCOP-UOS. For example, 32 of the 35 countries in the Americas (see 5.1.1.5) have jointly expressed their views as one voice through the Group of 77+ China, which usually is represented by one state in the Committee and its subsidiary bodies.

5.1.1.2 Space Treaties

Under the leadership of the LSC, five international space treaties were adopted between 1967 and 1979. The treaties are commonly referred to as the "five United Nations treaties on outer space". The foundation for the legal framework for all activities beyond the upper limits of national airspace jurisdiction is the Treaty of Principles Governing the Activities of States in the Exploration and Use of Outer Space, including the Moon and Other Celestial Bodies (the Outer Space Treaty, OST), while the other four treaties detail several specialized issues stipulated in this foundational treaty.

The space treaties are the following:

- The "Outer Space Treaty" (OST)

    - Treaty on Principles Governing the Activities of States in the Exploration and Use of Outer Space, including the Moon and Other Celestial Bodies, opened for signature on 27 January 1967 and entered into force on 10 October 1967.

- The "Rescue Agreement" (ARRA)

    - Agreement on the Rescue of Astronauts, the Return of Astronauts and the Return of Objects Launched into Outer Space, opened for signature on 22 April 1968 and entered into force on 3 December 1968.

- The "Liability Convention" (LC)

    - Convention on International Liability for Damage Caused by Space Objects, opened for signature on 29 March 1972 and entered into force on 1 September 1972.

- The "Registration Convention" (RC)

    - Convention on Registration of Objects Launched into Outer Space, opened for signature on 14 January 1975 and entered into force on 15 September 1976.

- The "Moon Agreement" (MOON)

    - Agreement Governing the Activities of states on the Moon and Other Celestial Bodies, opened for signature on 18 December 1979 and entered into force on 11 July 1984.

UNGA decided that the Dominican Republic shall become member of the Committee. See UNGA A/RES/74/82, 26 December 2019, p. 7, paragraph 38.

Table 5.2 Latin America countries engagement in UN Outer Space Treaties[a] (R = ratification, acceptance, approval, accession or succession; S = signature only)

Country	1967 OST	1968 ARRA	1972 LC	1975 RC	1979 MOON
Antigua and Barbuda	R	R	R	R	
Argentina	R	R	R	R	
Bahamas	R	R			
Barbados	R	R			
Belize					
Bolivia	S	S			
Brazil	R	R	R	R	
Chile	R	R	R	R	R
Colombia	S	S	R	R	
Costa Rica		S	S	R	
Cuba	R	R	R	R	
Dominica					
Dominican Republic	R	S	R		
Ecuador	R	R	R		
El Salvador	R	R	R		
Grenada					
Guatemala			S		S
Guyana	S	R			
Haiti	S	S	S		
Honduras	S		S		
Jamaica	R	S			
Mexico	R	R	R	R	R
Nicaragua	R	R	R	R	
Panama	S		R		
Paraguay	R				
Peru	R	R	R	R	R
St. Kitts and Nevis					
St. Lucia					
St. Vincent and the Grenadines	R	R	R	R	
Suriname					
Trinidad and Tobago	S		R		
Uruguay	R	R	R	R	R
Venezuela	R	S	R	R	R

[a]Status of International Agreements relating to activities in outer space as at 1 January 2019, UNOOSA Doc, A/AC.105/C.2/2019/CRP.3, http://www.unoosa.org/documents/pdf/spacelaw/treatystatus/AC105_C2_2019_CRP03E.pdf (accessed 2 May 2019)

Despite the fact that in 1979 there were only seven Latin American countries participating in the work of UNCOPUOS, it is remarkable that some states of the region actively participated in the discussions and drafting of space treaties. For example, Argentina and Mexico were very proactive in the travaux préparatoires of the five treaties, followed by Brazil and Chile, who participated in the drafting of the OST, the LC, the RC and MOON. Moreover, all the aforementioned countries together with Colombia, Ecuador and Venezuela participated prominently in the elaboration of the Moon Agreement,[5] showing the high interest of Latin American countries in the development of international space law (Table 5.2).

5.1.1.3 UNGA Resolutions and Declarations of Principles

Together with the UN Space Treaties, the work of the LSC led UNGA to adopt a series of declarations of principles on space-related activities. Since these principles were adopted in the form of UNGA Resolution, they do not possess a binding character, although states may agree with them. Due to the fact that some Latin America countries are members of UNCOPUOS and/or UNGA, they have supported the adoption of these principles.

The declarations of principles are the following:

- The "Declaration of Legal Principles"

  Declaration of Legal Principles Governing the Activities of states in the Exploration and Uses of Outer Space, adopted by UNGA Resolution 1962 (XVIII) of 13 December 1963.

- The "Broadcasting Principles"

  The Principles Governing the Use by states of Artificial Earth Satellites for International Direct Television Broadcasting, adopted by UNGA Resolution 37/92 of 10 December 1982.

- The "Remote Sensing Principles"

  The Principles Relating to Remote Sensing of the Earth from Outer Space, adopted by UNGA Resolution 41/65 of 3 December 1986.

- The "Nuclear Power Sources" Principles

  The Principles Relevant to the Use of Nuclear Power Sources in Outer Space, adopted by UNGA Resolution 47/68 of 14 December 1992.

---

[5]United Nations Treaties and Principles on Outer Space: Travaux Préparatoires, UNOOSA, http://www.unoosa.org/oosa/en/ourwork/spacelaw/treaties/travaux-preparatoires.html (accessed 4 May 2019).

- The "Benefits Declaration"

The Declaration on International Cooperation in the Exploration and Use of Outer Space for the Benefit and in the Interest of All states, Taking into Particular Account the Needs of Developing Countries, adopted by UNGA Resolution 51/122 of 13 December 1996.

Online records of the travaux préparatoires of the declarations of principles are scarcely available.[6] However, according to Lyall and Larsen, all declarations were adopted without vote by UNGA, except for the Broadcasting Principles, which were adopted by a majority vote.[7] That means that UNCOPUOS members of Latin America supported and agreed in general with all declarations of principles.[8]

5.1.1.4   Other International Agreements Related to Activities in Outer Space

The important role of the UN in formulating the rules of international space law has been one of the most fruitful aspects of the organization's universal character. The broad participation of states in the adoption of space treaties and declarations of principles demonstrates the harmony of the global nature of space activities, and the interest of all countries in space activities, regardless of the actual extent of their participation in space-related activities in that era.

Apart from the aforementioned treaties and declarations of principles, there are some international agreements concerning specific space activities that have been ratified by many Latin America countries, some of which took part in their negotiation.

The international space-related agreements are the following:

- 1963

Treaty Banning Nuclear Weapon Tests in the Atmosphere, in Outer Space and under Water, opened for signature on 5 August 1963 and entered into force on 10 October 1963 (NTB).

- 1974

Convention relating to the Distribution of Programme-Carrying Signals Transmitted by Satellite, opened for signature on 21 May 1974 and entered into force on 25 August 1979 (BRS).

---

[6]UNOOSA's website contains only the documents concerning the travaux préparatoires of the five UN space treaties and the Declaration of Legal Principles. See http://www.unoosa.org/oosa/en/ourwork/spacelaw/treaties/travaux-preparatoires.html (accessed 4 May 2019).

[7]Lyall Francis et al., Space Law: a Treatise, Ashgate, Dorchester, 2009, pp. 45–46.

[8]For example, the Declaration of Legal Principles was adopted by consensus by the Committee and UNGA. See Agenda Item 28, International cooperation in the peaceful uses of outer space: report of the Committee on the peaceful uses of outer space, UNGA 1280th Plenary Meeting, 13 December 1963, http://www.unoosa.org/pdf/garecords/A_PV1280E.pdf.

- 1971

  Agreement relating to the International Telecommunications Satellite Organization, opened for signature on 20 August 1971 and entered into force on 12 February 1973 (ITSO).

- 1971

  Agreement on the Establishment of the INTERSPUTNIK International System and Organization of Space Communications, opened for signature on 15 November 1971 and entered into force on 12 July 1972 (INTR).

- 1976

  Agreement on Cooperation in the Exploration and Use of Outer Space for Peaceful Purposes (INTERCOSMOS), opened for signature on 13 July 1976 and entered into force on 25 March 1977 (INTC).

- 1976

  Convention on the International Mobile Satellite Organization, opened for signature on 3 September 1976 and entered into force on 16 July 1979 (IMSO).

- 1992

  International Telecommunication Constitution and Convention, opened for signature on 22 December 1992 and entered into force on 1 July 1994 (ITU).

Table 5.3 shows that all Latin American countries have ratified the ITU. Most of them have ratified the NTB, IMSO, ITSO and BRS; INTR has been ratified by Cuba and Nicaragua, while the INTC has been ratified by Cuba only.

Table 5.3 Latin American countries' engagement in other agreements related to space and telecommunications[a] (R = ratification, acceptance, approval, accession or succession; S = signature only)

Country	1963 NTB	1974 BRS	1971 ITSO	1971 INTR	1976 INTC	1976 IMSO	1992 ITU
Antigua and Barbuda	R					R	R
Argentina	R	S	R			R	R
Bahamas	R		R			R	R
Barbados			R				R
Belize							R
Bolivia	R		R			R	R
Brazil	R	S	R			R	R
Chile	R	R	R			R	R

(continued)

Table 5.3 (continued)

Country	1963 NTB	1974 BRS	1971 ITSO	1971 INTR	1976 INTC	1976 IMSO	1992 ITU
Colombia	R	R	R			R	R
Costa Rica	R	R	R			R	R
Cuba			R	R	R	R	R
Dominica							R
Dominican Republic	R		R				R
Ecuador	R		R			R	R
El Salvador	R	R	R				R
Grenada							R
Guatemala	R		R				R
Guyana							R
Haiti	S		R				R
Honduras	R	R	R				R
Jamaica	R	R	R				R
Mexico	R	R	R			R	R
Nicaragua	R	R	R	R			R
Panama	R	R	R			R	R
Paraguay	S		R				R
Peru	R	R	R			R	R
St. Kitts and Nevis							R
St. Lucia							R
St. Vincent and the Grenadines							R
Suriname	R						R
Trinidad and Tobago	R	R	R				R
Uruguay	R		R				R
Venezuela	R		R			R	R

[a]"Status of International Agreements relating to activities in outer space as at 1 January 2019", UNOOSA Doc, A/AC.105/C.2/2019/CRP.3, http://www.unoosa.org/documents/pdf/spacelaw/treatystatus/AC105_C2_2019_CRP03E.pdf (accessed 2 May 2019)

5.1.1.5 Latin America's General Views on International Space Law

The existence of the UNCOPUOS LSC as an international forum has allowed Latin American countries to continuously express their views on international space law issues, some of which refer to legal problems that arose in the beginning of the space era, while others concern current or future legal problems. Every year, the LSC and its working groups deal with these issues through discussions, exchange of views, and debates, where practically all Latin America countries participate either individually or as a member of a group represented by a state member of UNCOPUOS.

Latin America's views expressed in the 58th LSC session held in Vienna in April 2019, as stated below, show the most recent discussions on international space legal problems. In this session, the G77 + China was represented by Egypt,[9] and the common opinions of a group of Latin America countries were expressed by Costa Rica or Chile. Moreover, several Latin America countries expressed their views in their individual capacity.

G77 + China

On one hand, the G77 + China is convinced that the use and exploration of outer space shall be carried out exclusively for peaceful purposes, for the benefit and in the interests of all countries, irrespective of their degree of economic or scientific development, and in conformity with applicable international law. The group strictly adheres to the space principles, namely the universal and equal access to outer space for all countries without discrimination; the equitable and rational use of outer space for the benefit and in the interests of all humankind; the principle of non-appropriation; the non-militarization of outer space, and international cooperation in the development of space activities.

On the other hand, the G77 + China affirms the importance of UNCOPUOS in the development of international space law. In this regard, it is of the view that UNCOPUOS, with its two subcommittees, is the only UN forum to discuss comprehensively all matters related to the peaceful uses of outer space, and that there should be greater interaction between both subcommittees. In addition, it considers that the discussions in the LSC should not lead to any measures that would limit access to outer space by nations with emerging space capabilities, especially developing countries. It also highlights that the international legal framework should take into account the concerns of all states, and that UNCOPUOS should promote legal capacity building in developing countries.

A Group of Latin American Countries

A group of Latin American countries composed of Argentina, Brazil, Bolivia, Chile, Costa Rica, Cuba, Dominican Republic, Ecuador, El Salvador, Mexico, Paraguay, Uruguay, and Venezuela, considering that the five space treaties are the fundamentals of international space law, supports the following principles: the peaceful use of outer space, universal access to outer space in equal conditions, the principle of common

---

[9]Latin America states that are currently members of the Group of 77 at the United Nations are: Antigua and Barbuda, Argentina, Bahamas, Barbados, Belize, Bolivia, Brazil, Chile, Colombia, Costa Rica, Cuba, Dominica, Dominican Republic, Ecuador, El Salvador, Grenada, Guatemala, Guyana, Haiti, Honduras, Jamaica, Nicaragua, Panama, Paraguay, Peru, Saint Kitts and Nevis, Saint Lucia, Saint Vincent and the Grenadines, Surinam, Trinidad and Tobago, Uruguay and Venezuela. See The Group of 77 at the United Nations, members, https://www.g77.org/doc/members.html (accessed 11 May 2019).

benefit of space activities, the non-appropriation principle, and international cooperation. These countries consider that the role of UNCOPUOS and its subcommittees is essential for the promotion of the benefits of space activities to all states.

Moreover, they are convinced that the exploration, use and exploitation of outer space exclusively for peaceful purposes are fundamental for the achievement of the UN 2030 Sustainable Development Goals (SDG). In this sense, the LSC should identify legal aspects to guarantee the sustainability of space activities bearing in mind the scientific and technical advances. For that reason, the group urges UNCOPUOS members to discuss this subject in order to continue the elaboration of guidelines concerning the long-term sustainability of space activities, and that would form part of the non-binding space instruments.

The Latin American group considers that the LSC should work together with the STC to discuss current space problems and update international space law. In this respect, the topic of the legal regime of outer space and global governance in space should utilise appropriate analysis that would pave the way to the elaboration of specific legal rules on the fields that could require them.[10]

Individual National Statements

Even when some states' opinions on the current legal discussions at UNCOPUOS' LSC have been presented together with the statements of the representatives of the G77 + China and/or a group of Latin American countries, Brazil, Chile, Cuba, Paraguay and Venezuela have expressed further reaching opinions in their individual capacity.

Brazil firmly considers that outer space must be used exclusively for peaceful purposes and to the benefit of all humankind, and in a sustainable manner. As modern life is unthinkable without space technology, international cooperation is essential for the sustainability of space activities, not only in the long term but also at present. In this regard, UNCOPUOS and its subcommittees are the appropriate fora to discuss all space issues because they reunite multiple space actors, discuss problems and propose solutions. Besides, the complexity of space matters makes necessary the adoption of joint commitments in order to achieve win-win partnerships. Brazil highlights that the conflicts and tensions on Earth must not be replicated in outer space because such a situation could harm all states with no exceptions. Thus, the work of the LSC should ensure the prevention of the use of outer space for military purposes. In addition, Brazil reaffirms its support and commitment to the principles enshrined in UN treaties and other international law instruments, in particular the principle that all countries are entitled to explore outer space in conditions of equality and that space should not be subject to appropriation.[11]

---

[10]Legal Subcommittee, 58th session, 1 April 2019, https://icms.unov.org/CarbonWeb/public/oosa/speakerslog/bb770449-b119-4018-a389-540c974c40ee (accessed 29 May 2019).

[11]Legal Subcommittee, 58th session, 3 April 2019, https://icms.unov.org/CarbonWeb/public/oosa/speakerslog/748c9f06-9f59-43c4-8b13-661c5cdaf741 (accessed 29 May 2019).

Chile highlights that the exploration and use of outer space shall be carried out in peaceful terms and to the benefit and interest of the international community, regardless of the economic or scientific level of development of its members, and in accordance with international law. Chile is also of the view that the exploration and use of outer space must be carried out in a sustainable manner to enable future generations to benefit from space operations. In this regard, Chile considers that it is essential to take measures to safeguard space operations, such as the creation of mechanisms for the elimination or mitigation of risks and unwanted effects for space navigation. Chile reaffirms the respect of the principles of universal and equitable access to space, non-appropriation, the use of outer space as common heritage of humankind, the non-militarization of outer space, and the sustainability of space activities. For Chile, international cooperation is fundamental for the control of spaceships and safe navigation, the identification and mitigation of space debris, the development of international law, and capacity building. It also manifests its will to contribute in the elaboration of measures concerning the sustainability of space operations and the safety of spaceships navigation, namely at the end of their useful life. Concerning space law education, Chile considers that the Education Curriculum on Space Law[12] elaborated by the LSC must be updated to incorporate new subjects, such as the definition and delimitation of outer space, the status and use of the GEO, the control and safety maintenance of spaceships in outer space, among other current legal problems.[13]

Cuba reaffirms the importance of the exclusively peaceful use of outer space in equal conditions for all states, without discrimination and according to the established technical and legal norms. Cuba considers that outer space must be preserved as the common heritage of humankind. International space legal norms should be strengthened and updated to address current practical problems and in a transparent manner. For Cuba, the LSC is the appropriate forum to discuss these issues and to promote the progressive development of international space law. Cuba expresses its concern on the threat of an arms race in outer space. Militarization of outer space and emplacement on it of all kind of arms must be prevented so as to guarantee its rational and peaceful use to the benefit of current and future generations. In this regard, Cuba proposes the adoption of an international treaty that prohibits the emplacement of arms in outer space, especially nuclear arms, as the only way to avoid the militarization of outer space. Cuba also highlights that the participation of developing countries in the work of UNCOPUOS and its subcommittees should be strengthened. Equal access to benefits of space technology and its applications must be promoted to attain the SGD's 2030.[14]

---

[12]Education Curriculum on Space Law, United Nations, March 2014, http://www.unoosa.org/res/oosadoc/data/documents/2014/stspace/stspace64_0_html/st_space_064E.pdf (accessed 29 May 2019).

[13]Legal Subcommittee, 58th session, 2 April 2019, https://icms.unov.org/CarbonWeb/public/oosa/speakerslog/16b21e6f-c01f-4d0f-bb51-80ea5b718498 (accessed 29 May 2019).

[14]Legal Subcommittee, 58th session, 3 April 2019, https://icms.unov.org/CarbonWeb/public/oosa/speakerslog/748c9f06-9f59-43c4-8b13-661c5cdaf741 (accessed 29 May 2019).

Paraguay recognizes the benefits derived from the exclusive peaceful use of outer space, as well as its sustainable use for the benefit of all humankind. For this country, international and regional cooperation is essential for the development of space capacities, and that the needs of developing countries should always be taken into account. Paraguay highlights the importance of space law as the fundamental pillar for the design of space policies that guarantee the responsible, safe and sustainable use of space technology applications, the strengthening of international cooperation, as well as the promotion of transparent and reliable relations.[15]

Venezuela considers that outer space is common heritage of humanity and that it belongs to all states in equal conditions, as affirmed by its national constitution. Venezuela underlines the necessity of interaction between the LSC and the STC to avoid out-dated legal norms. To this end, it is essential to assess and identify new potential legal norms on different space issues. Besides, the possibility of revision of the current space treaties should be considered. However, new space legislation must strengthen the core principles of space law, namely the peaceful use of outer space, equal access to outer space without discrimination and in equitable conditions, non-appropriation of outer space and celestial bodies, non-militarization of outer space, and international cooperation. Moreover, exploitation of outer space should be done with the sole purpose of achieving peaceful conditions for life on our planet. Capacity building in favour of developing countries should be also promoted.[16]

5.1.1.6 Latin America's Views on Specific Legal Space Issues

Apart from the views of Latin America's countries on international space law in its general aspects, they have also expressed their views at the LSC on specific space legal issues, such as the definition and delimitation of outer space, the character and utilization of the geostationary orbit, the use of nuclear power sources in outer space, legal mechanisms relating to space debris mitigation, legal aspects of space traffic management, application of international law to small-satellite activities, and potential legal models for activities in the exploration, exploitation and utilization of space resources.

The Definition and Delimitation of Outer Space

The G77 + China is of the view that the topic of the definition and delimitation of outer space must be kept on the agenda of the LSC and that more work should

---

[15]Legal Subcommittee, 58th session, 4 April 2019, https://icms.unov.org/CarbonWeb/public/oosa/speakerslog/d7e40e26-fc88-45b9-8da1-bceabe9f0137 (accessed 29 May 2019).
[16]Ibid.

be done in this direction to reach a consensus. The group highlights that the legal regimes governing airspace and outer space are different.[17]

More specifically, Argentina, Brazil, Bolivia, Chile, Costa Rica, Cuba, Dominican Republic, Ecuador, El Salvador, Paraguay, Uruguay and Venezuela consider that the issue of the definition and delimitation of outer space has created uncertainty regarding the applicability of space law and aeronautic law both nationally and internationally. Concepts such as territorial sovereignty and common heritage of mankind are present in the international debate and are some reasons why the subject must be kept on the UNCOPUOS agenda. The lack of progress to reach a consensus on the matter does not mean that the discussion must be interrupted. These countries consider as significant the participation of the International Civil Aviation Organisation (ICAO)[18] to find a solution to the matter.[19]

Colombia considers it important to safeguard the appropriate application of air law and space law activities both nationally and internationally. The definition would give clarity to states and other space actors, not only in terms of placing their own satellites into outer space but also in terms of scientific and commercial suborbital flights. Besides, this legal framework would help to clarify international responsibility and sovereignty of states. UNCOPUOS and its subcommittees must recognise that there is a grey area between air and space where a specific legal framework for suborbital flights must be implemented.[20]

Venezuela agrees that it is necessary to establish clear legal rules to distinguish the delimitation of airspace and outer space due to accelerated technological advancements, the increased activities of states and private actors in outer space, and the commercialisation of outer space. Although it recognises that it is difficult to reach a solution that satisfies all states on the matter, Venezuela considers that the subject must be kept on the agenda of the LSC.[21]

The Character and Utilisation of the Geostationary Orbit

Concerning the character and utilisation of the geostationary orbit (GEO), the G77 + China supports the work of the LSC for the development of adequate mechanisms to ensure equitable access to the GEO. It states that the GEO should be used

[17]G77 + China Statement during the 58th session of the Legal Subcommittee of the United Nations Committee on the Peaceful Uses of Outer Space, from 1 to 12 April 2019, delivered by H.E. Mr. Omar Amer Youssef, Ambassador, Permanent Representative of Egypt, https://www.g77.org/vienna/OOSAAPR19.htm (accessed 11 May 2019).

[18]ICAO is a UN specialized agency, established by states in 1944 to manage the administration and governance of the Convention on International Civil Aviation (Chicago Convention), See ICAO's official website at https://www.icao.int/about-icao/Pages/default.aspx.

[19]Legal Subcommittee, 58th session, 4 April 2019, https://icms.unov.org/CarbonWeb/public/oosa/speakerslog/d7e40e26-fc88-45b9-8da1-bceabe9f0137 (accessed 11 May 2019).

[20]Ibid.

[21]Legal Subcommittee, 58th session, 3 April 2019, https://icms.unov.org/CarbonWeb/public/oosa/speakerslog/3da65879-4237-4308-96cb-0c834270a5c0 (accessed 11 May 2019).

rationally and made available to all states, taking into consideration the needs and interests of developing countries and the geographical position of certain countries, and bearing in mind the processes of the ITU and other UN norms and instruments. Moreover, these countries highlight respect for the principle of non-appropriation of the GEO by any means, and that its utilisation should be governed by applicable international law and in accordance with the principle of non-appropriation of outer space.[22] The statement of the G77 + China is shared more specifically by Argentina, Brazil, Bolivia, Chile, Costa Rica, Cuba, Dominican Republic, Ecuador, El Salvador, Paraguay, Uruguay and Venezuela, who also consider that the issue should be kept on the LSC agenda.[23]

Colombia also stresses that the GEO is a limited natural space resource and that equal access must be guaranteed to this resource, bearing in mind the specific needs and interests of developing countries and their geographical situation. It is considered necessary that the ITU establishes a legal regime to guarantee equal access to states to orbital positions to address the needs of developing countries. Therefore, Colombia proposes that the ITU should modify the coordination procedures established in the Radiocommunications Regulations and highlights that the current legal regime on the exploitation and utilisation of the GEO, that gives priority to the *first come first serve* principle, is egalitarian but not equitable because it offers more opportunities to countries with higher financial and technical capacities.[24] The position of Colombia reflects to some extent its historical claims on the GEO expressed in the Declaration of Bogota of 1976 and other statements (see Table 5.4) regarding the necessity of recognizing certain rights in this orbit not only to the equatorial states but also to developing countries, and guaranteeing a rational and equitable access to the GEO.

Meanwhile, Mexico considers that the competent international body to address the issue is not UNCOPUOS but the ITU. Although Mexico recognises that the ITU let UNCOPUOS address the political issues of the utilisation of GEO, the subject cannot be addressed in this body because the competent authority on orbits and associated frequencies is the ITU.[25] In this sense, Mexico recommends that LSC members verify the content of the ITU's instruments concerning the rights of states

---

[22]G77 + China Statement during the 58th session of the Legal Subcommittee of the United Nations Committee on the Peaceful Uses of Outer Space, from 1 to 12 April 2019, delivered by H.E. Mr. Omar Amer Youssef, Ambassador, Permanent Representative of Egypt, https://www.g77.org/vienna/OOSAAPR19.htm (accessed 11 May 2019).

[23]Legal Subcommittee, 58th session, 4 April 2019, https://icms.unov.org/CarbonWeb/public/oosa/speakerslog/d7e40e26-fc88-45b9-8da1-bceabe9f0137 (accessed 11 May 2019).

[24]Ibid. Colombia's proposal takes into consideration three situations: (1) in case of comparable requests to access the orbit-spectrum resource by a country that has already accessed and a country that has not accessed to it, the country that has not accessed should have priority without implementing the coordination process; (2) in case of comparable requests to access the orbit-spectrum resource by a developing country and a developed country, the developing country should have priority without implementing the coordination process, and (3) in case of comparable requests made by two developed countries, the *first come first served* principle should be applied.

[25]Mexico declares that the concept of "equal access to outer space and celestial bodies by all states" stated in article 1 of the 1967 OST is different to the concept of "equity" stipulated in article 45

Table 5.4 Bogota Declaration of 1976

Bogota Declaration of 1976
In a conference held in Bogota, Colombia, in 1976, seven equatorial states: Colombia, Congo, Ecuador, Indonesia, Kenya, Uganda, and Zaire (now the Democratic Republic of the Congo), with Brazil as an observer, adopted a Declaration in which they asserted rights in the GEO above their respective territories. The Declaration affirms that the GEO is a limited natural resource over which the equatorial states exercise direct sovereignty. Accordingly, the placing of a satellite in a geostationary orbit above one of the territories of these states requires prior and express authorisation by the subjacent State. The Declaration also mentions that there is no definition of outer space, that the OST does not regulate the GEO, and that the non-appropriation principle does not apply to it.[a]

Signatories of the Bogota Declaration of 1976

The Bogota Declaration has no received support by the international community. Many members of UNCOPUOS rejected the declaration arguing that the GEO should be maintained in the benefit of all humankind and that the GEO is subject to the legal regime of the OST.[b] In addition, "no space-competent State has accepted its validity or complied with its requirements for permission to place a satellite in a geostationary orbital slot claimed by an equatorial state to be under its jurisdiction".[c]

Ecuador and Colombia, the only Latin America's signatories of the Declaration, have modified their respective claims over the years. Ecuador has now recently issued a statement on the subject, although it could be suggested that it shares the common view of the G77 + China expressed in UNCOPUOS. Regarding Colombia, this State currently claims the acknowledgment of certain rights but at the same time it agrees on the necessity of coordinating the use of the GEO between all states.[d]

[a]Cf. Lyall Francis et al., op. cit., pp. 253–256
[b]Constitutional Court of the Republic of Colombia, 23 March 2004, Judgment C-278/04 http://www.corteconstitucional.gov.co/relatoria/2004/C-278-04.htm (accessed 17 May 2019)
[c]Cf. Lyall Francis et al., op. cit., p. 255
[d]Constitutional Court of the Republic of Colombia, 23 March 2004, Judgment C-278/04 http://www.corteconstitucional.gov.co/relatoria/2004/C-278-04.htm (accessed 17 May 2019)

in the use of GEO to consider whether the subject must be kept on the LSC's agenda or be transferred to the ITU.[26]

The Use of Nuclear Power Sources in Outer Space

Chile is of the opinion that the use of nuclear power sources in outer space should be limited to the maximum and must be in conformity with international law, namely the OST, the Nuclear Non Proliferation Treaty, as well as the instruments and safeguard norms of the International Atomic Energy Agency (IAEA) in order to guarantee safety, harmlessness, and the sustainability of outer space. Chile agrees with UNGA Resolution 71/90 (2016) which considers that states should bear in mind the problem of collisions of space objects, in particular those with nuclear power sources, and asks for follow up on national investigations on the improvement of technology for the surveillance of space debris, and the necessity of compilation and diffusion of information on the matter. Chile does not pretend to be unaware that in some situations it might be necessary to use nuclear power sources when there are no other efficient power sources, for example, in the exploration of deep space. However, space activities that use nuclear power must be duly controlled and tracked. Chile also considers that international cooperation is the most appropriate tool for the diffusion of appropriate and affordable strategies to reduce to the minima the effects of space debris, in particular those containing nuclear power sources.[27]

For Mexico, the use of nuclear power sources is necessary in deep space missions. However, since the subject is currently being discussed by the STC, Mexico considers that the issue must be addressed by the LSC once the STC has completed its work.[28]

Legal Aspects of Space Traffic Management

Brazil highlights the importance of the LSC in negotiating the issue of space traffic management because it is a substantive contribution to the preservation of the long term sustainability of outer space activities. Besides, the exchange of information on space objects constitutes a transparency and confidence building measure. Thus, Brazil suggests that this subject must be kept on the LSC's agenda.[29]

Mexico manifests its concern about space traffic management in outer space, although this issue must not obviate airspace. Technology progress must go hand in

---

of the ITU Constitution, Legal Subcommittee, 58th session, 4 April 2019, https://icms.unov.org/CarbonWeb/public/oosa/speakerslog/d7e40e26-fc88-45b9-8da1-bceabe9f0137 (accessed 11 May 2019).

[26]Ibid.

[27]Legal Subcommittee, 58th session, 10 April 2019, https://icms.unov.org/CarbonWeb/public/oosa/speakerslog/c64b59c6-ff6e-4748-8d50-18ed22a6dda0 (accessed 13 May 2019).

[28]Ibid.

[29]Legal Subcommittee, 58th session, 2 April 2019, https://icms.unov.org/CarbonWeb/public/oosa/speakerslog/16b21e6f-c01f-4d0f-bb51-80ea5b718498 (accessed 13 May 2019).

hand with the respect of legal rules provided by international space related treaties, such as the rules concerning the registration of space objects and the provisions established in the ITU's Regulations. As sustainability of space activities is fundamental, the functioning of space operations must be done with the best conditions so as to avoid harmful interference due to the proliferation of space objects, namely in Low Earth Orbit (LEO).[30] Mexico considers that UNCOPUOS must identify the problematics related to the sustainability of space activities bearing in mind the impeding saturation of objects in LEO, the proliferation of satellite mega constellations, and the increase of space debris. Moreover, Mexico supports the continuation of UNCOPUOS' work related to the safety and security of space activities.[31]

Application of International Law to Small-Satellites Activities

The number of small-satellites launched into outer space has increased exponentially. The LSC decided to discuss the legal implications of small-satellites due to their particular characteristics. However, practically all Latin America countries consider that there should not be a legal framework different from the current international law applicable to all space objects.

For the G77 + China, the role of space objects, regardless of their size, in the socio-economic development of states is essential. For this reason, it is unnecessary for UNCOPUOS to create an ad hoc legal regime or any other mechanisms on small-satellites, which might impose limitations on designing, building, launching and use of space objects by developing countries.[32]

Mexico expresses its concern about the potential risk of the proliferation of space debris created by the use of small-satellites. For Mexico, perhaps the only legal issue related to small satellites is that of their registration. Mexico wants to know whether small-satellites that are part of mega constellations must be registered individually or jointly. Nevertheless, it considers that the LSC has not much to do on the matter, because international law does not distinguish among space objects, whereby small-satellites should be treated as any other space object. For Mexico, it is preferable that the STC addresses the issues of small-satellites.[33]

---

[30]Ibid.

[31]Legal Subcommittee, 58th session, 3 April 2019, https://icms.unov.org/CarbonWeb/public/oosa/speakerslog/748c9f06-9f59-43c4-8b13-661c5cdaf741 (accessed 13 May 2019).

[32]G77 + China Statement during the 58th session of the Legal Subcommittee of the United Nations Committee on the Peaceful Uses of Outer Space, from 1 to 12 April 2019, delivered by H.E. Mr. Omar Amer Youssef, Ambassador, Permanent Representative of Egypt, https://www.g77.org/vienna/OOSAAPR19.htm (accessed 14 May 2019).

[33]Legal Subcommittee, 58th session, 4 April 2019, https://icms.unov.org/CarbonWeb/public/oosa/speakerslog/d7e40e26-fc88-45b9-8da1-bceabe9f0137 (accessed 13 May 2019).

Legal Mechanisms Relating to Space Debris Mitigation

All Latin America countries express concern about the proliferation of space debris and the challenges posed by its removal. For the G77 + China, it is important that all states register their space objects launched into outer space. These countries consider that no object should be removed without prior consent or authorization of the Registering State. Moreover, the group encourages the work of UNCOPUOS Members to find a definition of space debris. Concerning the responsibility of removing space debris, the group considers that the actors that were responsible for creating space debris should be most involved in their removal. In other words, the group encourages states to take common but differentiated responsibilities in the removal activities. In addition, international cooperation should also be promoted to face this problem.[34]

Argentina, Brazil, Bolivia, Chile, Costa Rica, Cuba, Dominican Republic, Ecuador, El Salvador, Paraguay, Uruguay and Venezuela are of the view that the increase of new space actors has the consequence of the creation of more space debris, a phenomenon that puts in danger not only the continuity of satellite missions but also human life on Earth and in outer space. Regulation for space debris mitigation is a pressing need to guarantee the sustainability of space activities. The UN Space Debris Mitigation Guidelines are a significant progress in this field, but it is necessary to update and complete them by taking into account the current practice of states and international organizations. These countries consider that the actors that create the most space debris must become highly involved in its removal. Those actors should also share their scientific and legal knowledge with developing countries in order to avoid the creation of more space debris when carrying out their space activities.[35]

In its individual capacity, Brazil considers that the issue of space debris involves many legal questions, namely in terms of liability, registration, and ownership. Space debris threaten the sustainability of space activities, and affect space traffic management. Moreover, space debris removal has security implications. Although all states must contribute to the protection of the space environment, Brazil believes that the removal of space debris must take into account the historical responsibility of the creators of debris without imposing undue obligations on new space players, and that the costs of the removal should not be shared evenly among all countries. It is desirable that those who possess the capabilities to mitigate space debris facilitate its cooperation.[36] Mexico shares the same concerns as the G77 + China, and Brazil. It considers that the principle of equal access to outer space could disappear if there

---

[34]G77 + China Statement during the 58th session of the Legal Subcommittee of the United Nations Committee on the Peaceful Uses of Outer Space, from 1 to 12 April 2019, delivered by H.E. Mr. Omar Amer Youssef, Ambassador, Permanent Representative of Egypt, https://www.g77.org/vienna/OOSAAPR19.htm (accessed 14 May 2019).

[35]Legal Subcommittee, 58th session, 8 April 2019, https://icms.unov.org/CarbonWeb/public/oosa/speakerslog/9fbaf1fc-d017-4d5e-8b6c-800c723f4e28 (accessed 14 May 2019).

[36]Legal Subcommittee, 58th session, 9 April 2019, https://icms.unov.org/CarbonWeb/public/oosa/speakerslog/c5128d04-1986-4ec5-b2da-a0d6599fd23a (accessed 14 May 2019).

are no mechanisms to eliminate space debris. It also affirms that all countries have the obligation to clean outer space or to help those who do it for the benefit of the sustainability of space activities.[37]

Potential Legal Models for Activities in Exploration, Exploitation and Utilization of Space Resources

The G77 + China consider that the topic must be urgently discussed by UNCOPUOS due to the fact that some countries have begun to legislate on this matter. This would prevent gaps or contradictions in the international space legal framework and provide a clear understanding of the legal obligations of states in space exploration. In the view of this group, the topic is linked to future international cooperation in space exploration, and it will not depart from the principles of non-appropriation, egalitarian access, and the common province of humankind. For this reason, UNCOPUOS must work to develop a constructive and consensual approach for the utilization of space resources.[38]

Similarly, Argentina, Brazil, Bolivia, Chile, Costa Rica, Cuba, Dominican Republic, Ecuador, El Salvador, Paraguay, Uruguay and Venezuela express their concern that some states have adopted legislation concerning the exploration and exploitation of space resources. For these countries, states must comply with the space treaties and other appropriate regulations when regulating the exploration, exploitation and utilisation of space resources. They maintain that the principle of free exploration, use and exploitation of space resources is not absolute since it is limited by the OST's principles such as non-discrimination, equality of states, the maintenance of international peace and security, the promotion of international understanding and cooperation, and respect of international law. This group of Latin America's countries believes that any national legislation on the topic should be based on the following axes: the principle that the use and exploration is in the interest of humankind, and that these activities must be carried out in a sustainable manner and exclusively for the benefit of all countries, no matter their scientific and economic development level. The group considers that the clauses contained in several national legislations concerning the respect of international obligations[39] are insufficient to guarantee the fulfilment of those principles. They support the work of UNCOPUOS to analyse the

---

[37] Ibid.

[38] G77 + China Statement during the 58th session of the Legal Subcommittee of the United Nations Committee on the Peaceful Uses of Outer Space, from 1 to 12 April 2019, delivered by H.E. Mr. Omar Amer Youssef, Ambassador, Permanent Representative of Egypt, https://www.g77.org/vienna/OOSAAPR19.htm (accessed 14 May 2019).

[39] Although it is not explicitly mentioned by the group, it is possible that this declaration refers to some clauses contained in the U.S. Commercial Space Launch Competitiveness Act regarding the commercial exploration and commercial recovery of space resources. For example, paragraph 51302 states that "*(a) In General.- The President, acting through appropriate Federal agencies, shall-... (2) discourage government barriers to the development in the United States of economically viable, safe, and stable industries for commercial exploration for and commercial recovery of space resources in manners consistent with the international obligations of the United States.*" See U.S.

space treaties in order to prevent interpretations contrary to their object and purpose. In addition, the group proposes the creation of model clauses of national legislation on the subject. These clauses should reproduce exactly and explicitly the principles stated in space treaties. However, there must be institutional mechanisms to monitor their observance.[40]

In the view of Brazil, it is essential to develop innovation, foster progress and generate benefits to all humankind, whereby the discussion within UNCOPUOS on the exploration, exploitation and utilization of space resources should bear in mind the opinions of industry, academia and other organizations. Moreover, these discussions must be in line with the principles of non-appropriation and equitable conditions, and to the benefit of all humankind. UNCOPUOS should also take into account the works of The Hague Space Resources Governance Working Group and the building blocks for the negotiation of a legal regime on this matter (see Table 5.5).[41]

Colombia is the Latin America country that has expressed most concerns on this topic. Colombia deems that the principle of non-appropriation of outer space and celestial bodies could include the exploitation of natural space resources, which is not mentioned in that principle. It also suggests that the terms "use" and "exploration" should be defined, and to clarify the link between these terms and the concept of exploitation of space resources. For Colombia, commercial mining activities go beyond the exploration and use of outer space and celestial bodies; they are totally different from the exploration and use of outer space for scientific missions. Due to the fact that space resources are currently accessible to a limited number of space actors, the delegation of Colombia considers it important to assess the repercussion for global economy of the implementation of the *first come first served* doctrine, which could create a de facto monopoly contrary to the space treaties and UN Resolutions.[42]

Brazil, Colombia and Mexico support the creation within the LSC of a working group on the exploration, exploitation and utilization of space resources, an initiative proposed by Belgium and Greece in the 58th LSC session (2019). Finally, Mexico considers that the working group must take into account the work of The Hague Space Resources Governance Working Group, as well as the issues related to the interpretation of concepts such as equal access to outer space, equity, province of mankind, and common heritage of mankind.[43]

Commercial Space Launch Competitiveness Act Public Law 114-90, 114th Congress, 25 November 2015 (accessed 15 May 2019).

[40]Legal Subcommittee, 58th session, 5 April 2019, https://icms.unov.org/CarbonWeb/public/oosa/speakerslog/03806dc1-af87-4384-9be8-7e89f2c10840 (accessed 15 May 2019).

[41]Legal Subcommittee, 58th session, 8 April 2019, https://icms.unov.org/CarbonWeb/public/oosa/speakerslog/5002f7d0-a19f-4744-9790-2b9cad1e6999 (accessed 15 May 2019).

[42]Legal Subcommittee, 58th session, 5 April 2019, https://icms.unov.org/CarbonWeb/public/oosa/speakerslog/de86144b-3fa4-4ac3-a75c-17979298b18f (accessed 15 May 2019).

[43]Ibid.

Table 5.5   The Hague Space Resources Governance Working Group

The Hague Space Resources Governance Working Group[a]
The Hague Space Resources Governance Working Group is composed by different stakeholders of several countries and was established to assess the need for an international framework for space resource activities and to prepare the basis for such a framework. The final goal of the Working Group is to encourage states to engage in negotiations for an international agreement or non-legally binding instrument
The Working Group consists of members and observers and it is hosted by a Consortium of organizations from each continent. Members are representatives of governments, industries, universities, civil society and research centres, while observers are professionals directly involved in space resources
The Working Group is supported by a technical panel and a socio-economic panel, which provide input to the Working Group, namely on the validity of the building blocks
The activities of the Working Group develop in two phases: (1) from January 2016 to December 2017, and (2) from January 2018 to December 2019. In September 2017, the Working Group circulated the draft building blocks for the development of the international framework for space resources activities, and invited comments to improve the document[b]
In November 2019, the Working Group published the final version of the building blocks.
Representatives of Latin American Countries
Members[c]
Brazil: one person affiliated to the Catholic University of Santos
Mexico: one person affiliated to the Mexican Space Agency
Observers[d]
Mexico: one person affiliated to the National Institute of Astrophysics, Optics and Electronics
Mexico: one person affiliated to the Mexican Space Agency

[a]The Hague Space Resources Governance Working Group, Information provided by the Netherlands, UNCOPUOS, A/AC.105/C.2/2018/CRP.18, 12 April 2018, http://www.unoosa.org/res/oosadoc/data/documents/2018/aac_105c_22018crp/aac_105c_22018crp_18_0_html/AC105_C2_2018_CRP18E.pdf (accessed 16 May 2019). See the Working Group official website in The Hague International Space Resources Governance Working Group, International Institute of Air and Space Law, Leiden University, https://www.universiteitleiden.nl/en/law/institute-of-public-law/institute-for-air-space-law/the-hague-space-resources-governance-working-group# (accessed 16 May 2019)
[b]Draft Building Blocks for the Development of an International Framework on Space Resource Activities, Leiden University, September 2017, https://media.leidenuniv.nl/legacy/draft-building-blocks.pdf (accessed 16 May 2019)
[c]There is a Mexican national representing the United Nations Institute for Disarmament Research. List available at https://www.universiteitleiden.nl/binaries/content/assets/rechtsgeleerdheid/instituut-voor-publiekrecht/lucht–en-ruimterecht/space-resources/members-website-1-3.pdf (accessed 15 May 2019)
[d]List available at https://www.universiteitleiden.nl/binaries/content/assets/rechtsgeleerdheid/instituut-voor-publiekrecht/lucht–en-ruimterecht/space-resources/observers-april19.pdf (accessed 15 May 2019)

5.1.2   Implementation of Space Treaties

With a few exceptions, all Latin American countries have ratified at least one of the space treaties.[44] In general, international obligations stipulated in these treaties have been implemented by Latin American states, at least indirectly, such as the principles of peaceful use and exploration of outer space, and the non-appropriation of outer space, including the Moon and other celestial bodies.[45] Many international obligations have not been implemented because there has been no occasion to do it, such as the obligations concerning the rescue and return of astronauts according to ARRA, or the obligations to prevent harmful contamination of the Moon and other celestial bodies in conformity with the MOON. International cooperation, as mentioned, for example, in Articles 1, 3 and 10 of the OST, and Article 2 of the MOON has been respected and promoted by Latin American countries when adopting international cooperation agreements with other countries for the peaceful use and exploration of outer space.

However, there are specific international obligations that have been duly implemented by some Latin American countries, obligations which have been registered by UNOOSA and that are publicly available. These obligations include the registration of objects launched into outer space and the creation of a national registry of these objects and the respective notification to UNOOSA in conformity with the REG, and the notification of the discovery and rescue of space objects according to ARRA. Also, even though they do not constitute binding instruments, some space debris mitigation mechanisms have been accepted by a few Latin American states.

5.1.2.1   Registration of Objects Launched into Outer Space

In 1962 UNGA adopted Resolution 1721 B (XVI) that created a Register of Objects Launched into Outer Space to aid UNCOPUOS in its discussions on the political, legal and technical issues. UNOOSA keeps this register and, according to the resolution, UN members are invited to provide to the UN Secretary General information for the registration of launchings.[46] Due to the fact that UNGA resolutions are not legally binding, states submit the relevant information on a voluntary basis. To date, Brazil,

---

[44]Some states have only signed one or more space treaties. ITU's Convention and Constitution, and other related space treaties have been ratified by many Latin American states, see Tables 5.2 and 5.3.

[45]No Latin American state has threated international peace and security through its space activities by means of militarization or the emplacement of massive destruction weapons on Earth's orbit, and none of them have intended to claim sovereign rights, use or occupation over outer space, the Moon and other celestial bodies. Even though Colombia and Ecuador, at least theoretically, have claimed sovereign rights on the GEO through the Declaration of Bogota of 1976, these claims have not received international support; see Table 5.4.

[46]UNGA Res. 1721 B (XVI), International cooperation in the peaceful uses of outer space, 20 December 1961, http://www.unoosa.org/oosa/en/ourwork/spacelaw/treaties/resolutions/res_16_1721.html.

Bolivia and Venezuela are the only Latin American countries that have submitted information under Resolution 1721 B (XVI).

The necessity to have a register of objects launched into outer space has become more evident because it facilitates the identification of states that bear international responsibility and liability for damages caused by their space objects.[47] This situation prompted states to adopt the Convention on Registration of Objects Launched into Outer Space entered into force in 1976, according to which states and international organizations (IO) that agree to abide by it are required to establish their own national registers and to inform the Secretary General of the establishment of such registries (Article II), and to provide information on their space objects to the UN Secretary-General (Article IV) for inclusion in the UN Register (Article III).[48]

The fulfilment of international obligations by Latin American countries that are parties to the REG has been significant but imperfect. Concerning the creation of the national register, Chile, Mexico, Peru and Uruguay established their registers many years—even decades—after the ratification of the Convention, while Colombia and Venezuela have not yet created their respective registers, even when they have already launched objects into outer space (see Tables 5.6 and 5.7). The same can be said as to the notification to UNOOSA of the creation of Latin America's national registers.

Regarding the notification of objects launched into outer space, Latin American countries have properly fulfilled their obligations, although some countries have taken much time to do so (see Table 5.7); for example, Mexico took roughly 20 years to notify UNOOSA that it had launched two satellites in 1993 and 1994 (SOLIDARIDAD I and SOLIDARIDAD II, respectively).[49]

Nevertheless, the comprehensive response of Latin American countries to their obligations in conformity with the REG is very positive. It shows, on one hand, that Latin American countries are willing to participate in the application of international space law and to promote transparency concerning their space activities. On the other hand, it demonstrates an upward trend in the development of Latin America's space capacities, which can bring many benefits to the region and to their societies.

---

[47] United Nations Register of Objects Launched into Outer Space, UNOOSA, http://www.unoosa. org/oosa/en/spaceobjectregister/index.html (accessed 5 May 2019).

[48] Ibid.

[49] Information furnished in conformity with the Convention on Registration of Objects Launched into Outer Space, Note Verbale dated 16 April 2013 from the Permanent Mission of Mexico to the United Nations (Vienna) addressed to the Secretary-General, UNCOPUOS, ST/SG/SER.E/670, 30 May 2013, http://www.unoosa.org/documents/pdf/ser670E.pdf.

Table 5.6 Notification of establishment of a National Registry under Article II of the Registration Convention[a]

Country	Ratification of the Registration Convention	National Register Authority	Notification to UNOOSA of the creation of the National Registry
Argentina	1993	National Commission on Space Activities of Argentina (Comisión Nacional de Actividades Espaciales)	1996
Brazil	2006	Brazilian Space Agency (Agência Espacial Brasileira)	2007
Chile	1981	Directorate for International and Human Security of the Ministry of Foreign Affairs (Dirección de Seguridad Internacional y Humana del Ministerio de Relaciones Exteriores)	2016
Colombia	2014	–	–
Mexico	1977	General Coordination Office for Space-related Security and International Affairs of the Mexican Space Agency (Coordinación General de Asuntos Internacionales y de Seguridad en Materia Espacial de la Agencia Espacial Mexicana)	2013
Peru	1979	National Aerospace Research and Development Commission (Comisión Nacional de Investigación y Desarrollo Aeroespacial)	2016
Uruguay	1977	National Agency for Research and Innovation of Uruguay (Agencia Nacional de Investigación e Innovación)	2017
Venezuela	2016	–	–

[a]Index of Notifications by Member states and Organizations on the Establishment of National Registries of Objects Launched Into Outer Space, UNOOSA, http://www.unoosa.org/oosa/en/spaceobjectregister/national-registries/index.html (accessed 1 May 2019)

Table 5.7 United Nations Register of Objects Launched into Outer Space[a]

Country	Registration Submissions under Article IV of the Registration Convention[b]	Functional Space Objects registered with the SG	Registration Submissions under General Assembly resolution 1721B (XVI)	Notifications under Article XI of the Outer Space Treaty
Argentina	11 (1996–2018)	10	–	–
Bolivia	–	1	1 (2014)	–
Brazil	7 (2006–2019)	21	2 (1993 and 2000)	1 (2014)
Chile	4 (1998–2018)	4	–	–
Colombia	1 (2019)	1	–	–
Mexico	7 (1988–2016)	13	–	–
Peru	1 (2017)	3	–	–
Uruguay	2 (2017 and 2018)	5	–	–
Venezuela	–	2	2 (2009 and 2012)	–

[a]United Nations Register of Objects Launched into Outer Space, UNOOSA, http://www.unoosa.org/oosa/en/spaceobjectregister/submissions/states-organisations.html, and Online Index of Objects Launched into Outer Space, UNOOSA, http://www.unoosa.org/oosa/osoindex/search-ng.jspx?lf_id= (accessed 1 May 2019)

[b]A notification may include one or more space objects launched into outer space. For example, on 16 April 2013 Mexico transmitted to the Secretary General information concerning 7 space objects launched into outer space from 1993 to 2011, UNCOPUOS, ST/SG/SER.E/670, 30 May 2013, http://www.unoosa.org/documents/pdf/ser670E.pdf (accessed 1 May 2019)

5.1.2.2 Notification Under Article XI of the Outer Space Treaty

Article XI of the OST allows States Parties to provide information on space-related activities that do not fall under the provisions of other space treaties.[50] Under this article, UNOOSA also receives information of states relating to the conduct by states of pre-launch safety assessments of nuclear-powered space objects.[51] Moreover, some states have provided information under Principle IX of the Remote

---

[50]It specifically states that "*States Parties to the Treaty conducting activities in outer space, including the Moon and other celestial bodies, agree to inform the Secretary-General of the United Nations as well as the public and international scientific community, to the greatest extent feasible and practicable, of the nature, conduct, locations and results of such activities*", Art. XI OST.

[51]Index of Submissions by States under Article XI, OST, UNOOSA, http://www.unoosa.org/oosa/en/treatyimplementation/ost-art-xi/index.html (Accessed 6 May 2019).

Sensing Principles[52] and Principle IV, paragraph 3 of the Nuclear Power Sources Principles,[53] which refer directly to Article XI of the OST.

Chile is the only Latin American country that has provided information under Article XI of the OST. In 1996, the Government of Chile notified the UN Secretary-General of the discovery in the Pacific Ocean and in its territory of fragments of the Russian spaceship MARS-96 and expressed concern about the potential damages of the incident.[54] Later that year, the Government of Chile notified the UN Secretary-General that Russia had guaranteed Chile and other South American countries that the threat of radioactive contamination of the soil or ocean waters generated by the radioisotope power sources of the spacecraft was ruled out.[55]

5.1.2.3 Notification Under Article V of the Rescue Agreement

The 1968 Rescue Agreement complements and details the provisions of Article VIII of the OST, which stipulates that objects or component parts launched into outer space that have been found beyond the limits of a State Party to the Treaty on whose registry they are carried shall be returned to that State Party. Article V of the Rescue Agreement requires that should a space object or its component parts be discovered within a state's territory or on the high seas or in any other place not under the jurisdiction of any state, the object or its component parts should be returned to the launching authority.[56] Besides, states are required to notify the launching authority and the UN Secretary-General of the such recoveries.

A few incidents in Latin America have activated Article V of the Rescue Agreement. Argentina, Brazil, Mexico, Peru and Uruguay have notified the UN Secretary-General of the discoveries of multiple space objects in their respective territories since 1993 (see Table 5.8). Most of the space object's owners were identified at the

---

[52] Principle IX states that: "*In accordance with article IV of the Convention on Registration of Objects Launched into Outer Space and article XI of the Treaty on Principles Governing the Activities of States in the Exploration and Use of Outer Space, including the Moon and Other Celestial Bodies, a State carrying out a programme of remote sensing shall inform the Secretary-General of the United Nations*".

[53] According to which "*pursuant to article XI of the Treaty on Principles Governing the Activities of States in the Exploration and Use of Outer Space, including the Moon and Other Celestial Bodies, the results of this safety assessment, together with, to the extent feasible, an indication of the approximate intended time-frame of the launch, shall be made publicly available prior to each launch, and the Secretary-General of the United Nations shall be informed on how States may obtain such results of the safety assessment as soon as possible prior to each launch*".

[54] Nota verbal de fecha 29 de noviembre de 1996 dirigida a la Oficina de las Naciones Unidas en Viena por la Misión Permanente de Chile, UNGA A/AC.105/668, 3 December 1996, http://www.unoosa.org/pdf/reports/ac105/AC105_668S.pdf.

[55] Nota verbal de fecha 6 de diciembre de 1996 dirigida a la Oficina de las Naciones Unidas en Viena por la Misión Permanente de Chile, UNGA A/AC.105/669, 12 December 1996 http://www.unoosa.org/pdf/reports/ac105/AC105_669S.pdf.

[56] The obligation to return the space object or its component parts must be fulfilled upon request of the launching authority. See paragraphs 2 and 3 of Article V of the Rescue Agreement.

Table 5.8 Notifications under Article V of the Rescue Agreement[a]

Country	Date of discovery	Space objects discovered	Location of discovery	Remarks
Argentina	07/02/1991	7	Provinces of La Pampa, Santa Fe, and Entre Rios	Believed to debris from space station SALYUT 7
Argentina	20/01/2014	1	San Roque, Province of Corrientes	Confirmed as from DELTA-2 launcher
Brazil	26/04/2014	1	Uriandeua River, city of Salinópolis, 220 km from Belém, Pará	Identified as part of the payload fairing from an ARIANE launch vehicle
Mexico	13/06/2018	5	Mahahual, State of Quintana Roo	Identified as part of VEGA launch conducted by France
Paraguay	10/03/2018	1	Itanará, Canindeyú Department	Identified as part of Long MARCH 3B launch conducted by China. Notification provided under Article VIII of the OST
Peru	27/01/2018	5	Vicinity of the town of San José, and the communities of Catacora, Isla Huata, and Ninaruyo	–
Uruguay	23/04/2011	1	Third judicial section (administrative subdivision) of the Department of Artigas, North Uruguay	Confirmed as part of the Star 48 motor of the upper stage from a DELTA 7925 launch conducted by the USA

[a]More details can be found in List of reported space objects discovered by Member States, UNOOSA, http://www.unoosa.org/oosa/en/treatyimplementation/arra-art-v/unlfd.html (accessed 1 May 2019)

time of the notification, except for a space object found by Peru in 2018.[57]

A special case is that of Paraguay. Because Paraguay is not a State Party of the Rescue Agreement, in 2018 it notified the UN Secretary-General of the discovery of a Chinese space object in its territory under Article VIII of the OST.

5.1.2.4 Space Debris Mitigation Mechanisms

The proliferation of space debris in Earth orbits has led states, international organisations and other space-related actors to adopt international mechanisms containing standards, guidelines or recommendations on space debris mitigation. Although these mechanisms have a non-binding character,[58] states and international organisations may use them as a guide to conduct their space activities. In some cases, states have incorporated these standards as legal rules at the national level.

Some of the most important mechanisms on this matter are the Inter-Agency Space Debris Coordination Committee (IADC) Space Debris Mitigation Guidelines, adopted in 2002 and revised in 2007[59]; ITU Recommendation ITU-R S.1003.2 (12/2010) Environmental Protection of the GEO[60]; the European Code of Conduct for Space Debris Mitigation, adopted in 2004,[61] and the UNCOPUOS Space Debris Mitigation Guidelines, endorsed by UNGA in its Resolution 62/217 of 22 December 2007.[62]

To date, there are no Latin American countries with national space debris regulations or mechanisms. However, Argentina, Chile and Mexico have expressly agreed to some of the international mechanisms on space debris mitigation.

Argentina has stated that it fully adheres to the UNCOPUOS Space Debris Mitigation Guidelines, and supports IADC Space Debris Mitigation Guidelines. Moreover,

---

[57]Notifications to the Secretary-General do not usually mention whether the space objects have been returned to the launching authorities or will be kept by the discoverer states.

[58]Even though space debris mitigation guidelines and recommendations are voluntarily adopted by states (implying that the issue is not a matter of law), some argue that the creation of space debris is potentially harmful interference with the activities of other states contrary to Article IX of the Outer Space Treaty, Cf. Lyall Francis et al., op. cit., p. 301. Article IX of the Outer Space Treaty states that *"States Parties to the Treaty shall pursue studies of outer space, including the Moon and other celestial bodies, and conduct exploration of them so as to avoid their harmful contamination (…) and, where necessary, shall adopt appropriate measures for this purpose"*.

[59]See IADC Document Registration List, Inter-Agency Space Debris Coordination Committee, https://www.iadc-online.org/index.cgi?item=docs_pub.

[60]See S. 1003: Environmental protection of the geostationary-satellite orbit, ITU, https://www.itu.int/rec/R-REC-S.1003/_page.print.

[61]See European Code of Conduct for Space Debris Mitigation, Issue 1.0, 28 June 2004, http://www.unoosa.org/documents/pdf/spacelaw/sd/2004-B5-10.pdf.

[62]UNCOPUOS Guidelines are based on the IADC Space Debris Mitigation Guidelines, taking into consideration the UN treaties and principles on outer space. See Space Debris Mitigation Guidelines of the Committee on the Peaceful Uses of Outer Space, UNOOSA, Vienna, 2010, http://www.unoosa.org/pdf/publications/st_space_49E.pdf.

there are ongoing low profile discussions to develop national legislation on the matter.[63] Chile also adheres to the UNCOPUOS Space Debris Mitigation Guidelines, and supports IADC Guidelines, ISO Space Systems—Space Debris Mitigation Requirements (ISO 24113:2011), and ITU Recommendation ITU-R S.1003. Similarly, there are low profile discussions in progress to develop national space-related legislation in Chile.[64] Finally, Mexico also aligns with the UNCOPUOS Space Debris Mitigation Guidelines, and agrees with ITU Recommendation ITU-R S.1003, the standards of the European Code of Conduct for Space Debris Mitigation, and ISO 24113. Moreover, there are high level discussions within the Mexican competent bodies to develop national mechanisms in the short term.[65]

5.1.3   Space Law Education

Space Law education is vital for the development of national space capabilities, activities, legislation and policies. Moreover, it helps to understand not only the international space legal framework but also space science and technology. All over the world, national space agencies, universities, and IO's consider the promotion and understanding of space law to be essential. The 3rd UN Conference on the Exploration and Peaceful Uses of Outer Space (UNISPACE III) recognised the importance of indigenous capacity-building in space law and policy.[66] UNOOSA has also recognised that one of the pillars that supports the development of legal and policy frameworks at the national level is the availability of professionals able to provide services in the field.[67] Space law education has gained importance in some Latin America countries, although there is still a long way to go before this discipline is taught in a comprehensive way in the region.

5.1.3.1   National Space Agencies and the Promotion of Space Law

The role of space agencies and other governmental bodies in Latin America for the promotion of space law is essential for the understanding of the international space legal framework, and the rights and obligations of states and IO's in the area. Many of Latin America's space agencies have organised conferences, roundtables, colloquia,

---

[63] Compendium of space debris mitigation standards adopted by states and international organizations, UNCOPUOS Doc. A/AC.105/C.2/2014/CRP.15, 18 March 2014, p. 5, http://www.unoosa.org/pdf/limited/c2/AC105_C2_2014_CRP15E.pdf (accessed on 2 May 2019).

[64] Ibid, p. 18.

[65] Ibid, p. 31.

[66] Report of the Third United Nations Conference on the Exploration and Peaceful Uses of Outer Space (Vienna, 19–30 July 1999), UNGA A/CONF.184/6, 18 October 1999, http://www.unoosa.org/pdf/reports/unispace/ACONF184_6E.pdf, pp. 70–73.

[67] Space Law: Capacity Building, UNOOSA, http://www.unoosa.org/oosa/en/ourwork/spacelaw/capacitybuilding.html (accessed on 8 May 2019).

book's presentations, and other diffusion activities on space law. For example, in 2013 the National Commission of Space Activities of Argentina in conjunction with ESA and UNOOSA organized a colloquium on the Contribution of Space Law to Social and Economic Development[68]; the Bolivarian Agency for Space Activities included the topic of space law in the Venezuelan Congress on Space Technology held in 2017,[69] and the Mexican Space Agency has published various opinion articles on space legal issues in its magazine "Hacia el Espacio".[70]

5.1.3.2 Space Law Programmes

A few programmes on space law have been identified in the following Latin American countries: Argentina, Chile, Colombia and Mexico.[71] Table 5.9 shows that there is only one specialized programme on space law in the region. The other programmes are part of comprehensive programmes in Law or International Relations. Moreover, while some programmes focus on general issues of space law, other programmes concentrate on specific issues of space law.

5.1.3.3 Regional Centres for Space Science and Technology Education

Under the auspices of the UN Programme on Space Applications of UNOOSA some regional centres for space science and technology education affiliated to the UN were created to allow developing countries to build regional capacity in space science and technology and its applications. In 1997, a regional centre was established for the Latin America and the Caribbean region (CRECTEALC) through a cooperation agreement between the Governments of Mexico and Brazil, which act as the main coordinators and alternate headquarters.[72]

From 2007 to 2013, the UNOOSA Legal Subcommittee prepared a curriculum on space law to complement the activities of CRECTEALC. The curriculum, published

---

[68] Seminario Naciones Unidas/Argentina sobre Derecho Espacial, Buenos Aires, 5–8 November 2012, http://www.conae.gov.ar/index.php/espanol/2012/205-se-realizara-en-argentina-el-seminario-de-derecho-espacial-de-naciones-unidas-buenos-aires-5-al-8-de-noviembre-de-2012 (accessed 17 May 2019).

[69] II Congreso Venezolano de Tecnología Espacial, Agencia Bolivariana de Tecnología Espacial, Caracas, September 2017, http://2cvte.abae.gob.ve/ (accessed 17 May 2019).

[70] Available at https://haciaelespacio.aem.gob.mx/revistadigital/.

[71] UNOOSA keeps a directory of education opportunities in space law that is constantly updated thanks to the information received from national institutions and states. To date, there are only two Latin American institutions in the directory (both in Argentina). See Education Opportunities in Space Law: a Directory, UNCOPUOS, A/AC.105/C.2/2019/CRP.9, 1 April 2019 http://www.unoosa.org/res/oosadoc/data/documents/2019/aac_105c_22019crp/aac_105c_22019crp_9_0_html/AC105_C2_2019_CRP09E.pdf.

[72] Report of the Regional Centre for Space Science and Technology Education for Latin America and the Caribbean (CRECTEALC), UNCOPUOS, A/AC.105/2006/CRP.10, 12 June 2006 http://www.unoosa.org/pdf/limited/l/AC105_2006_CRP10E.pdf.

Table 5.9 Latin American Space Law Programmes

Country	Institution	Level	Programme	Duration	Characteristics
Argentina	Instituto Nacional de Derecho Aeronáutico y Espacial (INDAE)	Postgraduate	Air and Space Law	Two years	The programme focuses on all the branches of aeronautic, space and airport law[a]
Argentina	University of Belgrano	Postgraduate (Master's Degree)	International Relations	Three terms of eleven weeks each	The subjects "Public International Law & Human Rights", and "International Organizations & International Environmental Law", 50% of the syllabus, plus a moot court, deal with space law issues[b]
Chile	University of Chile	Bachelor's level	Space Law	One semester	The programme focuses on the relationship between space law and space policy; UNCOPUOS' work; Space conferences of the Americas, and UNISPACE; space technology and emergencies and natural disasters; and access to strategic information[c]

(continued)

Table 5.9 (continued)

Country	Institution	Level	Programme	Duration	Characteristics
Colombia	Universidad Libre	Bachelor's level	Space Law	One year (64 h)	The programme focuses on the generalities of outer space, and the use of space for telecommunications purposes[d]
Mexico	National Autonomous University of Mexico	Bachelor's level (with specialization in public international law)	Air and Space Law	One semester (48 h)	The programme focuses on the principles of space law, the legal regime of outer space, international liability, remote-sensing issues, peaceful uses of outer space, and militarization of outer space[e]

[a]Facultad de la Fuerza Aérea, Universidad de la Defensa Nacional, http://www.undef.edu.ar/academica/facultad-de-la-fuerza-aerea/ (accessed 8 May 2019)
[b]Education Opportunities in Space Law: a Directory, UNCOPUOS, op. cit., pp. 6–7. See also Plan de Estudios + Contenidos Mínimos, Universidad de Belgrano, Buenos Aires, Licenciatura en Relaciones Internacionales, Plan 2015, http://www.ub.edu.ar/sites/default/files/contenidos_minimos_Relaciones_Internacionales. pdf
[c]International Law Department, Universidad de Chile, Facultad de Derecho, http://www.derecho.uchile.cl/departamento-de-derecho-internacional/docencia/ 124640/asignaturas (accessed 20 May 2019). See the complete programme at http://web.derecho.uchile.cl/pregrado/2011_2/pro_d.php?recordID=1140 (accessed 20 May 2019)
[d]Universidad Libre, Facultad de Derecho, Contenido Programático de la Asignatura-Derecho Espacial http://www.unilibre.edu.co/derecho/images/stories/pdfs/ 2013/optativas/AREA%20DERECHO%20PUBLICO/DERECHO-ESPACIAL.pdf (accessed 8 May 2019). http://www.unilibre.edu.co/derecho/images/stories/ pdfs/2013/optativas/AREA%20DERECHO%20PUBLICO/DERECHO-ESPACIAL.pdf
[e]Universidad Nacional Autónoma de México, Facultad de Derecho, Derecho Aéreo y Ultraterrestre, https://www.derecho.unam.mx/oferta-educativa/licenciatura/ nuevoplan2011/optativas-eleccion/d_i-publico/Derecho_Aereo_y_Ultraterrestre.pdf (accessed 8 May 2019)

by UNOOSA in 2014, consists of four complementary modules: (1) basic concepts of international law and space law; (2) remote sensing/GIS, satellite meteorology and global climate + international law; (3) satellite communications + international law, and (4) global navigation satellite systems (GNSS) + international law.[73] It was proposed that the regional centre leads capacity-building and training courses in space law in order to enhance national and regional capabilities in the field. Therefore, it is expected that the curriculum will be used by instructors as a reference guide in their educational programme.

5.1.3.4 Manfred Lachs Space Law Moot Competition

The worldwide renowned Manfred Lachs Space Law Moot Court Competition created by the International Institute of Space Law (IISL) is an excellent opportunity for undergraduate and postgraduate students of all regions to put into practice their knowledge and oral and written skills in air law, space law, and public international law.

The Moot Court Competition has been organized since 1992. The competition is divided into regional rounds and the world finals. On the American continent there are two regional competitions: the North American Regional Competition,[74] and the Latin American Competition. Mexico is the only Latin American country that has participated in the North American Regional Competition. Mexico's unique participation took place in 2016 in the regional competition held at Georgetown University, Washington D.C., with a team of the National Autonomous University of Mexico.[75] In the same year, Mexico was host of the world finals held in the city of Guadalajara, which were part of the activities of the 67th International Astronautical Conference (IAC).[76] Although the North American Regional Competition welcomes teams from Latin American and Caribbean countries, no other country has participated in this competition.

The first part of the Latin American Regional Competition Test Round was held in May 2019 in Bogota, Colombia. The competition was organized by the Moot Court

[73]Education Curriculum on Space Law, United Nations, March 2014, http://www.unoosa.org/res/oosadoc/data/documents/2014/stspace/stspace64_0_html/st_space_064E.pdf (accessed 8 May 2019).

[74]Up to 2018, the North American Regional Competition also welcomed teams from Latin American and Caribbean countries.

[75]The Mexican team in collaboration with the Mexican Space Agency and the National Autonomous University of Mexico prepared a guide for the participation of future Mexican teams in the Moot Court. See D. Basurto Nickté et al., Manfred Lachs Space Law Moot Court Competition, Guía de Preparación, Mexican Space Agency, November 2016, https://www.gob.mx/cms/uploads/attachment/file/166123/Gu_a_de_preparaci_n_Manfred_Lachs_versi_n_final.pdf.

[76]The Mexican team did not classify to the world finals.

Committee of IISL, the ReLaCa-Espacio Network, and the Catholic University of Colombia. Colombia, Paraguay and Venezuela participated in the competition.[77]

5.1.3.5 ReLaCa-Espacio Network

The Latin American and Caribbean Network of Universities and Institutions that Investigate the Technology, Policy and Law of the Outer Space, also known as "ReLaCa-Espacio", was created under the impulse and support of the European Centre for Space Law (ECSL) of ESA during the 1st International Congress of ReLaCa-Espacio held at the University Sergio Arboleda, Colombia, in 2016. In the 2nd ReLaCa-Espacio's Congress held at the University of El Salvador, Argentina, in May 2017, a commitment document was adopted with the ReLaCa-Espacio's objectives, which are the following: to take advantage and complete the efforts made and the available resources in Latin America and the Caribbean in the research field related to space, and to promote the development of these efforts and resources, including documentation in a coordinated way; to encourage knowledge and support research activities, including the dissemination of information and the organization of workshops and professional exchanges; and to promote exchanges through the organization of national and international congresses and symposiums in all space related disciplines. The 3rd ReLaCa-Espacio Congress took place at the University of the Republic, Uruguay, in May 2018 in which the Montevideo Commitment was signed.[78] The 4th ReLaCa-Espacio Congress was held at the University of San Carlos, Paraguay, in May 2019.[79] To date, representatives of Argentina, Chile, Colombia, Ecuador, Uruguay, Spain, and ESA have participated in ReLaCa's activities.[80]

---

[77] See Latin America, International Institute of Space Law, 2019, http://iislweb.org/awards-and-competitions/manfred-lachs-space-law-moot-court-competition/participating-in-the-lachs-competition/latin-america/; Students from the Catholic University of Colombia have confirmed their participation, Universidad Católica de Colombia, 5 April 2019 https://www.ucatolica.edu.co/portal/universidad-realizara-la-ronda-de-pruebas-latinoamericana-del-manfred-lachs-moot-court/?fbclid=IwAR11o2bS7g4MGgelzkXnORIOzGc-aKgIZzr5zXOyxXl1yUbzEydLEBDKEg8 (accessed 8 May 2019); and Marcial Rafael, Paraguay, presente en competencia de derecho espacial, ABC Color, 19 May 2019, https://www.abc.com.py/nacionales/la-una-en-competencia-de-derecho-espacial-1815534.html (accessed 20 May 2019).

[78] All news and relevant information of ReLaCA-Espacio can be found on https://www.facebook.com/REDESPACIAL/.

[79] IV Encuentro de la Red Latinoamericana y del Caribe del Espacio, Call to Space, https://callto.space/event/iv-encuentro-internacional-red-latinoamericana-y-del-caribe-del-espacio/ (accessed 26 May 2019).

[80] See Programa, ReLaCA-Espacio, Encuentro de la Red Latinoamericana y del Caribe del Espacio, Montevideo, Uruguay, 25 May 2019, https://www.fder.edu.uy/sites/default/files/2018-05/Programa%20Invitacion%20RED%20Espacial%2025%20de%20mayo%202018%20salon%2026%20FD%20UDELAR.pdf?fbclid=IwAR0BWDyCRx_LQR6HL-adCfW3_iw6TcFRo2hfOFDHHptf8XPMPvFUYe6a0DM.

5.2 National Space Law

Latin American countries have not only contributed to the development of international space law, but also to the creation of national rules concerning space related activities. Even though several states do not have a strong national space sector, they have begun to pave the way in this direction by adopting national laws, decrees, regulations and other instruments regulating space, telecommunications and other related activities. Despite only few states having specialized legislation on the matter, some states have started to create their own space regulations, showing an upward trend in the interest in the space sector at the national level.

5.2.1 Argentina

As one of the space pioneers in Latin America, Argentina possesses several significant space regulations that aim to develop strong capacities and place the country in the international space arena. Laws and decrees are the most common Argentinian legal instruments on space related issues. Most of them concern the establishment, functioning and activities of the National Commission on Space Activities (CONAE), such as National Decree No. 995/91, which created the CONAE, and Decrees 1274/96, 2197/12 and 242/16, which transferred the CONAE to different administrative bodies in 1996, 2012 and 2016, respectively (nowadays CONAE is part of the Ministry of Education, Culture, Science and Technology).[81]

Other relevant instruments are NSP Decree 532/05 of 2005, which declares the development of the space activity as a national priority and a state policy, and recognises the National Space Plan (2004–2015) as a strategic plan, and the Decree N° 125/95 of 1995 that establishes the National Registry of Objects Launched into Outer Space.[82]

5.2.2 Bolivia

The two main space regulations of Bolivia are the Decree of the Establishment of the Bolivarian Space Agency (ABE) (DS N° 0423) and the Telecommunications

---

[81]National Decree No. 995/91, Creation of the National Commission on Space Activities (28 May 1991) http://www.conae.gov.ar/images/legislacion/l99591.pdf (accessed 20 May 2019).

[82]National Space Plan, Decree 532/2005, 26 May 2005, and National Decree No. 125/95, Establishment of the National Registry of Objects Launched into Outer Space (25 July 1995) http://www.conae.gov.ar/images/legislacion/l12595.pdf. These and other related regulations are available at https://www.argentina.gob.ar/ciencia/institucional/legislacion (accessed 20 May 2019).

Law.[83] DS N° 0423, adopted on 10 February 2016 and modified by Decree DS N° 0599 on 18 August 2010, creates the ABE, which is responsible for the management and implementation of the TUPAK KATARI Communications Satellite Programme, promotion of the development of new satellite and space projects, promotion of technology transfer and training of human resources, as well as implementation of satellite applications for use in social, productive, defense, environmental and other programmes.[84]

The Telecommunications Law adopted on 8 August 2011 establishes a general regime of telecommunications and ICTs. It guarantees the equitable distribution and efficient use of the radio electric spectrum, as well as the right of universal and equitable access to telecommunications and ICT's services. The Law also regulates issues concerning satellite communications. In this regard, it determines that the natural resource "orbit", which includes the spectrum and associated frequencies on behalf of Bolivia, will be assigned to the Bolivarian Space Agency (ABE) for its use in national satellite networks, namely the TUPAK KATARI Communications Satellite Programme.[85]

5.2.3   Brazil

Brazil's legislation concerning space activities consists mainly of laws, provisional measures and decrees. These instruments regulate space-related activities; the work and structure of national bodies linked to the space sector; and scientific, techno-logical, budgetary, administrative, and intellectual property issues, among others. There are also instruments related to the adoption of international space treaties and cooperation agreements with other states.[86]

The Law Establishing the Brazilian Space Agency adopted on 10 February 1994 is one of the most important laws in the field. It creates the Brazilian Space Agency (BSA) which is responsible for the implementation of the National Policy for the Development of Space Activities, the elaboration of the National Space Activities Programmes, the adoption of international space agreements, the promotion of the participation of private actors in space activities, and the promotion of scientific

---

[83]There are other administrative and internal regulations related to Bolivia's space activities. They are available at https://www.abe.bo/nosotros/transparencia/marco-normativo/ (accessed 20 May 2019).

[84]Supreme Decree N° 0423, Official Journal of Bolivia, 10 February 2010, and Supreme Decree N° 0599, Official Journal of Bolivia, 18 August 2010, both available at https://www.abe.bo/nosotros/transparencia/marco-normativo/ (accessed 20 May 2019).

[85]Chapter Three, Satellite Communications, Law N° 164, 8 August 2011, https://www.abe.bo/files/marco-normativo/LeyTelecomunicaciones.pdf (accessed 20 May 2016).

[86]Brazilian space-related legislation and instruments are available at Legislaçao, Agência Espacial Brasileira, http://www.aeb.gov.br/acesso-a-informacao/legislacao-2/.

and technological development in the space sector, among other functions.[87] Other relevant legislation is the Law Establishing the Programme for the Scientific and Technological Development of Space Sector, and the Decree of the Creation of the Inter-ministerial Group for the analysis, proposal and following of implementation of necessary actions for the strengthening of the National Space Activities Programme, both adopted in 2004.

5.2.4 Chile

In Chile, Decree N° 181 establishes the Presidential Advisory Commission entitled the Council of Ministers for Space Development. It stipulates the structure, functions and purposes of the Council of Ministers for Space Development. According to the Decree, the Council is responsible for advising the President of the Republic on the elaboration of public policies, plans, programmes and actions that contribute to the promotion of space activities and the use of space applications and technology. Moreover, the Council proposes the National Space Policy to the President; functions as the appropriate coordination point with public bodies that have space-related activities competences; identifies and proposes the elaboration of studies for the determination of national satellite needs and the model of use of satellite resources, and proposes the adoption of national and international cooperation agreements related to space science and technology.[88]

Another Chilean significant instrument is Decree N° 917, which promulgates an Agreement between the Republic of Chile and the Government of the USA on the use of the Mataveri Airport, Easter Island, as an emergency landing place, including for the rescue of space shuttles.[89]

5.2.5 Colombia

Colombian space legislation consists of Decree 2442/2006 that establishes the Colombian Space Commission (CSC), and Decree 2516/2013 that creates the Presidential Program for the Colombian Space Development under the Administrative

---

[87]Law No. 8.854 of 10 February 1994, Law Establishing the Brazilian Space Agency http://www.planalto.gov.br/ccivil_03/leis/L8854.htm (accessed 21 May 2019).

[88]Decree N° 181 that establishes the Presidential Advisory Commission entitled the Council of Ministers for Space Development, 28 October 2015, https://www.leychile.cl/Navegar?idNorma=1087964 (accessed 20 May 2019).

[89]Decree N° 917, Ministry of Foreign Affairs, Library of the National Congress of Chile, 6 November 1985.

Department of the Presidency of the Republic.[90] Moreover, the 2018–2022 National
Development Plan states that the national government, in the framework of the CSC,
will design a National Space Policy.[91]

5.2.6   Mexico

Mexican space legislation includes several laws, decrees or administrative instruments, and internal regulations. The two main laws are the Mexican Space Agency's
Act, and the Federal Telecommunications and Broadcasting Act. The first law,
adopted on 30 July 2010, creates the Mexican Space Agency (AEM), which is
responsible for the elaboration and implementation of the Mexican Space Policy
and the National Space Activities Programme; the promotion of the development of
space activities in multiple disciplines, as well as the development of space systems;
and the promotion of international cooperation in the space sector, among others.
The Act also enumerates the instruments of the Mexican Space Policy, and specifies
AEM's functions, budget, and structure.[92]

The second law is the Federal Telecommunications and Broadcasting Act, adopted
on 14 July 2014, which regulates the use and exploitation of the radio spectrum,
public telecommunication networks, orbital resources, and satellite communications.
It also determines how telecommunications and broadcasting public services must
be provided.[93]

Apart from the aforementioned laws, there are other relevant instruments related
to space activities, such as the General Guidelines of the Mexican Space Policy
adopted by the Ministry of Communications and Transportation, which defines this
policy as a state policy, and enumerates its strategic objectives in order to transform
Mexico into a regional space leader, and AEM's internal regulations, which specify
the functions of AEM's organs.[94]

---

[90]Decreto 2615 de 2013 por el que se crea el Programa Presidencial para el Desarrollo Espacial
Colombiano, 15 November 2013, https://www.funcionpublica.gov.co/eva/gestornormativo/norma_
pdf.php?i=67836 (accessed 22 September 2019).

[91]Bases del Plan Nacional de Desarrollo 2018-2022: Pacto por Colombia, pacto por la equidad,
Departamento Nacional de Planeación, Bogotá, 2019, p. 582, https://colaboracion.dnp.gov.co/CDT/
Prensa/PND-2018-2022.pdf (accessed 22 September 2019).

[92]Ley que crea la Agencia Espacial Mexicana, Official Journal of the Federation,
30   July   2010,   https://www.gob.mx/cms/uploads/attachment/file/73063/Ley_que_crea_la_
AgenciaEspacialMexicana.pdf (accessed 20 May 2019).

[93]Ley Federal de Telecomunicaciones y Radiodifusión, Cámara de Diputados del Congreso de la
Unión, 14 July 2014, http://www.diputados.gob.mx/LeyesBiblio/pdf/LFTR_140219.pdf (accessed
7 April 2019).

[94]Acuerdo mediante el cual se dan a conocer las Líneas Generales de la Política Espacial de
México, Ministry of Communications and Transportation, Official Journal of the Federation, 13
July   2011,   https://www.gob.mx/cms/uploads/attachment/file/73124/Lineas_Generalas_Politica_
Espacial_de_Mexico.pdf (accessed 20 May 2019), Acuerdo No 4/II/ORD./11.04.12/S adopting the

5.2.7 Paraguay

Paraguay's increasing interest in space activities has materialized in the recent adoption of Law N° 5151 that creates the Paraguayan Space Agency (AEP) which is responsible for the understanding, design, proposals and implementation of space and aerospace policies and programmes. The Law also stipulates that the AEP will promote and manage the development of the national space activities, and implement the Paraguayan Space Policy through the National Space Activities Programme. Besides, the Law enumerates the functions, structure, and financial sources of the Agency.[95]

Concerning the telecommunications sector, the Telecommunications Law or Law N° 642/95, adopted on 29 December 1995, creates the National Telecommunications Commission, which is responsible for the regulation of national telecommunications. The Law establishes the legal framework for the use of telecommunication and broadcasting services, and other services (private, value added, and basic services), and promotes radio ham activities. It also contains the conditions for obtaining licenses, concessions, and authorizations for the operation of those services.[96]

5.2.8 Peru

Peru can be considered as one of Latin America's pioneers in the space sector since the Law of the Establishment of the National Aerospace Research and Development Commission (CONIDA) or Law 20643, still in force, was adopted on 11 June 1974. The Law specifies the purposes, functions, structure, and financial resources of CONIDA, which is an organisation dependent on the Ministry of Aeronautics.[97]

Furthermore, Peru adopted the Telecommunications Law on 28 April 1993 that regulates all kind of telecommunications services, the competences of the Ministry

---

Organic Statute of the Mexican Space Agency (last reform on 10 April 2015), https://www.gob.mx/cms/uploads/attachment/file/73035/Estatuto_Organico_AEM.pdf (accessed 20 May 2019).

[95]Ley N° 5151, Agencia Espacial del Paraguay, Poder Legislativo, 24 March 2014, http://www.aep.gov.py/application/files/8115/1852/8075/LEY_5151_agencia_espacial.pdf (accessed 22 May 2019). There are some Decrees concerning the nomination of the authorities of the Agency, which are available at Legal Framework, Agencia Espacial del Paraguay, http://www.aep.gov.py/index.php/institucion/marco-legal (accessed 22 May 2019).

[96]Ley N° 642/95 de Telecomunicaciones, Agencia Espacial del Paraguay, https://www.conatel.gov.py/images/iprincipal/LEY%20642/Ley_N_642-95.pdf (accessed 22 May 2019).

[97]Decreto Ley de Creación de CONIDA, Decreto Ley 20643, 11 June 1974, http://www.conida.gob.pe/transparencia/datos_generales/PDF/Decreto%20Ley%20CONIDA.pdf (accessed 23 May 2019).

of Transportation, Communications, Housing, and Construction on telecommunications, the conditions and rights to use them, and several administrative procedures.[98] There is also the General Regulations of the Telecommunications Law, which regulates telecommunications services, radio spectrum management, and the market of services related to the sector, among others. For example, it regulates space services that are classified in three types: space research services, space operations services, and meteorological satellite services. Moreover, it determines the conditions to get licenses, concessions and authorizations for the use of the radio electric spectrum.[99]

5.2.9 Venezuela

The Law on the Establishment of the Bolivarian Agency for Space Activities of Venezuela was adopted on 9 August 2007 and entered into force on 1 January 2008. It creates the Agency as an institute responsible for the implementation of the policies and guidelines issued by the science and technology national body for the peaceful exploration and use of outer space and other areas that cannot be considered as common heritage of humankind, as well as everything related to space. The Law determines the functions, responsibilities, structure, assets, and other administrative arrangements of the Agency.[100]

The Organic Law of Telecommunications adopted on 7 February 2011 regulates the telecommunications sector. It stipulates that the telecommunications and radio electric spectrum integral regime is the responsibility of the National Public Power, and determines the functions of the National Commission on Telecommunications. It also includes the rights and duties of users and operators, the provision of services, and the establishment and exploitation of telecommunications networks, the development of the telecommunication sector, the participation of the government in the field, and several administrative, financial and procedural issues.[101]

---

[98]Texto Único Ordenado de la Ley de Telecomunicaciones, Decreto Supremo N° 013-93-TCC, 28 April 1993, Ministry of Justice, https://www.osiptel.gob.pe/repositorioaps/data/1/1/1/par/ds013-93-tcc-tuo-ley-de-telecomunicaciones/DS013-93-TCC-TUO-Ley-de-Telecomunicaciones.pdf (accessed 23 May 2019).

[99]Texto Único Ordenado del Reglamento General de la Ley de Telecomunicaciones, Decreto Supremo N° 020-2007-MTC, Ministry of Justice, 3 July 2007, https://www.osiptel.gob.pe/repositorioaps/data/1/1/1/par/ds020-2007-mtc/DS020-2007-MTC.pdf (accessed 23 May 2019). There are other telecommunications regulations that regulate specific issues related to the sector. They are available at https://www.osiptel.gob.pe/documentos/legislacion-en-telecomunicaciones.

[100]Ley de la Agencia Bolivariana para Actividades Espaciales, La Asamblea Nacional de la República Bolivariana de Venezuela, 9 August 2007, http://www.abae.gob.ve/web/leyes/SANC-%20LEY-%20DE-%20LA-%20AGENCIA-%20BOLIVARIANA-PARA-%20ACTIVIDADES-%20ESPACIALES-09-08-07.pdf (accessed 24 May 2019).

[101]Ley Orgánica de Telecomunicaciones, Official Journal N° 39.610, 7 February 2011, http://www.conatel.gob.ve/ley-organica-de-telecomunicaciones-2/ (accessed 24 May 2019).

Part II
Country Reports

Chapter 6
Argentina

Contents

Abstract Argentina's space activities go from the formation of human capital to the development of telecommunications satellites. In the early years of the space race, Argentina started to develop its own national space sector, namely through the development and launch of experimental rockets, becoming nowadays an international and regional reference in the sector. This chapter presents a general overview of Argentina's space sector. The first sections focus on general issues of the country, the history of Argentina's space activities, and the economic perspective of the

© Springer Nature Switzerland AG 2020
A. Froehlich et al., *Space Supporting Latin America*, Studies in Space Policy 25,
https://doi.org/10.1007/978-3-030-38520-0_6

sector. Next, there is a description of the main characteristics of the national space plan and legal framework, as well as the space actors responsible to implement them. International space cooperation agreements between Argentina and other parties are also noted. The last sections focus on issues concerning Argentina's satellite capabilities, in particular the current satellite programs and the national geostationary satellite system. In addition, there are several sections concerning Argentina's space technology, its launching capabilities, and its capacity building activities.

6.1 General Country Overview

Argentina is the second largest country in Latin America. It has borders with Bolivia and Paraguay in the north, with Chile and the Atlantic Ocean in the south, with Brazil, Uruguay and the Atlantic Ocean in the east, and with Chile in the west. Argentina's territory extends to 3,761,274 km^2, and it possesses multiple landscapes, such as ice fields, plains, arid zones, and mountains. There are four types of climate: warm, temperate, arid, and cold.[1]

Argentina has a population of 40,117,096. Spanish is the official language, which makes Argentina the largest Spanish speaking country in the world. However, there are also Amerindian languages, such as Mapuche, Guarani, and Quechua. The official currency is the Peso.[2]

6.1.1 Political System

The Republic of Argentina is a federal state composed of 23 provinces and an autonomous city, Buenos Aires, which is also the capital (Fig. 6.1). Argentina has a representative, federal and republican government, and it possess a presidential and democratic regime. Argentina's constitution was adopted in 1853, and was subsequently amended in 1860, 1898, 1957 and 1994. At the federal level there are three powers: the executive, the legislative and the judicial power. The head of the executive is the President (currently Alberto Fernández), who is elected for a four-year term with the possibility to be consecutively re-elected once. There is also a vice president, who is elected for the same term as the president. The legislature, also known as the Congress of the Nation, is composed of the Chamber of Deputies (257 members) and the Chamber of Senators (72 members, three for each province and

---

[1] Acerca de Argentina, Casa Rosada, Presidencia de la Nación, https://www.casarosada.gob.ar/nuestro-pais/acerca-de-argentina (accessed 5 July 2019).

[2] Ibid.

Fig. 6.1 Argentina's
political division (2019)

three for the autonomous city). The vice-president is the president of the Senate. The judicial power is exercised by the Supreme Court of Justice, as well as other inferior tribunals.[3]

---

[3]Organización, Casa Rosada, Presidencia de la Nación, https://www.casarosada.gob.ar/nuestro-pais/organizacion (accessed 5 July 2019).

6.1.2 Economic Situation

Argentina's economy has been unstable in the last decades. Weak policies have not been sufficient to increase productivity growth over the last 20 years. Although a few years ago some reforms were undertaken to strengthen growth and well-being, the 2018 economic crisis kept the national economy in recession. According to an OECD study, it is expected that the economic growth will shrink to 1.5% in 2019 before getting back to 2.3% in 2020.[4] The economic situation is difficult due to a fall in the national currency, the fall of stock values and country risk.[5]

Argentina is currently facing problems to integrate into the world economy, mainly because of the custom barriers that protect national enterprises from international competition, preventing Argentina's inclusion in global value chains.[6] The lack of structural reforms in the economy to enable greater insertion in the global economy has generated stagnation in employment, particularly in productivity and salary increases.[7]

6.2 History of Space Activities

Argentina's involvement in the space sector started in the early years of the space race when it developed its own space program. In just a few years, Argentina attained important breakthroughs such as the launching of rockets for experimental and military purposes, even from the Antarctic region, and the creation of the National Commission on Space Research (CNIE).[8] However, Argentina's space activities diminished from 1973 to 1981, the year in which the last scientific rocket was launched, although the CONDOR military space project continued to exist for several years. This project was finally cancelled in 1993. Two years before, in 1991, the National Commission on Space Activities (CONAE)[9] was created, a government body responsible for the implementation of the national space plan, which has now been renovated (Table 6.1).[10]

---

[4]"La OCDE afirma que Argentina necesita más reformas para conseguir una economía más fuerte e inclusiva", OECD, http://www.oecd.org/newsroom/la-ocde-afirma-que-argentina-necesita-mas-reformas-para-conseguir-una-economia-mas-fuerte-e-inclusiva.htm (accessed 5 July 2019).

[5]Argentina se soma al abismo económico entre dudas sobre la gestión de Macri, El País Internacional, 26 April 2019, https://elpais.com/internacional/2019/04/25/argentina/1556215894_662283.html (accessed 5 July 2019).

[6]"La OCDE afirma que Argentina necesita más reformas para conseguir una economía más fuerte e inclusiva", OECD, http://www.oecd.org/newsroom/la-ocde-afirma-que-argentina-necesita-mas-reformas-para-conseguir-una-economia-mas-fuerte-e-inclusiva.htm (accessed 5 July 2019).

[7]Ibid.

[8]Comisión Nacional de Investigaciones Espaciales (CNIE).

[9]Comisión Nacional de Actividades Espaciales (CONAE).

Table 6.1 Argentina's space history timeline

Year	Argentina's space history
1950s	First research and development of rocket engines
1961	Creation of the CNIE as part of the Argentina's Air Force
1961	Use of space platforms in Chamical for experiments with the Beta and Gamma Rocket-probes and stratospheric balloons
1965	Operation "MATIENZO" consisted of the launching of GAMMA CENTAURO rockets, and balloons from the Matienzo base (Antarctic)
1967	First rocket launch for biological experiments
1969	Launching of a monkey in a RIGEL rocket
1969	Construction of the Balcarce ground station for satellite communications
1972	Agreement between CNIE and the Max Planck Institute (Germany) to carry out the German-Argentinian Experience with Ionized Clouds (EGANI),[a] which was later transformed into the Argentinian Experience with Ionized Clouds (EANI)[b]
1980s	Development of the ALACRAN and CONDOR missiles by Argentina's Air Force
1980s	Management of orbital slots of the geostationary orbit (GEO) at the International Telecommunication Union (ITU)
1981	Last experience of the century with a national TAURO series rocket
1981	First satellite communication between Argentina and the Antarctic
1991	Establishment of the National Commission on Space Activities (CONAE)[c] as a body dependent on the Presidency
1993	Cancellation of the CONDOR II project and delivery of its missiles and components to Spain
1990s	Cooperation between Argentina and NASA for Argentinian satellite scientific missions
1990s	Launching of the first Satellites of Scientific Applications (SAC) in collaboration with the USA
1994	First National Space Plan to be implemented by CONAE
1996	CONAE is moved to the Ministry of Culture and Education
1996	CONAE is transferred to the Ministry of Foreign Affairs, International Trade and Worship
1997	Inauguration of the Space Centre of Falda del Carmen, Córdoba
1997	Satellite NAHUEL 1A of the Nahuelsat S.A. private company is launched into GEO
2003	Argentina becomes a member of the Group on Earth Observations (EO)
2006	Creation of the Argentinian Company of Satellite Solutions[d] (ARSAT)
2010	Establishment of the Centre of High Technology Tests (CEATSA)[e]
2011	Launching of the SAC-D satellite
2012	CONAE is moved to the Ministry of Federal Planning, Public Investment and Service
2013	Development of the Centenario Experience by Argentina's Air Force
2014	Launching of satellite ARSAT-1 to the GEO
2015	Launching of satellite ARSAT-2 to the GEO
2016	CONAE is transferred to the Ministry of Science, Technology and Productive Innovation

[a]Experiencia Germano-Argentina con Nubes Ionizadas (EGANI)
[b]Experiencia Argentina con Nubes Ionizadas (EANI)
[c]Comisión Nacional de Actividades Espaciales (CONAE)
[d]Empresa Argentina de Soluciones Satelitales (ARSAT)
[e]Centro de Ensayos de Alta Tecnología Sociedad Anónima (CEATSA)

6.3   Economic Perspective of Space Programs

Among its high technology activities, Argentina's satellite industry is perhaps the most important part of its national space sector and is a key element of its space economy.[11] Argentina belongs to the small group of countries that possess the capacity to develop telecommunication and observation satellites, and is currently developing a launch industry. The country also possesses recognised human capital in the aerospace sector.[12] However, Argentina's high technology competitiveness capabilities in several niches of the space economy have been inhibited due to the lack of government support.[13]

Moreover, Argentina's economic crisis and national political decisions have had a negative impact on the national space sector in recent years. The institutional budget allocated to the Minister of Science, Technology and Productive Innovation (MINCYT)[14] and its two decentralised organisms, the National Scientific and Technical Research Council (CONICET)[15] and CONAE (see Sect. 6.4.2), has seen progressive cuts since 2015.[16] For instance, the CONAE's projected budget for 2018 fell to the level of the period 2011–2012, and the 2019 budget suffered an important cut.[17]

Furthermore, in recent years, there has been a constant decrease in investments for the development, launching and operation of satellites. For example, in 2019, there was a major cut to the New Generation Space Vehicle's budget (VENG SA),[18] the executive body of CONAE, which has lost 60 job positions in the past year.

---

[10]See CONAE: la Argentina en el espacio, Educar, https://www.educ.ar/noticias/108996/conae-la-argentina-en-el-espacio; Adelanto de "Historia de la actividad espacial en Argentina", de Pablo de León, Periodismo.com, https://www.periodismo.com/2018/11/18/historia-de-la-actividad-espacial-en-argentina-de-pablo-de-leon/, and Manfredi Alberto, Argentina y la Conquista del Espacio, Fundación Histarmar, https://www.histarmar.com.ar/AVIACION/ArgylaConquistadelEspacio.htm (accessed 8 September 2019).

[11]López Andrés et al., Al infinito y más allá: una exploración sobre la economía especial en Argentina, Instituto Interdisciplinario de Economía Política de Buenos Aires (IIEP-BAIRES), March 2017, pp. 4–5.

[12]Ibid.

[13]Ibid., p. 37.

[14]Ministerio de Ciencia, Tecnología e Innovación Productiva (MINCYT).

[15]Consejo Nacional de Investigaciones Científicas y Técnicas (CONICET).

[16]In 2017, the total budget for these bodies was 18% lower than their 2015 budget. Cf. D. Stefani Fernando, "Evolución del presupuesto del Ministerio de Ciencia, Tecnología e Innovación Productiva (MINCYT), y de la función Ciencia y Técnica del presupuesto nacional", 1 February 2017, https://cibion.conicet.gov.ar/wp-content/uploads/sites/22/2017/10/Evolucion-de-presupuesto-MINCYT-y-f-CyT.pdf (accessed 8 September 2019).

[17]"CONAE, despidos y recortes entre festejos por SAOCOM", 26 October 2019, http://latamsatelital.com/conae-despidos-recortes-festejos-saocom/ (accessed 8 September 2019).

[18]Vehículo Espacial de Nueva Generación (VENG SA).

Moreover, there will be no investments for the company according to the new plan.[19] Between 2016 and 2017, the government suspended the development of the SARE, SARA-UAV and ARSAT-3 satellites, forcing the country to acquire a foreign satellite for €7 million to occupy the 81° W geostationary orbital position to avoid its loss by the ITU.[20] In addition, in the first trimester of 2019, INVAP, a national technology company of Rio Negro province, lost 36% of its budget compared to the same period of the previous year.[21] INVAP also suffered from the suspension of ARSAT-3. Besides, the company had to deal with payment delays by the government, forcing it to take on debt from the financial system.[22]

According to some sources, the current difficult situation of the Argentinian space sector, especially the satellite branch, is due to state policies that can be financially altered in a discretionary way. The aforementioned consequences and other situations, such as the closure of the STI technology company, the lack of new space projects and the authorisation of 25 foreign satellites to compete with ARSAT (see Sect. 6.4.2.2), are direct consequences of these policies.[23] Therefore, it seems that the Argentinian space sector will continue to have important challenges in its development in the following years.

6.4 Space Development Policies and Emerging Programs

Argentina's space activities are organised in a National Space Plan and Space Policy. CONAE and other actors with space activities and responsibilities contribute to developing the space sector through this policy.

6.4.1 Space Policies and Main Priorities

Since 1995, Argentina has implemented a series of National Space Plans (NSP) in different strategic areas. The government is currently preparing the NSP for the coming years.

---

[19]Martínez Gabriel et al., "Cae un 55% la inversión en el sector espacial argentino", Motor económico, http://motoreconomico.com.ar/cruda-realidad/cae-un-55-la-inversin-en-el-sector-espacial-argentino (accessed 8 September 2019).

[20]Ibid.

[21]"No despega la actividad espacial en INVAP", LATAM Satelital, 31 May 2019, http://latamsatelital.com/no-despega-la-actividad-espacial-invap/ (accessed 8 September 2019).

[22]Ibid.

[23]"CONAE, despidos y recortes entre festejos por SAOCOM", 26 October 2019, http://latamsatelital.com/conae-despidos-recortes-festejos-saocom/ (accessed 8 September 2019).

6.4.1.1 National Space Plan

The National Space Plan (NSP) or Strategic Plan for Space Activities is a state policy of national priority for the development of space knowledge and technology. The NSP is proposed and implemented by CONAE[24] and it comprises three elements: EO, peaceful exploration and use of outer space, and development of technology for space uses.[25]

Other objectives of the NSP are the strengthening of the national scientific and technological sector by providing advance knowledge and new education and work opportunities in space matters, such as the creation of specific careers and space specialities. Moreover, the NSP provides user training in space information and broad training programs, specific geospace development, and fieldwork for educational and labour purposes.[26]

The first version of the NSP covered the years 1995 to 2006. Succeeding NSPs covered the 1997 to 2008 and 2004 to 2015 terms. Currently, the NSP covers the 2016 to 2027 period, which has been approved by the Directorate of CONAE and transmitted to the national executive power.[27] The implementation and management of the NSP are carried out by a series of action courses (Table 6.2).

Earth Observation

EO is an important element of the NSP. It consists of the use of space data for the development of applications required and demanded by society. The data can be obtained from own missions or third party missions through data download from the national territory or by means of cooperation agreements. Own satellite missions are

Table 6.2 National space plan's action courses

National space plan's action courses
Use and management of space data
Satellite systems
Access to space
Integration, tests and assessment
Peaceful use and exploration of outer space
Education and training
National connection and international incorporation

---

[24]Comisión Nacional de Actividades Espaciales (CONAE).

[25]Plan Espacial Nacional, Comisión Nacional de Actividades Espaciales, https://www.argentina.gob.ar/ciencia/conae/plan-espacial (accessed 4 July 2019).

[26]Ibid.

[27]Ibid.

Table 6.3 EO's strategic areas and information sectors

Strategic areas	Information sectors
Environmental	Waters
	Land cover
	Atmosphere and climate
Production	Agriculture, livestock and forestry
	Fishing
	Mining
	Energy
Social	Health
	Emergencies
	Territorial ordering and integrity
	Security

This table was obtained from Observación de la Tierra, Comisión Nacional de Actividades Espaciales, https://www.argentina.gob.ar/ciencia/conae/plan-espacial/observacion-de-la-tierra (accessed 9 July 2019)

designed according to local user needs, and comprise the design, construction and putting into orbit of satellites through national or foreign launching services.[28]

CONAE has identified information sectors that correspond to specific societal demands and needs. These sectors have been assembled in interconnected strategic areas related to environmental issues, the production sector, and social aspects (Table 6.3).[29]

Peaceful Exploration and Use of Outer Space

Argentina has long been very interested in the peaceful exploration and use of outer space.[30] These activities allow the country to participate in interplanetary missions and scientific and technological research programs on outer space. To this end, for instance, the NSP envisages the development of global navigation technology and international cooperation in robotic and human missions to other planets. For this purpose, Argentina participates through CONAE in the Deep Space Networks of the European Space Agency (ESA) and the Launch and Control Agency of China. The

---

[28]Observación de la Tierra, Comisión Nacional de Actividades Espaciales, https://www.argentina.gob.ar/ciencia/conae/plan-espacial/observacion-de-la-tierra (accessed 9 July 2019).

[29]Ibid.

[30]Belonging to the United Nations Committee on the Peaceful Uses of Outer Space (UNCOPUOS) and the ratification of four of the five UN Space treaties is a strong commitment to this goal.

country also coordinates the available observation time of the monitoring antennas of interplanetary missions installed in Argentinian territory.[31]

Development of Technology for Space Uses

The creation of space technology is vital for national development. Moreover, the use and creation of own applications derived from space activities contribute to the independence and development of the country. In this sense, the work of CONAE in fostering the production of knowledge and the creation of space technology is essential. For example, the establishment and strengthening of new companies, some of which are CONAE providers, can create new technology and high added value exports. Additionally, the impulse to develop a national industry in the space area contributes to positioning Argentina in the international competition arena, as well as to creating specific careers in this field.[32]

6.4.2   Space Actors

The two institutions that develop Argentina's space missions are CONAE and ARSAT. CONICET, with its multiple centres and institutions, also contributes to the development of the national space sector. In addition, space related companies add value to this sector.

6.4.2.1   National Commission on Space Activities

CONAE, successor of CNIE, is the national body responsible for acting in a private and public capacity on the scientific, technical, industrial, commercial, administrative and financial issues of the space sector. It also proposes nationwide policies for the promotion and execution of peaceful space activities.[33] Pursuant to Decree N° 995/91, CONAE is the only competent state body to design, execute, control and administer outer space projects, endeavours and activities, including the implementation of certain satellite missions.

CONAE's mission aims to receive space data and develop applications on Argentina's sovereign territory; to deliver appropriate and timely information to the

---

[31] Uso del espacio ultraterrestre, Comisión Nacional de Actividades Espaciales, https://www.argentina.gob.ar/ciencia/conae/plan-espacial/uso-del-espacio-ultraterrestre (accessed 9 July 2019).

[32] Desarrollos tecnológicos para uso espacial, Comisión Nacional de Actividades Espaciales, https://www.argentina.gob.ar/ciencia/conae/plan-espacial/desarrollos-tecnologicos (accessed 9 July 2019).

[33] Misión, Comisión Nacional de Actividades Espaciales, https://www.argentina.gob.ar/institucional/mision (accessed 7 July 2019).

economic and productive sectors of the country in order to increase their competitiveness and productivity nationally and internationally; to promote the development of the national industry, namely new companies that innovate technologies, and to extend the sphere of their participation at the international level; to participate internationally in the peaceful exploration and use of outer space; and to make contributions to the national scientific and technological sector, as well as fostering new education and jobs opportunities.[34]

CONAE also proposes and implements the NSP (see Sect. 6.4.1.1) to use and take advantage of space science and technology, as well as to provide information to the state to collaborate in government management.[35] In addition, CONAE receives, processes and stocks mission data through the Teófilo Tabanera Space Centre.

6.4.2.2 The Argentinian Company of Satellite Solutions

ARSAT is a state company created in 2006 by Federal Law 26.092. It is responsible for the domestic design, development and construction of the launching or putting into service of geostationary telecommunications satellites in orbital slots resulting from coordination procedures in the ITU, and their associated frequencies, as well as the respective exploitation, use, and the provision of satellite facilities or commercialization of satellite services.[36] More specifically, ARSAT is responsible for the use of the geostationary orbital slots 72° W and 81° W and associated frequencies. It currently includes the satellites ARSAT-1 and ARSAT-2. Other ARSAT's services include open digital television, the Federal Fiber Optics Network, a national data centre, rural schools, and an infrastructure of space data.[37]

ARSAT is constituted as a public limited company in which 98% of the share capital of the company falls to the Ministry of Planning, Public Investment and Services, and the remaining 2% belongs to the Ministry of Economy and Production.[38]

6.4.2.3 National Scientific and Technical Research Council

CONICET is the main agency that fosters science and technology in Argentina. CONICET also encourages the development of the national economy and improvement of the quality of life of Argentinian society, promotes exchange and scientific and technological cooperation in the country and abroad, provides grants for research projects,

---

[34]Ibid.

[35]Ibid.

[36]Art. 5, ARSAT Statut, Ley 26.092, Créase la Empresa Argentina de Soluciones Satelitales Sociedad Anónima AR-SAT. Estatuto social, 26 April 2006, http://servicios.infoleg.gob.ar/infolegInternet/anexos/115000-119999/115886/texact.htm (accessed 7 July 2019).

[37]ARSAT, official website https://www.arsat.com.ar (accessed 7 July 2019).

[38]Ley 26.092, Créase la Empresa Argentina de Soluciones Satelitales Sociedad Anónima AR-SAT. Estatuto social. Otórgase a dicha empresa la autorización de uso de la posición orbital 81° de Longitud Oeste y sus bandas de frecuencias asociadas, 26 April 2006, art. 8.

fellowships for training and further education of university graduates, and organizes and funds institutes, laboratories and research centres. There are four large areas of knowledge over which CONICET has competence: agrarian, engineering and material sciences, biological and health sciences, exact and natural sciences, and social sciences and humanities. The area of exact and natural sciences includes physics, astronomy, and mathematics, among others. Around 23% of CONICET researchers and 21% of the fellows are part of this area.[39] Some of CONICET's Scientific and Technological Centres are related to this area, such as the Astronomical Complex "El Leoncito", the Institute of Astronomy and Space Physics, the Astrophysical Institute "La Plata", the Argentinian Institute of Radio astronomy, the Institute of Theoretical and Experimental Astronomy, the Institute of Astronomic, Earth and Space Sciences, and the Institute of Detection Technology and Astro particles.[40]

6.4.2.4   Other Actors

Apart from the aforementioned entities, Argentina has other actors that contribute to the development of its space sector, namely space and technology companies. Some of these companies are INVAP, CEATSA and VENG S.A.[41]

6.5   Perspectives on the Legal Framework for Space Activities

The development of the Argentinian space and telecommunication sectors could not be possible without the multiple national regulations that create the relevant bodies and mechanisms responsible for fostering these sectors. In addition, Argentina has ratified almost all the main international space and telecommunications treaties supporting the codification and progressive development of international space law.

---

[39] Description, Consejo Nacional de Investigaciones Científicas y Técnicas (CONICET), https://www.conicet.gov.ar/about-the-conicet/?lan=en (accessed 13 July 2019).

[40] Complejo Astronómico "El Leoncito" (CASLEO), Instituto de Astronomía y Física del Espacio (IAFE), Instituto de Astrofísica "La Plata" (IALP), Instituto Argentino de Radioastronomía (IAR), Instituto de Astronomía Teórica y Experimental (IATE), Instituto de Ciencias Astronómicas, de la Tierra y del Espacio (ICATE), and Instituto de Tecnología en Detección y Astropartículas (ITEDA).

[41] See Sect. 6.9 for a deep analysis of these actors.

6.5.1 National Space Regulations and Instruments

Legislation of space activities in Argentina comprises a series of laws, decrees and resolutions.[42] Most of these regulations pertain to the objectives, functions and administrative structure of CONAE, and determine the characteristics of the different NSPs adopted since 1991. Other national regulations govern specific issues on space and telecommunications such as the register of space objects and the creation of the Argentinian Geostationary Satellite System.

6.5.1.1 National Commission on Space Activities Regulations

The creation of CONAE was instituted by National Decree N° 995/91 adopted in 1991.[43] The decree regulates the functions, structure, assets and propriety of the Agency. The decree also replaces Decree N° 1164/60 and dissolves the National Commission on Space Research, the precursor of CONAE.[44]

Originally placed under the competence of the Presidency, CONAE was later transferred to different ministries pursuant to a series of decrees in the last thirty years. Decree N° 660/96 (1996) transferred CONAE to the Ministry of Culture and Education through the Secretary of Science and Technology[45]; Decree N° 1274/96 (art. 9) (1996) transferred the Agency to the Ministry of Foreign Affairs and Worship[46]; Decree N° 2197/12 (2012) transferred CONAE to the Ministry of Federal Planning, Public Investment and Services,[47] and Decree N° 242/16 (2016) put CONAE under the orbit of the Ministry of Science, Technology and Productive Innovation.[48]

The NSP, which is planned and implemented by CONAE, has been periodically approved for different periods. The first NSP was adopted by Decree N° 2076/94 for the period 1995–2006[49]; Decree N° 1330/99 approved the NSP for the period

---

[42]Most of these regulations are available at https://www.argentina.gob.ar/ciencia/institucional/legislacion (accessed 7 July 2019).

[43]Which was ratified by the Laws N° 24.061 (1992) and N° 11.672 (art. 32).

[44]Decreto N° 995/91, Creación de la Comisión Nacional de Actividades Espaciales, Buenos Aires, 3 June 1991.

[45]Decreto N° 660/96, Modificación de la actual estructura de la Administración Nacional, 24 June 1996.

[46]Decreto N° 1274/96, Apruébase la estructura organizativa del citado Departamento de Estado, Ministerio de Cultura y Educación, 7 November 1996.

[47]Decreto N° 2197/12, Transfiérese la Comisión Nacional de Actividades Espaciales (CONAE) a la Órbita del Ministerio de Planificación Federal, Inversión Pública y Servicios, Official Journal, 15 November 2012.

[48]Decreto N° 242/16, Dispónese el funcionamiento de la Comisión Nacional de Actividades Espaciales (CONAE) bajo la órbita del Ministerio de Ciencia, Tecnología e Innovación Productiva, Official Journal, 26 January 2016.

[49]Decreto N° 2076/94, Apruébase el Plan Espacial Nacional y sus acciones previstas para el período 1995–2006, Official Journal, 1 December 1994.

1997–2008,[50] which also determined that its content defined it as a strategic plan. Then, Decree N° 532/05 approved the 2004–2015 NSP,[51] which declared that space activity is a state policy and a national priority. Currently, the NSP for the period 2016 to 2027 has already been approved by CONAE and sent to the National Executive for its approval.[52]

6.5.1.2 Other Space Related Regulations

Other space regulations include Decree N° 125/95 (1995) that created the National Register of Space Objects Launched into Outer Space; Federal Law 26.092 that creates ARSAT and grants it the 81° W geostationary orbital position and associated frequencies[53]; Federal Law 27.208 for the Development of a Satellite Industry, that creates the Argentinian Geostationary Satellite Plan for the period 2015–2035[54]; Decree N° 1552/2010 which creates the National Telecommunications Plan "Argentina conectada" (Argentina connected), which has among its strategic axes the optimization of the use of the radioelectric spectre[55]; Decree N° 626/2007, which granted ARSAT the orbital position 72° W,[56] and Resolution N° 79/2007 of the ex-Ministry of Communications, which authorized ARSAT to provide satellite services in this orbital position; Resolution N° 581/11 adopted by the Ministry of Security for the establishment of the Department of Images Intelligence, which use space images for national security issues,[57] and Resolution N° 1141/2011 that created the Joint Course of Interpretation of Satellite Images to be implement by the Ministry of Security (CCIISAT).[58]

---

[50]Decreto N° 1330/99, Establécese que el Plan Espacial Nacional 1997–2008 y las acciones previstas para dicho período reviste el carácter de plan estratégico, Official Journal, 17 November 1999.

[51]Decreto N° 532/05, Declárese al desarrollo de la actividad espacial como política de Estado y de prioridad nacional. Apruébase el Plan Espacial Nacional 2004–2015, Official Journal, 26 May 2005.

[52]Plan Espacial Nacional, Comisión Nacional de Actividades Espaciales, https://www.argentina.gob.ar/ciencia/conae/plan-espacial (accessed 8 July 2019).

[53]Ley 26.092 Empresas Argentinas de Soluciones Satelitales, Creación, Official Journal, 27 April 2006.

[54]Ley 25.208 de Desarrollo de la Industria Satelital, Plan Satelital Geoestacionario Argentino, 9 November 2015.

[55]Decreto N° 1552/2010, Créase el Plan Nacional de Telecomunicaciones "Argentina Conectada", 21 October 2010.

[56]Decreto N° 626/2007, Poder Ejecutivo Nacional, Official Journal, 30 May 2007.

[57]Resolución N° 1141/2011, Ministerio de Seguridad, 15 November 2011.

[58]Curso Conjunto de Interpretación de Imágenes Satelitales (CCIISAT). Resolución 1141/2011, Ministerio de Seguridad, 23 November 2011.

6.5.2   Argentina and International Space Instruments

Argentina's commitment to respect international space law is materialized through the ratification of several of the most important treaties on space and telecommunications issues. The creation of a national register of space objects launched into outer space is an example of this commitment. Argentina has also committed unilaterally to respect various multilateral guidelines on space debris mitigation.

6.5.2.1   Space Treaties and Other Instruments

The long record of Argentina's space activities is reflected in the engagement of this South American country in respect of international legal rules governing space activities. Argentina is a party to several international treaties on space related issues (Table 6.4), including four of the five UN Space Treaties, the Convention and Constitution of ITU, and the Treaty Banning Nuclear Weapon Tests in the Atmosphere, in Outer Space and under Water (NTB).[59] Argentina has also signed, but not ratified, the Convention Relating to the Distribution of Programme-Carrying Signals Transmitted by Satellite (BRS) of 1974. Moreover, even though Argentina has not signed nor ratified the Moon Agreement of 1979, it actively participated in the negotiations

Table 6.4 International space related treaties ratified by Argentina

International space treaties ratified by Argentina		
Treaty	Adoption	Ratification (R)/signature (S)
OST	1967	R
ARRA	1968	R
LIAB	1972	R
REG	1975	R
MOON	1979	–
NTB	1963	R
BRS	1974	S
ITSO[a]	1971	R
IMSO[b]	1976	R
ITU	1992	R

[a]Agreement Relating to the International Telecommunications Satellite Organization, 20 August 1971
[b]Convention on the International Mobile Satellite Organization, 3 September 1976

---

[59]Status of International Agreements relating to activities in outer space as at 1 January 2019, UNCOPUOS, A/AC.105/C.2/2019/CRP.3, 1 April 2019, http://www.unoosa.org/documents/pdf/spacelaw/treatystatus/AC105_C2_2019_CRP03E.pdf (accessed 5 July 2019).

of this treaty as documented by the travaux préparatoires.[60] In this sense, Argentina had an important role in supporting the inclusion in the Moon Agreement of the common heritage of mankind concept relating to the use of the natural resources of the Moon.[61]

6.5.2.2   National Registry

As state party to the Registration Convention, Argentina has created a national registry to enter its space objects launched into outer space.[62] On 30 December 1996, Argentina informed the UN Secretary-General of the establishment of its national register, which is kept by CONAE.[63] In addition, Argentina continuously informs the UN Secretary-General on all its space objects launched into outer space in accordance with Article IV of the Registration Convention. Accordingly, Argentina has submitted eleven notifications to UNOOSA to date.[64]

6.5.2.3   Space Debris Mitigation Instruments

Although Argentina has not yet adopted a national regulatory framework on space debris mitigation, it fully adheres to the UNCOPUOS Space Debris Mitigation Guidelines, and it supports the IACD Space Debris Mitigation Guidelines.[65]

---

[60] Agreement Governing the Activities of States on the Moon and Other Celestial Bodies (Moon Agreement), UNOOSA, http://www.unoosa.org/oosa/en/ourwork/spacelaw/treaties/travaux-preparatoires/moon-agreement.html (accessed 5 July 2019).

[61] Jakhu Ram, 20 years of the Moon Agreement: Space Law Challenges for Returning to the Moon, in Disseminating and Developing International and National Space Law: the Latin America and Caribbean Perspective, Proceedings, United Nations/Brazil Workshop on Space Law, ST/SPACE/28, UNOOSA, New York, 2005, p. 346.

[62] Art. II, paragraph 1 REG: "When a space object is launched into Earth orbit or beyond, the launching state shall register the space object by means of an entry in an appropriate registry which it shall maintain. Each launching state shall inform the Secretary-General of the United Nations of the establishment of such a registry".

[63] Note verbale dated 30 December 1996 from the Permanent Mission of Argentina to the United Nations (Vienna) addressed to the Secretary-General, UN Secretary-General, ST/SG/SER.E/INF.13, 15 January 1997, http://www.unoosa.org/pdf/reports/regdocs/SER_INF_013E.pdf (accessed 5 July 2019). National regulations concerning the Argentinian registry are available at Registro de satélites, Comisión Nacional de Actividades Espaciales, https://www.argentina.gob.ar/ciencia/conae/institucional/registro-de-satelites.

[64] Notifications from States & Organizations: Argentina, UNOOSA, http://www.unoosa.org/oosa/en/spaceobjectregister/submissions/argentina.html (accessed 5 July 2019).

[65] These instruments have a recommendatory character. Compendium of space debris mitigation standards adopted by States and international organizations, Argentina, UNCOPUOS A/AC.105/C.2/2014/CRP.15, 18 March 2014, p. 5, http://www.unoosa.org/pdf/limited/c2/AC105_C2_2014_CRP15E.pdf.

6.6 Cooperation Programs

The diversification of Argentina's space activities has been possible thanks to the treaties concluded between Argentina and other countries, and the cooperation agreements adopted between CONAE and other institutions. Argentina also participates through CONAE in many international committees and initiatives of coordination and integration related to the application of space science and technology.[66]

6.6.1 Argentina and Latin America

Argentina has taken advantage of its geographical situation by concluding specific space cooperation agreements with its Latin American neighbours, namely Brazil, Chile, Colombia, Ecuador, Mexico, and Venezuela.

Bilateral relations between Argentina and Brazil concerning space cooperation have had a long trajectory. The Joint Declaration on the Peaceful Uses of Outer Space of 1989 marks the beginning of these relations.[67] In 1996, the two countries signed a Framework Cooperation Agreement on the Peaceful Applications on Space Science and Technologies. This agreement was followed by specific cooperation programs between both countries, such as the Cooperation Program concerning the SAC-D/AQUARIUS project and the Cooperation Program concerning the SAOCOM Project, adopted in 2005 and 2006, respectively. More recently, in 2014 Argentina and Brazil adopted the Protocol of Intentions for the joint development of the SABIA-MAR Space Mission, which consists of the development of a satellite for taking ocean measurements for the understanding and analysis of its biosphere and the ways it is affected by anthropogenic activities (see Sect. 6.7.1.2).[68]

Space cooperation between Argentina and Chile is materialized by a cooperation agreement signed in 2008,[69] and a cooperation agreement between CONAE and the Chilean Information Centre of Natural Resources (CIREN),[70] adopted in 2009, concerning the exchange of optical satellite and radar images, joint research efforts on the natural and productive heritage, and the development of early warning and response for emergencies.[71]

---

[66] Cooperación internacional, Comisión Nacional de Actividades Espaciales, https://www.argentina. gob.ar/ciencia/conae/institucional/cooperacion-internacional (accessed 12 July 2019).

[67] Argentina, Agência Espacial Brasileira, 22 April 2018, http://www.aeb.gov.br/programa-espacial-brasileiro/cooperacao-internacional/argentina/ (accessed 12 July 2019).

[68] Ibid.

[69] Firma de acuerdos bilaterales en Chile, Casa Rosada, presidencia de la Nación, 5 December 2008, https://www.casarosada.gob.ar/informacion/archivo/20313-blank-49596941 (accessed 12 July 2019).

[70] Centro de Información de Recursos Naturales (CIREN).

[71] Chile y Argentina firman convenio de cooperación espacial, El Ranco, Diario, http://www.diarioelranco.cl/2009/08/07/chile-y-argentina-firman-convenio-de-cooperacion-espacial/ (accessed 12 July 2019).

Colombia and Argentina adopted a cooperation agreement for the peaceful use and exploration of outer space in 2008. The same year, the Geographic Institut Agustín Codazzi and the Hydrology Institute of Colombia signed an agreement with CONAE.[72]

Argentina and Ecuador signed a space cooperation agreement in 2007. Since then, the two countries have carried out many bilateral actions, especially between their technical institutions. For instance, negotiations on space cooperation issues, a workshop on the use of satellite images and many conferences on space policy and space technology applications for risk management were held during a visit of CONAE's Secretary General to Ecuador in 2016.[73]

Argentina concluded a framework agreement with the Mexican Space Agency (AEM) concerning space cooperation for peaceful purposes in 2016. In particular, the agreement concerns cooperation on satellite data applications related to natural emergencies, health and productive development, as well as the use of capacities of ground segments and the joint development of satellite technology.[74] There is also a Memorandum of Understanding (MoU) adopted in 2018 between AEM and the Sur Technological Entrepreneurs, S.R.L.[75]

Two cooperation agreements have been signed between Argentina and the Bolivarian Republic of Venezuela: a Framework Cooperation Agreement on the peaceful uses of outer space, space science, technology and applications between the Government of the Bolivarian Republic of Venezuela and the Government of the Republic of Argentina, adopted in 2011, and a Specific Cooperation Agreement in the satellite field between the Ministry of Popular Power for Science, Technology and Innovation through the Bolivarian Agency for Space Activities (ABAE),[76] and the Ministry of Federal, Planning, Public Investment and Services[77] of the Argentinian Republic through CONAE, signed in 2013.[78]

---

[72]La Argentina firmó con Colombia un acuerdo de cooperación espacial, infobae, 23 August 2008, https://www.infobae.com/2008/08/23/399144-la-argentina-firmo-colombia-un-acuerdo-cooperacion-espacial/ (accessed 12 July 2019).

[73]Ecuador y Argentina promueven cooperación en el ámbito espacial, Ministerio de Relaciones Exteriores y Movilidad Humana, Quito, 9 August 2016, https://www.cancilleria.gob.ec/ecuador-y-argentina-promueven-cooperacion-en-el-ambito-espacial/ (accessed 12 July 2019).

[74]Firma de Acuerdo de Cooperación Espacial con México, Comisión Nacional de Actividades Espaciales, 29 July 2016, http://www.conae.gov.ar/index.php/espanol/2016/849-acuerdo-con-mexico (accessed 12 July 2019).

[75]Instrumentos Internacionales, Acuerdos Vigentes firmados con Agencias Espaciales y Organismos Internacionales, Agencia Espacia Mexicana, 15 September 2017, https://www.aem.gob.mx/downloads/mapa_acuerdos_internacionales_AEM.pdf (accessed 12 July 2019).

[76]Ministerio del Poder Popular para Ciencia, Tecnología e Innovación de la República Bolivariana de Venezuela, and Agencia Bolivariana para Actividades Espaciales (ABAE).

[77]Ministerio de Planificación Federal, Inversión Pública y Servicios de la República Argentina.

[78]Cooperación Internacional de la Agencia Bolivariana para Actividades Espaciales (ABAE), http://www.abae.gob.ve/web/ConveniosInt.php (accessed 12 July 2019).

At the multilateral level, Argentina participates in regional fora related to space issues such as the America's Space Conference and the Latin American Society on Remote Sensing and Space Information Systems (Selper Internacional).[79]

6.6.2 Argentina and Europe

Argentina and several European countries and institutions have worked together in different space programs, particularly in the development of space capacities and technology. France and Italy have strong relations with Argentina concerning the space sector. At the multilateral level, Argentina and ESA continue to strengthen their relations with several scientific space projects.

Argentina and France have been working together on space matters since 1996, when a space cooperation agreement was adopted between CONAE and the Centre national d'études spatiales (CNES). Then, several bilateral MoU made possible the participation of CNES in Argentine satellite missions. In May 2019, CNES and CONAE adopted a Framework Agreement on peaceful space cooperation to further strengthen their relations in the sector.[80]

International space cooperation between Argentina and Italy has been very productive. For example, Italy participated through a cooperation agreement in the development of the Argentine satellite SAC-B by providing the solar panels and the Imaging Particle Spectrometer for Energetic Neutral Atoms (ISENA).[81] Furthermore, both countries have created the Italian-Argentine Satellite System for Emergency Management (SIASGE),[82] which possesses six SAOCOM satellites, two provided by CONAE and four by the Agenzia Spatiale Italiana (ASI). The satellites take images of fires, floods, eruptions, earthquakes, avalanches, and landslides. Each year, representatives of ASI visit Argentina to discuss perspectives for the following year.[83]

Argentina and ESA collaborate in the operation, reception and distribution of data of diverse scientific missions through a cooperation agreement for the installation of the DEEP SPACE 3 station—Maragüe in the province of Mendoza, Argentina, which

---

[79]Cooperación internacional, CONAE, https://www.argentina.gob.ar/ciencia/conae/institucional/otros-ambitos-multilaterales (8 September 2019); see also Sociedad Latinoamericana en Percepción Remota y Sistemas de Información Espacial (Selpel Internacional) at https://selper.info/.

[80]Acuerdo de Cooperación con CNES de Francia, Comisión Nacional de Actividades Espaciales, 9 May 2019, https://www.argentina.gob.ar/noticias/acuerdo-de-cooperacion-con-cnes-de-francia (accessed 12 July 2019).

[81]Participación internacional, Comisión Nacional de Actividades Espaciales, http://www.conae.gov.ar/index.php/espanol/2012-09-13-17-12-35 (accessed 12 July 2019).

[82]Sistema Italo-Argentino de Satélites para la Gestión de Emergencias (SIASGE).

[83]Cooperación internacional entre la UNC, la CONAE y la Agencia Espacial de Italia, Universidad Nacional de Córdoba, 7 September 2018, https://www.unc.edu.ar/comunicaci%C3%B3n/cooperaci%C3%B3n-internacional-entre-la-unc-la-conae-y-la-agencia-espacial-de-italia (accessed 12 July 2019).

was inaugurated in 2012. The station is the most modern tracking antenna for deep space exploration missions of ESA, such as BEPICOLOMBO, EXOMARS, and GAIA. Moreover, Argentina has available time to use the antenna for radio astronomy research and other applications.[84] ESA also tracks the Argentinian SAOCOM-1A satellite through ESTRACK, the ESA's global network of ground antennas. In addition, ESA and CONAE have organized different courses and scholarships for Argentinian students, as well as workshops for the study of EO data applications, namely in hydrology, natural disasters surveillance, and radar applications.[85]

6.6.3 Argentina and North America

Since 2011, Argentina and the U.S. have signed different cooperation agreement on space issues.[86] Currently, NASA and CONAE work together in different space programs. NASA is the main partner of CONAE's SAC-D/AQUARIUS mission, and supports with technical expertise the launch operations of the Argentinian radar satellites SAOCOM 1A and SAOCOM IB (see Sect. 6.7.1.2). In this regard, NASA has supported the ground segment of the SAOCOM mission through NASA's ground station of Fairbanks, Alaska, and the NASA-NEN stations network. Furthermore, both agencies have carried out joint validation campaigns and data calibration actions of CONAE's SAOCOM mission and NASA's SMAP mission (Soil Moisture Active Passive) using the NASA's Uninhabited Aerial Vehicle SAR in calibration sites of the SAOCOM mission. Moreover, NASA proposed that CONAE should participate in scientific experiments on board the International Space Station (ISS).[87] In 2015, NASA and CONAE signed a cooperation agreement to support the Van Allen Probes NASA's mission[88] through the CONAE ground station of the Teófilo Tabanera Space Centre.

---

[84] Estación DS3 Malargue, Comisión Nacional de Actividades Espaciales, https://www.argentina. gob.ar/ciencia/conae/centros-y-estaciones/estacion-ds3-malargue (accessed 11 July 2019).

[85] La ESA y Argentina estrechan su colaboración científica con una jornada espacial, European Space Agency, 4 April 2019, https://m.esa.int/esl/ESA_in_your_country/Spain/La_ESA_y_Argentina_estrechan_su_colaboracion_cientifica_con_una_jornada_espacial (accessed 13 July 2019).

[86] "La Argentina firmó un convenio con la NASA para la exploración satelital", Infobae, 24 October 2011, https://www.infobae.com/2011/10/24/613095-la-argentina-firmo-un-convenio-la-nasa-la-exploracion-satelital/ (accessed 8 September 2019); see also "La Argentina y la NASA acordaron investigaciones conjuntas, una de ellas sobre la radiación solar", Télam, 19 February 2015, telam.com.ar/notas/201502/95539-argentina-y-estados-unidos-acuerdan-investigar-la-radiacion-solar.php (accessed 8 September 2019).

[87] La cooperación especial entre CONAE y NASA: reunión técnica y firma de acuerdo, Comisión Nacional de Actividades Espaciales, 23 February 2015, http://www.conae.gov.ar/index. php/espanol/2015/741-la-cooperacion-espacial-entre-conae-y-nasa-reunion-tecnica-y-firma-de-acuerdo (accessed 13 July 2019).

[88] Van Allen Probes Mission Overview, NASA, https://www.nasa.gov/mission_pages/rbsp/mission/index.html (accessed 13 July 2019).

6.6.4 Argentina and the Asia-Pacific

International space cooperation between Argentina and other parties from the Asia-Pacific region has been strengthened in recent years, namely with China and Turkey. Moreover, Argentina and Russia have started discussions to cooperate in the space sector.

In the last two decades, Argentina and China have increased their mutual cooperation in space matters. In 2004, both countries signed a bilateral agreement under which China provided commercial launch services, satellite components and other space-related technology. A year later, the Argentinian University of San Juan collaborated with China National Astronomical Observatories and the China National Academy of Science to develop a satellite laser ranging facility.[89] In 2012, two bilateral agreements were adopted at the institutional level: a cooperation agreement under the Chinese Lunar Exploration Program framework between the China Satellite Launch and Tracking Control General (CLTC) and CONAE for the establishment of ground tracking, command and data acquisition facilities, including an antenna for deep space research, in the province of Neuquén, Argentina. According to the agreement, CONAE has 10% of the annual use time of the antenna; and there is a cooperation agreement between the aforementioned parties for the establishment of the conditions concerning the construction of the ground tracking, command and data acquisition facilities, including an antenna for deep space research, in Neuquén.[90] At the governmental level, in 2014 Argentina and China signed a cooperation agreement for the construction, establishment and operation of a Chinese deep space station in Neuquén, Argentina.[91]

Thanks to the support of the Argentinian Government, in 2019 the INVAP Company and Turkish Aerospace Industries created the "GSATCOM Space Technologies" company for the development of new generation geostationary communication satellites. This association will open new opportunities to the Argentinian space industry abroad, and consolidate INVAP as a strong space company in the international satellite market.[92] Meanwhile, CONAE has recently proposed to Russia that it establish some facilities in Argentinian territory for its global positioning system

---

[89]Michelle Julie, Journal of Latin American Geography, Vol. 17, N° 2, July 2018, p. 56.

[90]Comunicado de Prensa, Comisión Nacional de Actividades Espaciales, 8 September 2014, http://www.conae.gov.ar/index.php/espanol/2014/705-comunicado-prensa-20140909; see also Estación CLTC-CONAE-NEUQUEN, Comisión Nacional de Actividades Espaciales, https://www.argentina.gob.ar/ciencia/conae/centros-y-estaciones/estacion-cltc-conae-neuquen (accessed 12 July 2019).

[91]Comunicado de Prensa, CONAE, 8 September 2014. See also General Resolution AFIC N° 4365/2018 (Argentina) of 18 December 2018 which specifies certain clauses of these agreements, available at http://biblioteca.afip.gob.ar/dcp/REAG01004365_2018_12_18 (accessed on 12 July 2019).

[92]Cooperación satelital INVAP-Turkish Aerospace Industries, Comisión Nacional de Actividades Espaciales, Information for the press N° 187/19, 10 May 2019, https://www.cancilleria.gob.ar/es/actualidad/noticias/cooperacion-satelital-invap-turkish-aerospace-industries (accessed 12 July 2019).

GLONASS. It is also expected that Russia and Argentina will sign a Cooperation Protocol for the peaceful use and exploration of outer space in the second semester of 2019. Currently, international space cooperation between the two countries dates from the period of the former USSR.[93]

6.6.5 Argentina and the United Nations

Space cooperation between Argentina and the UN is mostly concentrated in the UN-SPIDER program. Argentina has also contributed to the development of the space and telecommunications sectors at the multilateral level in UN organs and specialised organisations related to these areas.

6.6.5.1 Bilateral Cooperation

Argentina signed an MoU with the UNOOSA to transform CONAE into the UN-SPIDER's thirteenth Regional Support Office to promote development and strengthening of national capacities in Latin America and the Caribbean, as well as to provide horizontal cooperation and technical assistance to countries and organizations involved in disaster prevention and mitigation. The CONAE's Mario Gulich Institute for Advance Space Studies contributes to the UN-SPIDER program by offering a two-year Masters course on Space Applications for Early Warning and Response to Emergencies, which is open to Latin American participants. CONAE's contributes to this program by carrying out several activities on remote sensing applications.[94]

6.6.5.2 Institutional Participation

As a member of UNCOPUOS and the ITU, Argentina participates actively in the work of both entities. Moreover, Argentina contributes to the activities of UNOOSA on different space issues.

UN Committee on the Peaceful Uses of Outer Space

Since the creation of UNCOPUOS on 13 December 1958, Argentina has actively participated in the work of the UN Committee and of its Technical and Scientific and

---

[93] Argentina dispuesta a acoger instalaciones del sistema de navegación ruso Glonass, Sputnik, 13 June 2019, https://mundo.sputniknews.com/america-latina/201906131087604087-argentina-dispuesta-a-acoger-instalaciones-del-sistema-de-navegacion-ruso-glonass/ (accessed 12 July 2019).

[94] Argentina Regional Support Office, UNOOSA, UN-SPIDER, http://www.un-spider.org/network/regional-support-offices/argentina-regional-support-office (accessed 15 July 2019).

Legal Subcommittees in an individual capacity or as a member of a group, such as the G77 + China or a Latin America group.[95] Argentina has had an important role in the development of the UN Space Treaties and other instruments.

United Nations Office for Outer Space Affairs

Argentina has given to UNOOSA information concerning its space activities to fulfil its international obligations or on a voluntary basis. For example, as part of its international obligations as the State of Registry according to the 1975 Registration Convention, Argentina has duly notified the Secretary-General of its space objects launched into outer space[96] as well as the creation of its national register.[97] On a voluntary basis, Argentina has responded to many UNOOSA questionnaires on space law and other space issues, such as a questionnaire concerning the definition and delimitation of outer space.[98]

UNOOSA and Argentina have jointly cooperated to carry out various space activities. For example, the UN, together with the Government of Argentina and CONAE, and with the support of ESA, organized the UN/Argentina Workshop on Space Law on the theme "Contribution of Space Law to Economic and Social Development", held in Buenos Aires in November 2012.[99] More recently, in March 2018, UNOOSA and CONAE organized a workshop on Global Navigation Satellite Systems (GNSS) operations and development, in which participants formulated a common plan of action for Latin America and several recommendations on the use of GNSS, such as improving GNSS training and education, strengthening regional GNSS networks and partnerships, and enhancing capacity-building on geo-referencing.[100]

International Telecommunication Union

Argentina has been a member of the ITU since 1997. Argentina's engagement in ITU´s instruments includes the ratification of ITU's Constitution and Convention, and

---

[95]Question of the peaceful use of outer space, UNGA RES 1348 (XIII), 13 December 1958, http://www.unoosa.org/pdf/gares/ARES_13_1348E.pdf.

[96]Notifications from States & Organizations: Argentina, UNOOSA, http://www.unoosa.org/oosa/en/spaceobjectregister/submissions/argentina.html (accessed 15 July 2019).

[97]Note verbale dated 30 December 1996 from the Permanent Mission of Argentina to the United Nations (Vienna) addressed to the Secretary-General, Information furnished in conformity with the Convention on Registration of Objects Launched into Outer Space, UNCOPUOS, ST/SG/SER.E/INF.13, 15 January 1997. See Sect. 6.5.2.2.

[98]Preguntas relativas a la definición y delimitación del espacio ultraterrestre: respuestas recibidas de los Estados Miembros, UNCOPUOS, A/AC.105/889/Add.14, 18 December 2013, http://www.unoosa.org/pdf/reports/ac105/AC105_889Add14S.pdf (accessed 15 July 2019).

[99]Report on the United Nation/Argentina Workshop on Space Law on the theme "Contribution of space law to economic and social development", UNCOPUOS, A/AC.105/1037, 15 January 2013.

[100]Annual Report 2018, UNOOSA, Vienna, June 2019, p. 47.

the approval ipso facto of the International Telecommunications Regulations (Melbourne, 1988), Radio Regulations (Geneva, 1979), and many World Administrative Radio Conferences and World Radiocommunication Conferences.[101]

6.7 Satellite Capacities

Argentina's satellite capacities mainly consist of satellite missions implemented by CONAE, and the Argentinian Geostationary Telecommunications Satellite System operated by ARSAT.

6.7.1 Satellite Missions of the National Commission on Space Activities

CONAE designs, builds, calibrates, integrates, tests and puts in orbit satellites for different uses and applications with own capacities or through cooperation agreements with other space agencies. Since its inception, CONAE has accomplished several satellite missions and it is currently on the way to planning, developing and putting into orbit new satellite missions for different purposes (Table 6.5).

6.7.1.1 Accomplished Missions

To date, CONAE has carried out four satellite missions that consist of joint cooperation projects with the National Aeronautics and Space Administration (NASA-USA), the Agenzia Spaziale Italiana (ASI), the Centre national d'études spatiales (CNES-France), the Danish Space Research Institute (DSRI), and the Canadian Space Agency (CSA). All satellites were conceived and designed by CONAE and built in Argentina with the participation of the national scientific-technological system and the private sector, namely the INVAP company. The operation of the satellites was sited from the Cordoba ground station of the Teófilo Tabanera Space Centre.[102]

The four satellite missions, SAC-A, SAC-B, SAC-C and SAC-D/Aquarius, were part of the SAC series with optical instruments in which CONAE collaborated with other space agencies. Except for satellite SAC-B, an astrophysical satellite that did not fulfil its mission due to a problem with its launcher, the other satellites achieved multiple results. Satellite SAC Λ, a joint technological mission between NASA and CONAE, tested material and human infrastructure of telemetry, remote control, and

---

[101]For a complete list of agreements see https://www.itu.int/online/mm/scripts/gensel26?ctryid= 1000100530.

[102]Misiones cumplidas, Comisión Nacional de Actividades Espaciales, https://www.argentina.gob. ar/ciencia/conae/misiones-espaciales/misiones-cumplidas (accessed 8 July 2019).

Table 6.5 Argentina's satellites launched into outer space (1996–2018)

Argentina's satellites launched into outer space				
Satellite	Year	Orbit	Features	Focus
μSAT-1	1996	• Nodal period: 98.88 min • Inclination: 62.8° • Apogee: 1183 km • Perigee: 239 km	• Launch vehicle: Rocket MOLNIYA • Launch site: Plesetsk, Russian Federation • Launching organization: NPO Lavochkin	• Owner: Coratec SE and AIT • Status: decayed • General functions: experimental platform capable of taking and sending images of the national territory and of receiving, storing and retransmitting messages between low-cost ground stations (PC-type)
SAC-B	1996	• Nodal period: 95.7 min • Inclination: 38° • Apogee: 550 ± 20 km • Perigee: 510 ± 91 km	• Launch vehicle: PEGASUS XL • Launch site: Wallops NASA Flight Facility, USA • Launching organization: NASA, USA • Operator of object launched: CONAE	• Status: decayed (the satellite could not accomplish its mission due to a failure of the launch vehicle) • General functions: scientific applications satellite: hard and soft solar X-ray observation; detection of background non-solar X-ray levels; detection of neutral particles in orbital altitudes; and technological demonstration
NAHUEL 1 A	1997	Geostationary −71.8° E • Nodal period: 23 h and 56 min • Inclination: 0° • Apogee: 42,164 km from the centre of the Earth Perigee: 42,164 km	• Launch vehicle: ARIANE 4 • Launch site: Kourou, French Guiana • Launching organization: Arianespace	• Owner: NahuelSat S.A. • Status: inactive • General functions: telecommunications
SAC-A	1998	• Nodal period: 92 min • Inclination: 51.6° • Apogee: 384 km • Perigee: 362 km	• Launch vehicle: Space shuttle ENDEAVOUR (Mission STS-88) • Launch site: Kennedy Space Center • Launching agency: NASA, USA • Operator of the object launched: CONAE	• Status: decayed • General functions: technology

(continued)

Table 6.5 (continued)

Argentina's satellites launched into outer space

Satellite	Year	Orbit	Features	Focus
SAC-C	2000	• Nodal period: 98.3 min • Inclination: 98.28° • Apogee: 703 km • Perigee: 671 km	• Launch vehicle: DELTA II, Boeing Corporation • Launch site: Vandenberg AFB, California, USA • Launching organization: NASA, USA • Operator of the object launched: CONAE	• Status: decayed • General functions: scientific and EO satellite. It formed part of the "Matutina Constellation" with NASA satellites LANDSAT-7, EO-1 and TERRA
SAC-D Aquarius	2011	• Nodal period: 88 min • Inclination: 98.008° • Apogee: 668.4 km • Perigee: 651.5 km	• Launch vehicle: Boeing DELTA II • Launch site: Vandenberg Air Force Base, California, USA • Launching authority: NASA, USA • Operator of space object: CONAE	• Status: decayed • General functions: scientific and EO satellite
CubeBug-1	2013	• Nodal period: 97.6 min • Inclination: 98.06° • Apogee: 660 km • Perigee: 636 km	• Launch vehicle: LONG MARCH 2D carrier rocket • Launch site: Jiuquan Satellite Launch Centre, China • Launching authority: China National Space Administration • Operator of space object: Radio Club Bariloche	• General functions: technology demonstration to test in space components of low-cost nano-satellites intended for scientific and education applications
BugSat-1	2014	• Nodal period: 5800.24 s • Inclination: 97.99° • Apogee: 620 km • Perigee: 578 km • Argument of perigee: 6.03°	• Launch vehicle: DNEPR • Launch site: Dombarovsky Cosmodrome, Yasny, Orenburg Province, Russian Federation • Launching organization: International Space Company, Kosmotras • Operator of space object: Satellogic S.A.	• General functions: in-flight assessment of design and components for building microsatellites, in-orbit testing of flight software and medium-resolution imaging, communication test with a low-cost ground station and operator training activities, and in-flight testing of components for use in future satellites and experiments designed by the National Institute of Industrial Technology of Argentina

(continued)

Table 6.5 (continued)

Argentina's satellites launched into outer space

Satellite	Year	Orbit	Features	Focus
ARSAT 1	2014	Geostationary 72° W	• Launch vehicle: ARIANE 5 ECA VA 220 • Launch site: Guiana Space Centre, Kourou, French Guiana • Launching authority: Arianespace	• Owner: Empresa Argentina de Soluciones Satelitales S.A. (ARSAT S.A.) • Status: active • Expected life: 15 years • General functions: communications satellite
ARSAT 2	2015	Geostationary 81° W	• Launch vehicle: ARIANE 5 VA226 • Launch site: Guiana Space Centre, Kourou, French Guiana • Launching authority: Arianespace • Operator of space object: ARSAT S.A.	• Owner: ARSAT S.A. • Status: active • General functions: geostationary telecommunication satellite
SAOCOM 1A	2018	Heliosynchronous Altitude 620 km	• Launch vehicle: FALCON-9 • Launch site: Vandenberg Air Force Base, California, USA • Launching organization: SpaceX • Operator of space object: CONAE	• Status: active • General functions: to map moisture and monitor Earth resources, agriculture and urban development, for cartography and to assist in management and response to natural and man-made disasters

Information available at Online Index of Objects Launched into Outer Space, UNOOSA, http://www.unoosa.org/oosa/osoindex/search-ng.jspx?lf_id=. See also Graham William, SpaceX Falcon 9 launches with SAOCOM 1A and nails first West Coast landing, NASA Spaceflight.com, 7 October 2018, https://www.nasaspaceflight.com/2018/10/spacex-falcon-9-saocom-1a-launch-west-coast-landing/ (accessed 15 July 2019)

control equipment. The EO satellite SAC-C, a joint mission between NASA and CONAE, with the participation of France, Italy, Brazil and Denmark, took images of Argentinian territory, namely of the terrestrial and marine ecosystems. The mission also conducted studies of the geomagnetic field and the atmospheric structure. Finally, the satellite SAC-D/AQUARIUS, that formed part of a cooperation program between CONAE and the NASA Goddard Centre and Jet Propulsion Laboratory, got climate data from salinity measures and a new vision of the circulation and mixture processes in the ocean. The satellite also detected high temperature zones of the terrestrial surface to obtain risk maps of fires and soil humidity that are useful for early warning of floods.[103]

---

[103] Ibid.

6.7.1.2   Current and Future Missions

CONAE is currently developing four new satellite missions that will carry out a myriad of scientific and technological studies. There is the SABIA-Mar mission, an Argentinian-Brazilian satellite system that consists of two satellites equipped with national optical and thermal instruments and that will perform studies on the productivity of marine zones, including coasts and estuaries. While Brazil is responsible for the satellite SABIA-Mar 2, CONAE fully operates SABIA-Mar 1.[104]

SAOCOM mission, integrated by two satellites with Argentinian instruments in the active microwave range, will get data for emergencies management. Specifically, the EO satellite system will measure land humidity. SAOCOM 1A and SAOCOM 1B will be part of the Italian-Argentinian Satellite System for Emergencies Management (SIASGE),[105] together with the Italian satellites COSMO-SkyMed. SAOCOM 1A was launched in 2018 and SAOCOM 1B is planned to be launched in 2019.[106]

Another CONAE satellite mission is the Segmented Architecture program, which aims to promote the technological development of a new generation of satellite platforms and instruments, together with the development of launchers (mainly the TRONADOR launchers), in order to respond quickly to user demands and where the typical functions of the satellite instruments are distributed in platforms or segments that fly in formation.[107] In other words, each segment is a satellite with a platform that has its typical functions, but carries a single payload or a single resource of the system.[108]

Finally, there is a future mission called SARE, which consists of light satellites for EO applications that will be integrated by different instruments varying in function according to user demands. These satellites will be launched into Earth orbit by the Argentinian launchers TRONADOR II and TRONADOR III. SARE's satellites will also be part of the Segmented Architecture program. SARE comprises two types of satellites: SARE with a useful optical payload, a constellation of satellites with high resolution sensors, and SARE with a useful microwave payload, a constellation with synthetic aperture radar.[109]

---

[104] SABIA-Mar, Comisión Nacional de Actividades Espaciales, https://www.argentina.gob.ar/ciencia/conae/misiones-espaciales/sabia-mar (accessed 8 July 2019).

[105] Sistema Italo-Argentino de Satélites para la Gestión de Emergencias (SIASGE).

[106] SAOCOM, Comisión Nacional de Actividades Espaciales, https://www.argentina.gob.ar/ciencia/conae/misiones-espaciales/saocom (accessed 8 July 2019).

[107] Misiones satelitales, Comisión Nacional de Actividades Espaciales, https://www.argentina.gob.ar/ciencia/conae/misiones-espaciales (accessed 8 July 2019).

[108] Arquitectura segmentada, Comisión Nacional de Actividades Espaciales, https://www.argentina.gob.ar/ciencia/conae/misiones-espaciales/arquitectura-segmentada (accessed 8 July 2019).

[109] SARE, Comisión Nacional de Actividades Espaciales, https://www.argentina.gob.ar/ciencia/conae/misiones-espaciales/sare (accessed 8 July 2019).

6.7.2 Argentinian Geostationary Telecommunications Satellite System

To strengthen its national telecommunications sector, Argentina decided to create the Argentinian Geostationary Telecommunications Satellite System (SSGAT),[110] a program aimed to design, build, put into orbit and operate its own telecommunications satellites. Currently, there are three satellites that are part of SSGAT: ARSAT-1, ARSAT-2 and ASTRA-1H. These satellites are operated by the national company ARSAT, created in 2006 (see Sect. 6.4.2.2), although the management, development, design, construction, integration, and functional and environmental tests are conducted by the INVAP company.[111] To do this, in 2007 Argentina transferred to ARSAT the assets of the Nahuelsat S.A. company, which exploited the 72° W Geostationary Orbital Position (GOP) through the satellite NAHUEL-1. In 2014, four years after the decay of NAHUEL-1, the satellite ARSAT-1 occupied the 72° W GOP. In 2015, the satellite ARSAT-2 was launched into orbit to occupy the 81° W GOP (Table 6.6).[112]

Satellite ARSAT-1 aims to send radiofrequency signals in Ku band for telecommunication services such as direct television, access to Internet through VAST antennas, and IP data services and telephony to Argentina and bordering countries .[113] Satellite

Table 6.6 Argentina's geostationary orbital resources

Argentina's geostationary orbital resources					
GOP and associated spectrum	Satellite	Services	Extension	Operator	Launch
72° W Ku-band	ARSAT-1	Radiofrequency for telecommunication services	Regional	ARSAT S.A.	2014
81° W Ku and C bands	ARSAT-2	Radiofrequency for telecommunication services	Regional	ARSAT S.A.	2015
81° W Ka, Ku and C bands	ASTRA-1H	Broadband Internet services	National	ARSAT S.A.	1999

---

[110]Sistema Satelital Geoestacionario Argentino de Telecomunicaciones (SSGAT).

[111]The environmental tests of ARSAT-1 were conducted by the High Technology Test Centre (Centro de Ensayos de Alta Tecnología—CEATSA), an Argentinian company created in 2010 by an agreement between ARSAT and INVAP, in ARSAT 1—Argentina lanza un satellite al espacio, Embajada en Tailandia, https://etail.cancilleria.gob.ar/es/content/arsat-1-argentina-lanza-un-sat.%C3%A9lite-al-espacio (accessed 9 July 2019).

[112]Satélites ARSAT, INVAP, http://www.invap.com.ar/es/espacial-y-gobierno/proyectos-espaciales/satelite-arsat.html (accessed 9 July 2019).

ARSAT-2 complements ARSAT-1. It transports radiofrequency signals in Ku and C bands for telecommunications, and extends its coverage to the rest of the American continent.[114]

In 2015, Argentina promulgated Law 27.208 to develop the Argentinian Geostationary Satellite Plan for the period 2015–2035, which contemplated the creation of four new satellites in addition to ARSAT-1 and ARSAT-2. The first of these satellites was ARSAT-3, which should have been launched in 2019 to occupy the Ka-band of the 81° W GOP. However, the ARSAT-3 project was cancelled in 2016. To avoid losing the Ka-band of the 81° W GOP by the ITU, ARSAT decided to rent for €7 million the satellite ASTRA-1H, whose useful life finished in 2015, through an agreement with the New Skies Satellites B.V. (SES) company in 2018. In January 2019, Argentina notified ITU of the relocation of ASTRA-1H to the 81° W GOP.[115]

6.8   Launch Capability and Sites

Argentina's early space history (see Sect. 6.2) is marked by several experimental rocket launches for scientific and military purposes. Argentina's decision to invest in a national space launch industry came with the launching of the TRONADOR I and Ib rocket probes in 2007 and 2008, respectively. After the failure of a test launch in 2011, Argentina decided to transform the design of its launch system so as to develop the TRONADOR project, for which CONAE is mainly responsible.[116] Currently, Argentina is developing several launch projects that include launchers and launch facilities.

6.8.1   Launch Projects

One of the better examples of Argentina's ambition to become a spacefaring nation is the planning and development of launchers capable to put own satellites into Earth orbit. CONAE is currently working on the development of national technology for future launches through the Program on Research and Development of Means to Access to Space, which has four projects[117]:

---

[113]Ibid.

[114]Ibid.

[115]Krakowiak Fernando, El costo de haber abandonado el Arsat-3, p. 12, 11 February 2019, https://www.pagina12.com.ar/174238-el-costo-de-haber-abandonado-el-arsat-3 (accessed 9 July 2019); see also ARSAT ubica el ASTRA-1H en la posición de 81°, Latam Satellital, 11 February 2019, http://latamsatelital.com/arsat-ubica-astra-1h-la-posicion-81/ (accessed 9 July 2019).

[116]Tronador, Comisión Nacional de Actividades Espaciales, https://www.argentina.gob.ar/ciencia/conae/acceso-al-espacio/tronador (accessed 13 July 2019).

[117]Acceso al espacio, Comisión Nacional de Actividades Espaciales, https://www.argentina.gob.ar/ciencia/conae/acceso-al-espacio (accessed 10 July 2019).

(a) The Light Payloads Satellite Launcher (ISCUL)[118] consists of the elaboration of all the necessary steps to build the TRONADOR II launcher. This launcher will be capable of sending payloads of up to 250 kg into polar orbit, at 600 km of altitude.

(b) The development of the TRONADOR III launcher that will be capable of putting payloads of up to 1000 kg into the polar orbit, at 600 km of altitude.

(c) The launch through third parties of satellites SAOCOM and SABIA-Mar by providing the interface between the satellite project and the injector vehicle.

(d) The launch of SARE satellites, as part of the segmented architecture, through third parties or with the TRONADOR II launcher.

6.8.2 Launching Sites

There are three space centres and one ground station in the territory of Argentina managed by CONAE[119]:

(a) The Teófilo Tabanera Space Centre (CETT) located in Falda de Cañete, in the province of Cordoba. The centre has a ground station, a mission control centre, two laboratories, a unit for superior training, and several development departments concerning access to space. The centre also has the headquarters of the Gulich Institute (see Sect. 6.10.2).[120]

(b) The Punta Indio Space Centre (CEPI) located in Punta Indio, in the province of Buenos Aires. This centre is conceived for manufacture, integration and tests of big structural elements. The zone has also a launching support for experimental vehicles, such as the VEX1A/B and VEX5A/B.[121]

(c) The Manuel Belgrano Space Centre (CEMB) located in the Naval Zone of Puerto Belgrano, in the province of Buenos Aires. This centre will have its own launching platform for the TRONADOR II and III launchers, together with the final integration zone of the launcher and the respective satellite.[122]

(d) The Ground Station of Tierra del Fuego located near the Tolhuin population, in the province of Tierra del Fuego. The station has two systems of parabolic

---

[118]Proyecto Inyector Satelital para Cargas Útiles Livianas (ISCUL).

[119]Centro y Estaciones, Comisión Nacional de Actividades Espaciales, https://www.argentina.gob.ar/ciencia/conae/centros-y-estaciones (accessed 11 July 2019).

[120]Centro Espacial Teófilo Tabanera, Comisión Nacional de Actividades Espaciales, https://www.argentina.gob.ar/ciencia/conae/centros-y-estaciones/centro-espacial-teofilo-tabanera (accessed 11 July 2019).

[121]Centro Espacial Punta Indio, Comisión Nacional de Actividades Espaciales, https://www.argentina.gob.ar/ciencia/conae/centros-y-estaciones/centro-espacial-punta-indio (accessed 11 July 2019).

[122]Centro Espacial Manuel Belgrano, Comisión Nacional de Actividades Espaciales, https://www.argentina.gob.ar/ciencia/conae/centros-y-estaciones/centro-espacial-manuel-belgrano (accessed 11 July 2019).

reflector satellite antennas that will receive data from different EO satellites, and support the SAOCOM mission. The station will also support the space access zone for the TRONADOR launcher.[123]

6.9   Space Technology Development

As noted in Sect. 6.4.1.1.3, one of the priorities of the NSP is to develop space technology and applications for the benefit of the country. Since the beginning of its space history, Argentina has developed technology for its space projects, from rocket probes and atmospheric balloons to missiles and satellites.[124] Currently, CONAE and other national institutions and organisms develop space technology for their programs. Besides, many space and technology industries in Argentina participate in this sector.

All CONAE's space programs contribute to the development of space technology. The TRONADOR program, for example, aims to demonstrate the technological capacities of the propulsion, avionic and structures components necessary to satisfy the launcher requirements,[125] and the SAOCOM series (see Sect. 6.7.1.2) requires the development of active instruments that operate in the microwave range to be developed by national scientific and technological organisms and companies, such as the National Atomic Energy Commission (CNEA),[126] and the VENG and INVAP companies.[127]

Other institutions that contribute to the development of space technology are the CETT, IAR, and the Institute of Astronomy and Space Physics (IAFE), to name a few. The CETT develops liquid fuel engines and associated systems and propellants, and carries out micro engines, avionic, and fair tests, among other experiments.[128] IAR develops and implements technological projects aimed at generating strategic value to the public and private sectors. To this end, IAR strongly supports the connection and transfer of technology generated by its programs.[129] The technological services

---

[123]Estación Terrena de Tierra del Fuego, Comisión Nacional de Actividades Espaciales, https://www.argentina.gob.ar/ciencia/conae/centros-y-estaciones/estacion-terrena-de-tierra-del-fuego (accessed 11 July 2019).

[124]Viera Cristian, El desarrollo de la tecnología espacial en Argentina y sus posibles aportes a la región, Letras Internacionales, Universidad ORT, Uruguay, N° 180-8, 22 May 2014.

[125]Tronador II, Comisión Nacional de Actividades Espaciales, https://www.argentina.gob.ar/ciencia/conae/acceso-al-espacio/tronador2 (accessed 13 July 2019).

[126]Comisión Nacional de Energía Atómica (CNEA).

[127]SAOCOM, Comisión Nacional de Actividades Espaciales, https://www.argentina.gob.ar/ciencia/conae/misiones-espaciales/saocom (accessed 13 July 2019).

[128]Desarrollo de proyectos vinculados con Acceso al Espacio, Comisión Nacional de Actividades Espaciales, https://www.argentina.gob.ar/desarrollo-de-proyectos-vinculados-con-acceso-al-espacio (accessed 13 July 2019).

[129]Transferencia de tecnología y vinculación tecnológica, Instituto Argentino de Radioastronomía, https://www.iar.unlp.edu.ar/slider/transferencia-de-tecnologia/ (accessed 13 July 2019).

offered by IAFE consist of the use of satellite images for the creation of an early warning system for the prevention of forest deforestation.[130]

Many Argentinian companies also develop space technology. In this area, ARSAT, the national company responsible for the operation of the Argentinian Geostationary Satellite System (see Sect. 6.4.2.2) advances the implementation of national telecommunication projects. The company also carries out satellite control, monitors the Federal Fiber Optic Network (REFEFO), the National Centre Data, and the technological platform for the Argentinian Digital Television System (TDA).[131]

INVAP is an Argentinian company that designs and build complex technological systems. It develops advanced technology in different industry, science and applied research fields, as well as high added value "technological products". The company was created by a convention signed between the CNEA and the Government of the province of Río Negro. INVAP, the only Latin American company capable of implementing full satellite projects, has designed and built several satellites for CONAE, such as the SAC and SAOCOM series. Besides, INVAP, the main contractor for ARSAT, has a prominent role in developing the Argentinian Geostationary Satellite System.[132] It also develops, constructs and installs the radars for the National Aerospace Surveillance and Control System (SINVICA),[133] which aims to control the national air space for civil and defence purposes.[134]

CEATSA is a company that carries out environmental tests linked to the satellite industry. The company, which was created in 2010 and began operations in 2012, is owned by ARSAT (80%) and INVAP (20%). CEATSA is a regional reference for all kinds of industries for the development of new technologies and the assessment of new products.[135]

VENG S.A. is a CONAE's contractor for the access to space department. The company develops the TRONADOR project and is currently working on the construction of the TRONADOR II rocket.[136]

---

[130] Servicios tecnológicos, Instituto de Astronomía y Física del Espacio, CONICET, https://www.conicet.gov.ar/new_scp/detalle.php?keywords=&id=05512&inst=yes&ofertas=yes (accessed 13 July 2019).

[131] Red Federal de Fibra Óptica (REFEFO); Centro Nacional de Datos, and Sistema Argentino de Televisión Digital (TDA). See Drewes Lorena, El Sector Espacial Argentino: instituciones, empresas y desafíos, Benavidez: ARSAT—Empresa Argentina de Soluciones Satelitales, 1st edition, 2014, pp. 30–31.

[132] Aerospace and Government, INVAP, http://www.invap.com.ar/en/aerospace-and-government/aerospace-and-government-area/introduction-aerospace/940-introduction.html (accessed 13 July 2019).

[133] Sistema Nacional de Vigilancia y Control Aerospacial (SINVICA).

[134] Drewes Lorena, El Sector Espacial Argentino: instituciones, empresas y desafíos, op. cit., pp. 30–31.

[135] Ibid, p. 32.

[136] Ibid, p. 35.

Other companies that currently develop space technology in Argentina (some of which are CONAE and INVAT's providers) are Arsultra, Ascentio, Satellogic, SpaceSur, Latam Sat, DTA, SADE, Mecánica 14, STI, the Laboratorio de Investigación Espacial, Theia Technologies SA, and Kohlenia.[137]

6.10   Capacity Building and Outreach

According to its NSP, Argentina must provide new education and job opportunities in the space sector.[138] Capacity building is being materialized through specific space education programs and training programs and the professional activities of research centres, some of which are strongly supported by CONAE. In addition, dissemination of space knowledge and other related disciplines is also possible thanks to the work of several NGO's.

6.10.1   Space Education

Space education in Argentina mainly consists of high education programs in different space related disciplines. CONAE also has specific education programs, activities and projects that contribute to the formation and training of students and professionals in the space sector.

6.10.1.1   Space Related Programs

As the country increases the development of its space activities, the training of human resources specialized in space issues becomes a necessity. In this regard, many Argentinian higher institutions and universities have created undergraduate and postgraduate programs related to space, telecommunications, aeronautics, astronomy and other related disciplines. At the bachelor's level, most of the programs last five years, while at the postgraduate level, the programs last from two to five years (Table 6.7).

---

[137]Ibid, pp. 47–58. See also "La industria espacial argentina depende de los profesionales de hoy", Aeroespacio, 10 December 2018, http://www.aeroespacio.com.ar/sc/la-industria-espacial-argentina-depende-de-los-profesionales-de-hoy/ (accessed 13 July 2019).

[138]Plan Espacial Nacional, Comisión Nacional de Actividades Espaciales, https://www.argentina.gob.ar/ciencia/conae/plan-espacial (accessed 13 July 2019).

Table 6.7 Space and telecommunications programs in Argentina (2019)

Space and telecommunications programs			
University	Program	Education level	Duration
Instituto Universitario Aeronáutico (Centro Regional Universitario Córdoba)	Aeronautical Engineering	Undergraduate	5 years
Instituto Universitario Aeronáutico (Centro Regional Universitario Córdoba)	Telecommunications Engineering	Undergraduate	5 years
Universidad Nacional de Córdoba and Instituto Universitario Aeronáutico	Engineering Sciences—Specialization in Aerospace (master)	Postgraduate	1 year
Universidad Nacional de La Plata	Astronomy	Undergraduate	5 years
Universidad Nacional de La Plata	Astronomy (Ph.D.)	Postgraduate	–
Universidad Nacional de La Plata	Aeronautical Engineering	Undergraduate	5 years
Universidad Nacional de La Plata	Telecommunications Engineering	Undergraduate	5 years
Universidad Nacional de San Martín	Telecommunications Engineering	Undergraduate	5 years and a half
Universidad Nacional de San Martín	Astrophysics	Postgraduate	–
Universidad Tecnológica Nacional (Regional Faculty Haedo)	Aeronautic Engineering	Undergraduate	–
Universidad Nacional de Córdoba	Aeronautical Engineering	Undergraduate	5 years
Universidad Nacional de Córdoba	Astronomy	Undergraduate	5 years
Universidad Nacional de Córdoba	Space Data Applications (master)	Postgraduate	2 years
Universidad Nacional de Córdoba	Astronomy (Ph.D.)	Postgraduate	5 years
Universidad Nacional de San Juan	Astronomy	Undergraduate	5 years
Universidad Nacional de San Juan	Astronomy (Ph.D.)	Postgraduate	–
Universidad de Blas Pascal	Telecommunications Engineering	Undergraduate	5 years

(continued)

Table 6.7   (continued)

Space and telecommunications programs

University	Program	Education level	Duration
Universidad Nacional de Cuyo (Balseiro Institute)	Telecommunications Engineering	Undergraduate	5 years and a half
Universidad Argentina de la Empresa	Telecommunications Engineering	Undergraduate	5 years
Universidad Católica de Salta	Telecommunications Engineering	Undergraduate	5 years
Universidad Nacional de Río Cuarto	Telecommunications Engineering	Undergraduate	5 years
Universidad de Buenos Aires	Telecommunications Engineering (masters)	Postgraduate	2 years
Instituto de Altos Estudios Espaciales "Mario Gulich" and Universidad Nacional de Cordoba	Space Data Applications (master)	Postgraduate	2 years
Instituto de Altos Estudios Espaciales "Mario Gulich" and Universidad Nacional de Cordoba	Geomatics and Space Systems (Ph.D.)	Postgraduate	–
Instituto Nacional de Derecho Aeronáutico y Espacial	Specialization in Aeronautical and Space Law	Postgraduate	2 years
Universidad de Belgrano	International Relations (courses focused on Space Law)	Undergraduate	33 weeks

6.10.1.2   Space Education and the National Commission on Space Activities

One of CONAE's functions is to disseminate space data to society. Apart from a great array of educational materials freely available online by CONAE,[139] the Agency has an Education Unity that is responsible for sharing space information and training people on the use of satellite images with the aim of improving and optimizing education and socio-productive activities. The Unity organizes courses, conferences and other activities to bring people from Argentina and Latin America closer to the use of satellite images and geospace technology. There are two work lines developed for this purpose: one for teachers and students for teaching, and one for technicians and professionals of different disciplines interested in the use of satellite images.[140]

---

[139] Materiales educativos, Comisión Nacional de Actividades Espaciales, https://www.argentina.gob.ar/ciencia/conae/unidad-educacion/software/2mp (accessed 10 July 2019).

[140] Aula Virtual de la Unidad Educación de la Comisión Nacional de Actividades Espaciales, CONAE, https://educacion.conae.gov.ar/aulavirtual/ (accessed 10 July 2019).

In addition, CONAE has created the Educational Program 2Mp (from "2 million pibes and pibas"—boys and girls) that incorporates the use of satellite technology in the national educational system, enabling eight-year-old students or older to become familiar with the use of space data. The program aims to develop teaching, to provide technological tools to favour inclusion in the workplace, and to promote the democratization of access to satellite data.[141]

To this end, CONAE has developed a digital tool called "Software 2Mp" for the analysis and development of cases based on the application of satellite images. The tool, which is addressed to teachers and students, includes thematic modules on different topics that are capable of being analysed through the use of satellite data.[142]

CONAE also organizes visits to the Téofilo Tabanera Space Centre located in Falda del Carmen, Cordoba, where students of all educational levels from the fourth grade of elementary school hear a dissemination talk on Argentina's space activities and the main concepts linked to the satellite technology, as well as an introduction to the development of satellite launchers in the country.[143]

6.10.2 Space Research

Argentina has invested important resources in space research in order to become a more self-sufficient country in the use of space technology and knowledge, and to generate high added value services for the sustainable development of the country. Several research centres and institutes work in this field for the benefit of Argentinian society.

CONAE and the National University of Cordoba (UNC) created the Institute of High Space Studies "Mario Gulich" in 1997 to do research, develop and transfer of knowledge, and train human resources in remote control of land, oceans and the atmosphere. The vision of the Institute, which receives support from the Agenzia Spaziale Italiana, is to become an interdisciplinary centre of excellence on space data applications in Latin America.[144] The Institute, whose headquarters were inaugurated in 2001, has several research lines related to the development of solutions to different social, productive and environmental problems.[145]

---

[141]Programa 2Mp, Comisión Nacional de Actividades Espaciales, https://www.argentina.gob.ar/ciencia/conae/unidad-educacion/acerca-de/programa-2mp (accessed 10 July 2019).

[142]Software 2Mp, Comisión Nacional de Actividades Espaciales, https://www.argentina.gob.ar/ciencia/conae/unidad-educacion/software/2mp (accessed 10 July 2019).

[143]Visitas al CETT, Comisión Nacional de Actividades Espaciales, https://www.argentina.gob.ar/ciencia/conae/unidad-educacion/acerca-de/programa-2mp/visitas-al-cett (accessed 10 July 2019).

[144]Visión y misión, Instituto de Altos Estudios Espaciales "Mario Gulich", http://ig.conae.unc.edu.ar/institutcional/; see also Instituto Gulich, CONAE, https://www.argentina.gob.ar/ciencia/conae/gullich (accessed 10 July 2019).

[145]The current research lines are: panoramic epidemiology, monitoring and modelling of environmental quality indicators, agriculture production and food safety, advanced processing, biodiversity and ecosystemic services, and space applications of early warning and response to emergencies,

Other research centres on space related issues are the Colomb Institute of the National University of San Martin, which organizes activities related to satellite communications, such as the use of ground stations, communications and antennas[146]; IAR created by CONICET and the Commission of Scientific Research of the Province of Buenos Aires (CIC),[147] which promotes and coordinates radio astronomy research and technical development, and collaborates in its teaching[148]; the Institute on Astronomy and Space Physics (IAFE), dependent on CONICET and the University of Buenos Aires, which does research on observational and theoretical sciences related to the universe,[149] and the Institute of Theoretical and Experimental Astronomy of CONICET and the UNC, which does astrophysical research and characterizes sites for the installation of astronomical structures in Argentinian territory.[150] The Astronomical Complex "El Leoncito", dependent on CONICET and the National Universities of La Plata, Cordoba, and San Juan, serves an astronomical centre for the national scientific community, which is located in the Province of San Juan.[151] The Civil Institute of Space Technology (ICTE) also has different space projects such as a suborbital exploration probe in collaboration with the Qatari ADGS company.[152] The Pierre Auger Observatory, located in the province of Mendoza, does research on the exploration of cosmic rays.[153] Finally, the University of the National Defence has a research line on aerospace interests and national defence.[154]

6.10.3   Argentina and Non-governmental Space Institutions

The role of civil society has been essential for the development of Argentina's space sector. International and national space NGO's and associations have contributed

---

see Líneas de Investigación, Instituto de Altos Estudios Espaciales "Mario Gulich", http://ig.conae. unc.edu.ar/lineas-de-investigacion/ (accessed 10 July 2019).

[146]Instituto Colomb, Universidad Nacional de San Martín, http://www.unsam.edu.ar/institutos/colomb/ (accessed 10 July 2019).

[147]Comisión de Investigaciones Científicas de la Provincia de Buenos Aires (CIC).

[148]Finalidad, Instituto Argentino de Radioastronomía, https://www.iar.unlp.edu.ar/institucional/finalidad/ (accessed 10 July 2019).

[149]Instituto de Astronomía y Física del Espacio (IAFE), http://www.iafe.uba.ar/ (accessed 11 July 2019).

[150]Instituto de Astronomía Teórica y Experimental (IATE), http://iate.oac.uncor.edu/ (accessed 11 July 2019).

[151]Complejo Astronómico El Leoncito, Consejo Nacional de Investigaciones Científicas y Técnicas, https://casleo.conicet.gov.ar/ (accessed 11 July 2019).

[152]Proyectos, Instituto Civil de Tecnología Espacial (ICTE), https://icte.com.ar/proyectos.html (accessed 13 July 2019).

[153]Pierre Auger Observatory, https://www.auger.org/index.php?start=30 (accessed 13 July 2019).

[154]Líneas de investigación, Universidad de la Defensa Nacional, https://www.undef.edu.ar/investigacion/lineas-de-investigacion/ (accessed 10 July 2019).

to this development through different activities that go from the organizations of conferences, workshops and seminars to the publication of space related works.

6.10.3.1   Space Generation Advisory Council

The Space Generation Advisory Council (SGAC), a global non-profit NGO and network that aims to represent young space students and professionals to the UN, space agencies, industry, and academia, has motivated Argentinian youth to become space sector leaders.[155] One of its most important activities carried out in South America was the first South American Space Generation Workshop SA-SGW held in Buenos Aires, Argentina in 2015, a two-day event where students, professionals and leaders discussed several topics related to space, such as simulation missions on Mars, the creation of a South American Space Agency, technological and research advances in South America, and the education and dissemination of programs in the region. The objectives of the workshop were the strengthening of the regional community of young and professional students of the space sector in the region, analysis and assessment of the key issues of the space community in South America, and personal interactions between future space leaders.[156]

6.10.3.2   Other Space Related Institutions

Apart from the Argentinian government's attention on space activities, there is also a high interest in the space sector by civil society. In this regard, different national associations have been constituted to disseminate and teach space sciences and technology knowledge to the general public. The Argentinian Association on Space Technology, the Argentinian Association of Astronomy, and the Latin American Association on Aeronautic and Space Law are some examples of space related NGO's.[157] The Robotic Club of the National Technological University of Cordoba, the Argentinian

---

[155] One of the two current SGAC Regional Coordinators of South America is the Argentinian Santiago Enriquez, see South America, Space Generation Advisory Council, https://spacegeneration.org/regions/south-america (accessed 11 July 2019).

[156] The workshop was organized together with the 8th Argentinian Congress on Space Technology, see the event brochure at http://aate.org/CATE_2015/CATE2015.HTML (accessed 11 July 2019).

[157] See Asociación Argentina de Tecnología Espacial (AATE), https://www.aate.org/, Asociación Argentina de Astronomía, http://www.astronomiaargentina.org.ar/, and Asociación Latinoamericana de Derecho Aeronáutico y Espacial (ALADA), https://alada.org/inicio/.

Radio Club, and Mars Society Argentina also contributes to this aim. Moreover, there exist space magazines created by Argentinians, such as the LATAM satelital magazine[158] and the Space Law Review,[159] which have contributions from authors from different countries.

Appendix

See Table 6.8.

Table 6.8 Institutional contact information in Argentina (2019)

Institution	Details
Comisión Nacional de Actividades Espaciales (CONAE)	751 Paseo Colón Avenue, 1063, Buenos Aires, Argentina Website: http://conae.gov.ar/ Email: info@conae.gov.ar
Empresa Argentina de Soluciones Satelitales S.A. (ARSAT)	7934 Juan D. Perón Avenue, B1621BGY, Benavidez PBA, Argentina Website: https://www.arsat.com.ar/ Email: contacto@arsat.com.ar
Centro Espacial Teófilo Tabanera	Route C45, km 8, Falda de Cañete, 5187, Córdoba, Argentina Website: http://conae.gov.ar/ Email: info@conae.gov.ar
Consejo Nacional de Investigaciones Científicas y Técnicas (CONICET)	Godoy Cruz 2290, C1425FQB, Buenos Aires, Argentina Tel: +5411 4899-5400 Email: info@conicet.gov.ar

---

[158]See Latam satelital, http://latamsatelital.com/, and a Latam's director interview at Desarrollo Productivo y Tecnológico Empresarial de la Argentina http://latamsatelital.com/ (accessed 11 July 2019).

[159]Revista de Derecho Espacial, Editores Argentina, http://ar.ijeditores.com/index.php?option= publicacion&idpublicacion=173&hash_t=730d02efaa9d59ba9d29d2dc804d1115&hash_g=pub. The Argentinian Air Force also publishes a magazine on space related articles called the Aeroespacio magazine, see http://www.aeroespacio.com.ar/.

Chapter 7
Bolivia

Contents

Abstract One decade ago, Bolivia decided to develop its national space sector by establishing a national space agency. A few years later, Bolivia created its first institutional strategic plan to foster its space sector in the following years. Since then, Bolivia has increased little by little its space activities thanks to the work, jointly or separately, of the government and the academia. This chapter briefly presents a general overview of Bolivia's current space activities. Special attention is put on Bolivia's space actors and programs, the legal space and telecommunications framework, the cooperation programs adopted by Bolivia and other international partners, national satellite capacities, and capacity building and outreach on space issues.

7.1 General Country Overview

The Plurinational State of Bolivia (Bolivia) is a landlocked country located in western-central South America. With 1,098,581 km^2 of territory, Bolivia borders Brazil to the north and northeast, Argentina to the south, and Chile to the west and southwest. It has a population of 11,307,000 inhabitants of which 49.6% are women

© Springer Nature Switzerland AG 2020

A. Froehlich et al., *Space Supporting Latin America*, Studies in Space Policy 25,
https://doi.org/10.1007/978-3-030-38520-0_7

and 50.4% are men (2018).[1] Bolivia's official languages are Spanish and all the national and indigenous languages of the country (36 in total). Bolivia's currency is the Peso Boliviano.[2]

Geographically, Bolivia is divided into three regions: the Andean or Western Region, a mountainous region with twelve summits higher than 6000 m; the sub-Andean region, which has many valleys and agricultural zones, and the Eastern Tropical Plains, a low lands zone that covers roughly 60% of Bolivia's territory.[3]

7.1.1 Political System

Bolivia is a unitary social state of plurinational, community-based law, free, independent, sovereign, democratic, intercultural, decentralized, and with indigenous autonomies.[4] Its public power is organised into legislative, executive, judicial and electoral bodies. The Plurinational Legislative Assembly is composed of two chambers: the Chamber of Deputies and the Chamber of Departmental Representatives[5]; the executive body is composed of the President,[6] the Vice-president and the state ministers[7]; the judicial body is the Supreme Court of Justice, the Departmental Courts of Justice, and other jurisdictions[8]; and the electoral body is the Plurinational Electoral Council.[9]

The territory of Bolivia is organised in nine departments (see Fig. 7.1), which are divided into 113 provinces and more than 300 municipalities and indigenous origin rural territories.[10] Departmental governments are constituted by a governor and a department assembly. Bolivia's constitutional capital is Sucre, and the administrative capital and seat of government and the parliament is La Paz.

---

[1] Bolivia cuenta con más de 11 millones de habitantes a 2018, Instituto Nacional de Estadística, Estado Plurinacional de Bolivia, https://www.ine.gob.bo/index.php/notas-de-prensa-y-monitoreo/itemlist/tag/Poblaci%C3%B3n (accessed 5 October 2019).

[2] Datos generales, Estado Plurinacional de Bolivia, Ministerio de Relaciones Exteriores, Embajada de Bolivia en Buenos Aires – Argentina, https://www.embajadadebolivia.com.ar/bolivia/el-pais/datos-generales/ (accessed 5 October 2019).

[3] Bolivia, Ficha país, Oficina de Información Diplomática del Ministerio de Asuntos Exteriores, Unión Europea y Cooperación, España, September 2019, p. 1, http://www.exteriores.gob.es/Documents/FichasPais/BOLIVIA_FICHA%20PAIS.pdf (accessed 5 October 2019).

[4] Art 1, Constitución Política del Estado, H. Congreso Nacional, 7 February 2009.

[5] Ibid., Art. 146.

[6] On 10 November 2019, president Evo Morales stepped down and flew to Mexico three days later. On 12 November, Jeanine Áñez became the intern president of the Republic of Bolivia.

[7] Ibid., Art. 166.

[8] Ibid., Art. 180.

[9] Ibid., Art. 206.

[10] Bolivia, Ficha país, Oficina de Información Diplomática del Ministerio de Asuntos Exteriores, Unión Europea y Cooperación, España, September 2019, p. 1, http://www.exteriores.gob.es/Documents/FichasPais/BOLIVIA_FICHA%20PAIS.pdf (accessed 5 October 2019).

Fig. 7.1 Bolivia's
Departments (2019)

7.1.2   Economic Situation

Bolivia has one of the highest growth rates in Latin America thanks to the increase
in exports, namely hydrocarbons and mining. Bolivia's GDP growth has been stable
in the last three years (4.3% in 2016, 4.2% in 2017 and 4.3% in 2018), its inflation
rate in 2017 was the lowest in recent years (2.8%), and Bolivia's unemployment rate
decreased to 4.2% in 2018. In addition, the Bolivian Government has planned public
investments of up to USD 48.5 billion for the period 2016–2020,[11] which shows the
growth rate of the country.

7.2   Space Development Policies and Emerging Programs

In order to develop and strengthen its space capacities, Bolivia has established several
national space and telecommunications bodies. Moreover, Bolivia's government has
recently created an institutional strategic plan that defines the short-term objectives
for Bolivia's space development.

---

[11]Ibid., p. 2.

7.2.1 Space Actors

The national bodies responsible for the promotion and development of Bolivia's space activities, including telecommunication activities related to space issues, are the Bolivian Space Agency and the telecommunications authorities.

7.2.1.1 The Bolivian Space Agency

The Bolivian Space Agency (ABE)[12] is the national main space body that was created on 10 February 2010 (see Sect. 7.3.1). ABE's objective is to manage and implement the TUPAK KATARI Communications Satellite Project[13] and other space projects.[14]
  ABE's main functions are the following[15]:

- Manage and implement the TUPAK KATARI Communications Satellite Project;
- Promote the development of new space and satellite projects;
- Promote technology transfer and human resources training on space technology, and
- Promote the implementation of satellite applications for use in social, productive, defence, environmental and other programs.

  Moreover, ABE commercialises several satellite services: rent of the satellite space segment, connectivity service, access to Internet, and radio and television broadcasting services.[16]
  ABE is headed by a General Executive Director appointed by Bolivia's president, and a Board of Directors serves as the highest body of control and approval of ABE's plans and regulations. The Board of Directors is composed of five members of different national ministries.[17]

---

[12] Agencia Boliviana Espacial (ABE).

[13] The natural resource orbit-spectrum and its associated frequencies registered in the name of Bolivia according to the ITU's Radiocommunications Regulations are assigned to ABE for its use in Bolivian satellite networks, namely the Communications Satellite Program TUPAK KATARI. In case ABE does not use the available orbit-spectrum and associated frequencies natural resources, the use of this resources will be defined according to the integral policies of the Telecommunications and TIC's sector. See Art. 17 of Law N° 164—General Law on Telecommunications and TIC's.

[14] Ibid., Art. 3.

[15] Historia, Agencia Boliviana Espacial, https://www.abe.bo/nosotros/historia/ (accessed 18 September 2019).

[16] Plan Estratégico ABE 2016–2020, p. 6.

[17] Art. 7, Decreto Supremo N° 0423, Gaceta Oficial de Bolivia, 10 February 2010, https://www.abe.bo/files/marco-normativo/DS0423.pdf (accessed 17 September 2019).

7.2.1.2   The Telecommunications Authorities

The Vice-Ministry of Telecommunications that is part of the Ministry of Public Works, Services and Housing, is the national body that proposes policies on telecommunications, information and communications technologies (TIC's) and postal services. Among its functions are the proposal and implementation of policies for the development of TIC's services by using and exploiting the electromagnetic spectrum.[18]

The Authority of Regulation and Control of Telecommunications and Transportation is the only authority responsible for the assignment, control, supervision, and management of the radio spectrum associated with satellite networks in the territory of Bolivia. It grants ABE the authorisation to use Bolivian's orbit-spectrum and associated frequencies natural resources.[19]

7.2.2   Bolivian Space Agency's Strategic Plan

ABE's Strategic Plan 2016–2020[20] is a mid-term planning instrument that articulates the national, sectoral and corporate planning of the space and satellite sectors. It contains ABE's policy approach, a diagnostic of internal and external variables, as well as the objectives and strategic actions to be implemented by the Agency. There are five strategic objectives and 14 actions defined by the Strategic Plan to be achieved in the 2016–2020 period (see Table 7.1).

7.3   Current and Future Perspectives on the Legal Framework Related to Space Activities

Bolivia's space related regulations are scarce. However, they constitute the foundations of the national space and telecommunications sectors. At the international level, Bolivia has signed and ratified several space related treaties showing the country's engagement in respecting the international legal framework on space.

---

[18]Marco legal, Viceministerio de Telecomunicaciones, https://vmtel.oopp.gob.bo/index.php/informacion_institucional/Marco-Legal,981.html (accessed 18 September 2019).

[19]Autoridad de Regulación y Fiscalización de Telecomunicaciones y Transportes. See art. 17, Law N° 164—General Law on Telecommunications and TIC's.

[20]Plan Estratégico 2016–2020, Agencia Boliviana Espacial, Empresa Pública Nacional Estratégica, La Paz, Bolivia, December 2016, https://www.abe.bo/files/planificacion/PlanEstrategicoABE2016-2020.pdf (accessed 18 September 2019).

Table 7.1 Strategic objectives and actions of ABE's Strategic Plan 2016–2020

N°	Strategic objectives	N°	Strategic actions
1	Reach the full capacity use of the satellite TUPAC KATARI and establish ABE as the operator of rural telecommunications for social inclusion	1.1	Manage the assignment of social inclusion projects
		1.2	Diversify the satellite services portfolio
		1.3	Position ABE in the rural market telecommunications services without affecting ABE's clients
2	Consolidate Bolivia's presence in outer space	2.1	Formulate a project for the launch of a new telecommunications satellite
		2.2	Shape and formulate new satellite applications projects and programs
		2.3	Increase relationships with the international space community
		2.4	Promote satellite applications and scientific and technological space development
3	Manage ABE's human resources as the most valuable resource of the state company	3.1	Create and implement a personnel management plan
		3.2	Develop new forms of motivation and recognition of ABE's personnel
4	Transform ABE into a highly efficient modern company that utilises cutting-edge technology in all its processes	4.1	Develop and carry out business and operations support systems
		4.2	Reduce the use of paper in the company
		4.3	Implement a teleworking policy
5	Incorporate quality in ABE's organisational culture	5.1	Reach and maintain the ISO 9001 certificate
		5.2	Implement other international norms that increase the company's value

7.3.1   National Space Regulations and Instruments

Bolivia's space related regulations include various decrees concerning ABE's creation, competence and functions, and one law governing the telecommunications sector.

Bolivia's space related regulation includes Supreme Decree N° 0423 of 10 February 2010 that created the ABE.[21] It determines ABE's legal nature, assets, budget, legal address, structure, functions and competences. Two additional decrees adopted in 2010 expand ABE's role. Supreme Decree N° 0599[22] defines ABE as a National

---

[21] Decreto Supremo N° 0423, Gaceta Oficial de Bolivia, 10 February 2010, https://www.abe.bo/files/marco-normativo/DS0423.pdf (accessed 17 September 2019).

[22] Decreto Supremo N° 0599, Gaceta Oficial de Bolivia, 18 August 2010, https://www.abe.bo/files/marco-normativo/DS0599.pdf (accessed 17 September 2019).

Public Strategic Company[23] dependent on the Ministry of Public Works, Services and Housing,[24] and recognises ABE's legal personality, assets and administrative, financial, legal and technical management. This Decree also allows ABE to enter into the space industry.[25] Supreme Decree N° 0746 of 25 February 2011[26] authorises a loan subscription of USD 251.124 million with the China Development Bank (CDB) to finance the Bolivian satellite project (see Sect. 7.5).[27]

Concerning the telecommunications sector, Law N° 164 or the General Law on Telecommunications, Information and Communication Technologies[28] establishes the general regime on telecommunications and TIC's in Bolivia[29] and guarantees the equitable distribution and efficient use of the radioelectric spectrum.[30] It also declares that the electromagnetic spectrum, a part of the radioelectric spectrum, is a natural limited resource of strategic character and public interest owned by the Bolivian's people and administered by the State.[31]

7.3.2   *Bolivia and International Space Instruments*

Bolivia is one of the few Latin American countries that have not ratified any of the UN Space Treaties.[32] It has only signed the OST and ARRA treaties. However, the country has ratified other space and telecommunications international instruments such as the International Telecommunication Union's (ITU) Convention and Constitution, and the Agreement Relating to the International Telecommunications Satellite Organisation (ITSO) (see Table 7.2).

---

[23]Empresa Nacional Pública Estratégica.

[24]Ministerio de Obras Públicas, Servicios y Vivienda.

[25]Historia, Agencia Boliviana Espacial, https://www.abe.bo/nosotros/historia/ (accessed 17 September 2019).

[26]On 25 February 2011, the Bolivian Senate approved the binational convention for the construction of the satellite TUPAC KATARI, Ibid.

[27]Several ABE's internal regulations are available at Marco normativo, Agencia Boliviana Espacial, https://www.abe.bo/nosotros/transparencia/marco-normativo/ (accessed 17 September 2019).

[28]Ley N° 164, Ley General de Telecomunicaciones, Tecnologías de Información y Comunicación, 8 August 2011, https://www.abe.bo/files/marco-normativo/LeyTelecomunicaciones.pdf (accessed 17 September 2019).

[29]Ibid., Art. 1.

[30]Ibid., Art. 2.1.

[31]Ibid., Art. 3.1.

[32]In South America only Bolivia and Suriname have not ratified any of the UN Space Treaties. See Status of International Agreements relating to activities in outer space as at 1 January 2019, A/AC.105/C.2/2019/CRP.3, 1 April 2019, http://www.unoosa.org/documents/pdf/spacelaw/treatystatus/AC105_C2_2019_CRP03E.pdf.

Table 7.2 International space and telecommunications treaties ratified by Bolivia

International space treaties ratified by Bolivia[a]

Treaty[b]	Adoption	Ratification (R)/signature (S)
OST	1967	S
ARRA	1968	S
LIAB	1972	–
REG	1975	–
MOON	1979	–
NTB	1963	R
BRS	1974	–
ITSO	1971	R
IMSO	1976	R
ITU	1992	R

[a]See Status of International Agreements relating to activities in outer space as at 1 January 2019, A/AC.105/C.2/2019/CRP.3, 1 April 2019, http://www.unoosa.org/documents/pdf/spacelaw/treatystatus/AC105_C2_2019_CRP03E.pdf
[b]Treaty on Principles Governing the Activities of States in the Exploration and Use of Outer Space, including the Moon and Other Celestial Bodies (OST); Agreement on the Rescue of Astronauts, the Return of Astronauts and the Return of Objects Launched into Outer Space (ARRA); Convention on International Liability for Damage Caused by Space Objects (LIAB); Convention on Registration of Objects Launched into Outer Space (REG); Agreement Governing the Activities of States on the Moon and Other Celestial Bodies (MOON); Treaty Banning Nuclear Weapon Tests in the Atmosphere, in Outer Space and under Water (NTB); Convention Relating to the Distribution of Programme-Carrying Signals Transmitted by Satellite (BRS); Agreement Relating to the International Telecommunications Satellite Organisation (ITSO); Convention on the International Mobile Satellite Organization (IMSO); International Telecommunication Constitution and Convention (ITU)

7.4   Cooperation Programs

To develop its space sector, Bolivia has adopted several agreements with other countries, such as China and Argentina.

The bilateral relation between Bolivia and China on space matters began in 2010 with the adoption of an MoU for the creation of the TUPAC KATARI project.[33] A year later, ABE and the China Great Wall Industries Corporation signed a commercial contract for the construction, acquisition and implementation of the satellite TUPAC KATARI, which was followed by a credit agreement signed between the Bolivian Government and the CDB for the construction of the satellite.[34]

---

[33]B. de Selding Peter, China to Build, Launch Bolivian Telecom Satellite, SpaceNews, 2 April 2010, https://spacenews.com/china-build-launch-bolivian-telecom-satellite/ (accessed 19 September 2019).

[34]Historia, Agencia Boliviana Espacial, https://www.abe.bo/nosotros/historia/ (accessed 19 September 2019).

In 2014, Bolivia and Argentina entered into negotiations to give impulse to prospective satellites for the benefit of Bolivia.[35] In 2019, both countries signed a Bilateral Agreement for the Provision of Satellite Services and Facilities.[36]

7.5 Satellite Capacities

The Telecommunications Satellite TUPAC KATARI, also known as TKSAT-1, is the first Bolivian telecommunications satellite.[37] Built by China Great Wall Industries Corporation, TKSAT-1 was launched on 20 December 2013 into GEO and started operations for commercial purposes in April 2014. The operation and management of TKSAT-1 is ABE's main activity. The satellite is controlled from the ABE's ground stations of Amachuma, La Paz, and La Guardia, Santa Cruz.[38] In 2016, ABE obtained a licence to offer telecommunication services directly to Bolivia's rural areas.[39]

TKSAT-1 operates in the C, Ku FSS, Ku BSS and Ka satellite frequencies and offers many telecommunications services, in particular Internet and satellite television services (see Table 7.3). Since the beginning of its operation, TKSAT-1 has contributed to reducing the digital gap and giving equal opportunities for all Bolivians.[40]

Apart from satellite TKSAT-1, there is currently a plan to launch a second national telecommunications satellite in the coming years (the expected year is 2020, although the date could vary depending on financial and political factors). The objective of

---

[35]Bolivia y Argentina se reúnen para impulsar cooperación espacial en satélites de prospección, Ministerio de Comunicación, 2 September 2014, https://www.comunicacion.gob.bo/?q=20140902/16565 (accessed 19 September 2019).

[36]Bolivia y Argentina realizan la III Reunión Técnica Bilateral de Vicecancilleres, Ministerio de Relaciones Exteriores, La Paz, 25 February 2019, http://www.cancilleria.gob.bo/webmre/noticia/3159 (accessed 19 September 2019). In April 2019, both countries signed an Interinstitutional Agreement between the Ministry of Public Communication of the Republic of Argentina, and the Ministry of Public Works, Services and Housing of the Plurinational State of Bolivia on the Provision of Satellite Services and Facilities, see Argentina – Bolivia: Comunicado Conjunto, Información para la Prensa N° 155/19, Ministerio de Relaciones Exteriores y Culto, República Argentina, 22 April 2019, https://www.cancilleria.gob.ar/es/actualidad/noticias/argentina-bolivia-comunicado-conjunto (accessed 19 September 2019).

[37]After the successful launch of Tupac-Katari, Bolivia planned to launch into outer space an Earth-observation satellite called "Bartolina Sisa". However, the project was indefinitely posponed due to an unfavorable economic situation, see Plan Estratégico ABE 2016–2020, p. 11.

[38]Telecomunicaciones,   Agencia   Boliviana   Espacial,   https://www.abe.bo/actividades/telecomunicaciones/ (accessed 17 September 2019). In Amachuma, there is a teleport which offers telecommunications services to different clients, Ibid.

[39]Ibid.

[40]Ficha técnica TKSAT-1, Agencia Boliviana Espacial, https://www.abe.bo/ficha-tecnica-tksat-1/ (accessed 17 September 2019).

Table 7.3 Satellite TKSAT-1 technical information

Satellite TKSAT-1 technical information			
Type of satellite	Communications	**Number of channels**	30
Frequency bands	C/Ku FSS/Ku BSS/Ka	**Launching date**	December 2013
Estimated lifetime	15 years	**Platform**	DFH-4
Dimensions	2360 mm × 2100 mm × 3600 mm	**Weight**	5100 kg
Orbital position	87.2° West	**Orbital type**	Geostationary
Orbital altitude	36,000 km	**Launching vehicle**	LM.3BE

This table was obtained from Ficha técnica TKSAT-1, Agencia Boliviana Espacial, https://www.abe.bo/ficha-tecnica-tksat-1/ (accessed 17 September 2019)

this new satellite is to solve the communications problems that still exist in some rural areas of Bolivia's territory.[41]

7.6 Capacity Building and Outreach

Capacity building on space issues is mostly promoted and carried out by ABE and Bolivia's universities.[42] In this regard, ABE promotes the formation of human resources in space technology and its applications by building relations with national universities, research centres and specialized institutions that develop space related activities. ABE has adopted cooperation agreements with the Higher University of San Andres, the Bolivian Catholic University and the Military School of Engineering.[43] These agreements aim to develop research and training programs, data exchanges and university internships.[44]

Other examples of ABE's space capacity building include the offer made by the Agency in 2019 of three scholarships for a Masters program in space technology applications (MASTA) at the Regional Centre for Space Science and Technology

---

[41]Larocca Nicolás, "El satélite boliviano paga más impuestos que los extranjeros", TeleSemana.com, 7 March 2019, (accessed 18 September 2019).

[42]Civil associations also contribute to this aim, such as the Bolivian Astronomy Association. See Asociación Boliviana de Astronomía, http://www.astrosurf.com/astrobo/ (accessed 5 October 2019).

[43]Universidad Mayor de San Andrés, Universidad Católica Boliviana, and Escuela Militar de Ingeniería, respectively.

[44]Formación de Recursos Humanos, Agencia Boliviana Espacial, https://www.abe.bo/actividades/formacion-de-recursos-humanos/ (accessed 18 September 2019).

Education in Asia and the Pacific (RCSSTEAP), University of Beihang, China[45]; the 2019 call concerning the search for professionals in science, engineering and technology to join ABE's personnel,[46] and the five-month Certification Program on Satellite Communications Systems organized by ABE and the Institute of Applied Electronic of the Higher University of San Andres to acquire knowledge on the foundations, operation and design of these systems.[47]

Appendix

See Table 7.4.

Table 7.4 Institutional Contact Information in Bolivia (2019)

Institution	Details
Bolivian Space Agency (ABE)	Calacoto 14, N° 8164, La Paz, Bolivia Website: https://www.abe.bo/ E-mail: abe@abe.bo
Vice-ministry of Telecommunications of the Ministry of Public Works, Services and Housing	Avenue Mariscal Santa Cruz 591, La Paz, Bolivia Website: http://www.oopp.gob.bo
Authority of Regulation and Control of Telecommunications and Transportation	Calacoto 13 N° 8260, La Paz, Bolivia Website: https://www.att.gob.bo/ E-mail: webmaster@att.gob.bo

---

[45]This master offers three specialities: satellite communications and global navigation satellite systems, remote sensing and geo-information systems, and micro-satellite technology. See Becas MASTA 2019, Agencia Boliviana Espacial, https://www.abe.bo/estudiantes/becasmasta2019/ (accessed 18 September 2019). ABE's have channeled these scholarships since three years ago, see ABE lanza tres becas para maestrías en China, Agencia Boliviana Espacial, 8 January 2019, https://www.abe.bo/abe-lanza-tres-becas-maestrias-china/ (accessed 18 September 2019).

[46]Convocatoria, Búsqueda de Profesionales en Ciencias, Ingeniería y Tecnología, Agencia Boliviana Espacial, 2019, https://abe.bo/convocatoriaprofesionales.pdf (accessed 18 September 2019).

[47]Formación de Recursos Humanos, Agencia Boliviana Espacial, https://www.abe.bo/actividades/formacion-de-recursos-humanos/ (accessed 18 September 2019). See also Diplomado en Sistema de Comunicaciones Satelitales, 1era versión, Instituto de Electrónica Aplicada, Universidad Mayor de San Andrés, http://iea.umsa.bo/sistemas-de-comunicaciones-satelitales (accessed 18 September 2019).

Chapter 8
Brazil

Contents

Abstract In order to take stock of Brazil's progress in the space arena, and to determine how far advanced the space sector is across the country, this chapter provides the basis for the subsequent study of the Brazilian space ecosystem. The political context of the country is examined from the beginning of space activities in the 1960s until the creation of the Brazilian Space Agency in 1994, which is the key body. This is followed by a discussion of the investments made and of international cooperation, within a framework of the needs approach to evolving space activities. The priority areas of critical importance, which require further funding, if the country wants independence from other countries' satellites are: climate and biodiversity, broadband, education and capacity building related to space. Throughout the discussion, the focus is placed on Brazil's National Space Activities Program 2012–2021

© Springer Nature Switzerland AG 2020

A. Froehlich et al., *Space Supporting Latin America*, Studies in Space Policy 25,
https://doi.org/10.1007/978-3-030-38520-0_8

as the central document that will guide the country towards a future of sustainable development. The Brazilian space capabilities are described, as well as the project to develop new space technologies.

8.1   General Country Overview

Brazil is the largest country in both Latin America and the southern hemisphere with 8,515,770 km^2 and the world's fifth-largest country by area. With over 211 million people, it is the fifth most populated with 86.3% of urban population. Brazil is considered an advanced emerging economy with an emerging financial market; it is the ninth world economy[1] with high inequalities and the leading producer of sugar and coffee in the world.

Brazil is a regional power and sometimes considered a great or a middle power in international affairs. The country is a founding member of the UN, the G20, BRICS, MERCOSUR, Union of South American Nations (UNASUR),[2] Organization of American States (OAS), Organization of Iberic-American States (OEI) and the Community of Portuguese Language Countries (CPLP).

The political and administrative organization of Brazil comprises the federal government, the 26 states, a federal district and the municipalities. The states are autonomous sub-national entities with their own governments that, together with the other federal units, form the Federative Republic of Brazil. It is a federal presidential constitutional republic, based on representative democracy whereby the President is both head of state and head of government. The federal government has three independent branches: executive, legislative, and judicial (Fig. 8.1).

Throughout its history, Brazil has struggled to build a democratic and egalitarian society because of its origins as a Portuguese colony from the 16th to 19th century. After independence in 1822, Brazil became a parliamentary constitutional monarchy until 1889, succeeded by a Republic. During this period, the country had a military dictatorship that ruled from 1964 to 1985, until it became a democracy again, in 1988.

In recent years, Brazil experienced a period of economic and social progress between 2003 and 2014, when more than 29 million people left poverty and inequality was reduced significantly (the Gini coefficient dropped 6.6%—from 58.1 to 51.5%).

---

[1]Smith, R., 18 April 2018, World Economic Forum, https://www.weforum.org/agenda/2018/04/the-worlds-biggest-economies-in-2018 (accessed 2 February 2019).

[2]On 15 April 2019, the Brazilian Ministry of Foreign Affairs denounced the Constitutive Treaty of the UNASUR, formalizing its departure from the organization, just hours after receiving the Pro Tempore Presidency from Bolivia, http://www.itamaraty.gov.br/pt-BR/notas-a-imprensa/20291-denuncia-do-tratado-constitutivo-da-uniao-de-nacoes-sul-americanas-unasul (accessed 2 July 2019).

Fig. 8.1 Brazil's Political Division. Blank Map of Brazil, http://www.4geeksonly.com/countries/brazil/blank-map-of-brazil.html (accessed 22 September 2019)

Since 2015, however, the pace of poverty and inequality reduction seems to have stagnated.[3]

In the wake of a strong recession, after 2015, Brazil has been going through a phase of highly depressed economic activity. There was a significant contraction in economic activity in 2015 and 2016, with the GDP dropping by 3.6% and 3.4% (respectively). The economic crisis was a result of falling commodity prices and the country's limited ability to carry out necessary fiscal reforms at all levels of government, thus undermining consumer and investor confidence. The country's growth rate has been slowing since the beginning of the last decade, from an annual growth rate of 4.5% (between 2006 and 2010) to 2.1% (between 2011 and 2014). 2017 saw the beginning of a slow recovery in Brazil's economic activity, with 1% growth of GDP.

---

[3]The World Bank, https://www.worldbank.org/en/country/brazil/overview (accessed 2 February 2019).

8.2   History of Space Activities

Brazilian space history (see Table 8.1) began in the late 1950s. Since then, development of the space sector in Brazil has increased considerably in different areas, from satellite launches to launch vehicle projects.

Brazil was one of the pioneers, among developing countries, to lean toward space sciences. The Brazilian Space Agency (AEB)[4] was created in 1994, but the first space activities on Brazilian soil were conducted already around 1956, when, for four years, in Fernando de Noronha (archipelago that was federal territory at the time), the USA installed and operated, a rocket tracking station for missions launched from the then Cape Canaveral—today Cape Kennedy. In 1957, at the beginning of these operations at the Aerospace Technical Centre (CTA),[5] two students of the Aeronautical Technological Institute (ITA)[6] built a satellite reception station,[7] with which they were able to pick up signals from the Soviet satellite SPUTNIK and from the North American EXPLORER I.

In 1960, in Argentina, the first Inter-American Meeting on Space Research took place, where countries agreed that "each local group should encourage the formation of national governmental commissions or state support for a greater activity in space research." Thus, in 1961, Brazil created the Organizing Group of the National Commission of Space Activities (GOCNAE),[8] linked to the National Research Council (CNPq).[9]

As a result of the active military participation of the Ministry of Aeronautics since the initial phase of space activities in Brazil, a joint project with the National Commission of Space Activities (CNAE)[10] was initiated. Thus, in 1964, the Executive Group of Work and Study Space Projects (GETEPE)[11] was created,[12] subordinated to the General Staff of the Aeronautics Ministry, with the objective of (a) establishing a centre for rocket launching and preparing teams specialized in launchings; (b) establishing meteorological and ionospheric surveys in cooperation with foreign organizations; and (c) encouraging Brazilian private industry to improve the development of space technology.

---

[4]Agência Espacial Brasileira (AEB).

[5]Centro Técnico Aeroespacial (CTA).

[6]Instituto Tecnológico da Aeronáutica (ITA).

[7]INPE, Institutional, http://www.inpe.br/institucional/sobre_inpe/historia.php (accessed 2 February 2019).

[8]Grupo de Organização da Comissão Nacional de Atividades Espaciais (GOCNAE).

[9]Conselho Nacional de Desenvolvimento Científico e Tecnológico (CNPq).

[10]Comissão Nacional de Atividades Espaciais (CNAE).

[11]Grupo Executivo e de Trabalho e Estudos de Projetos Espaciais (GETEPE).

[12]Rollemberg, D., Pinheiro, A., Filho, D., Felipe, B., Lins, E., Silva, E., Martins, D., A Política Espacial—Parte 1, Câmara dos Deputados, 2010, p. 38, https://www2.camara.leg.br/a-camara/estruturaadm/altosestudos/arquivos/politica-espacial/a-politica-espacial-brasileira (accessed 2 February 2019).

Table 8.1 Brazil's space history

Year	Brazil's space history
1950	Foundation of the Aeronautics Technical Centre (CTA). Nowadays, called the Department of Aerospace Science and Technology (DCTA), under the Ministry of Aeronautics.
1957	ITA students set up the Minitrack Mark II station to receive signals from the U.S. Project Vanguard.
1961	Establishment of the National Space Activities Commission Organization Group (GOCNAE).
1963	GOCNAE became the National Commission for Space Activities (CNAE).
1964	Established the Working Group on Space Studies and Projects (GTEPE).
1965	Inauguration of Barreira do Inferno Launch Centre (CLBI), in Rio Grande do Norte State.
1966	Foundation of the Executive Working Group and Study of Space Projects (GETEPE).
1967	Launch of SONDA I rocket from CLBI.
1969	1. Launch of SONDA II rocket from CLBI; 2. Foundation of the Brazilian Aeronautics Company (Embraer).
1971	1. CNAE is replaced by the Space Research Institute; 2. Foundation of the Brazilian Space Activities Commission (COBAE).
1973	After USA and Canada, Brazil became the third country to have an operating station to receive satellite images, in Cuiabá, Mato Grosso State.
1976	Launch of SONDA III rocket from CLBI.
1979	The Brazilian Complete Space Mission (MECB) is released to develop satellites, launch vehicles and establish launching centres in Brazil.
1983	Inauguration of the Alcântara Launch Centre (CLA), in Maranhão State.
1984	Launch of SONDA IV rocket from CLBI.
1985	CLA begins operations with the launch of SONDA II rocket.
1988	Brazil and China sign cooperation agreement to develop CBERS satellites.
1993	1. Launched the first Brazilian satellite, Data Collection Satellite (SCD-1), with the mission of collecting environmental data; 2. First rocket VS-40 launch from CLA.
1994	Foundation of the Brazilian Space Agency (AEB).
1997	1. First VLS-1 launch test from CLA and first VS-30 rocket launch from CLA; 2. Brazil joins the International Space Station (ISS) program.
1998	1. Launch of the Brazilian Satellite SCD-2 in Florida, USA; 2. AEB selects the first Brazilian astronaut candidate.
1999	1. CBERS-1 Satellite is launched by the Chinese Long March IV rocket from Taiyuan Base, China; 2. VLS-1 second prototype launch with failure on second stage; 3. Launch of SACI 1 (failure) and 2 scientific microsatellites aboard the Long March IV and VLS rocket;
2000	Launch of the first VS-30 Orion sounding rocket prototype.

(continued)

Table 8.1 (continued)

2003	1. Treaty between Brazil and Ukraine on long-term cooperation in the use of the Cyclone-4 Launch Vehicle at CLA; 2. VLS-1 accident at CLA; 3. Launch of CBERS-2 from Taiwan Launch Centre.
2006	1. On 30 March 2006, Colonel Marcos Pontes became the first Brazilian astronaut to go to the ISS, aboard the Russian SOYUZ 8 spacecraft, with eight scientific experiments on microgravity environment execution; 2. Establishment of the binational (Brazil and Ukraine) company Alcântara Cyclone Space (ACS).
2007	Launch of the Brazilian Satellite CBERS-2B from China.
2012	AEB promotes first Workshop on Brazilian Space Industry: Challenges and Opportunities.
2013	1. First GLONASS Station outside Russia installed at University of Brasilia (UnB); 2. AEB release the National Space Activities Program (PNAE), which establishes the guidelines and actions of the Brazilian Space Program between 2012 and 2021.
2014	Brazil joins the nanosatellite segment and develops NanoSatC-BR1, by the Federal University of Santa Maria (UFSM-RS).
2017	1. Launch of the Geostationary Defence and Strategic Communications Satellite (SGDC) from Kourou Space Centre, in French Guiana. The satellite, for civil and military use, will bring broadband Internet to the entire Brazilian territory, promoting the digital and social connectivity of millions of citizens; 2. The first Space Technological Vocational Centre of Brazil (CVT-Espacial) is inaugurated in Parnamirim, Rio Grande do Norte State. The CVT-Espacial was conceived with the aim to encourage students to work with Space, Science & Technology areas, as well as making them aware of the importance of the Brazilian Space Program and its benefits for the country.
2019	1. Launch of CBERS 4-A, from Taiyuan Satellite Launch Center (TSLC) in China; 2. Launch of FloripaSat-1, a Brazilian Cubesat mission developed by university students of the Federal University of Santa Catarina (UFSC). The launch was from TSLC in China together with the CBERS-4A.

In practice, GETEPE became the operational arm of space activities in Brazil, which enabled its rapid development, especially in the field of launches and vehicles. In fact, one of the first activities of GETEPE was the selection of an area near the city of Natal, state of Rio Grande do Norte, to construct what was to become Barreira do Inferno Launch Centre (CLBI),[13] inaugurated in 1965, with the launch of a North American NIKE-APACHE rocket.[14] The success of this launch was also a result of the training received by the Brazilian teams at the Wallops Flight Centre and the Goddard Space Flight Centre, both of NASA. Following the inauguration of the CLBI in 1965, more than a hundred launches were followed up until the year 1970. Some of them, already in the national series of sounding-rockets since 1965, were based on an agreement between the CNPq/CNAE, NASA and the National

---

[13]Centro de Lançamento da Barreira do Inferno (CLBI).

[14]Harvey, B., Smid, H., Emerging Space Powers—The New Space Programs of Asia, the Middle East and South-America, Springer, 2010, p. 336.

Commission of Spatial Investigations of Argentina (CNIE),[15] such as part of Project EXAMETNET.[16]

The initial series of national rockets was designated SONDA and consisted of four models, through which successive gains were made in qualification for design, production and launching of rockets. The rockets of the SONDA family were the first rockets manufactured in Brazil, launched from the CLBI.[17]

SONDA I was a simple, two-stage rocket, with a take-off mass of only 59 kg, specified by GETEPE and ordered from a Brazilian company. Its peak was 65 km, with a payload capacity of 4 kg. In all, about 225 of these rockets were launched between 1967, the date of the flight of the first prototype, and 1977. Although several technologies were developed in Brazil to make SONDA I viable, the field to produce this technology was developed not only for the preparation of the SONDA I propellant wrap but also for the SBAT rockets, launched from military aircraft, later used in the production of valve guides for internal combustion engines. It is estimated that the foreign exchange savings generated by import substitution of this type of tube was higher than the total resources applied in the Brazilian space program until 1992, at which time the estimate was made.[18]

SONDA II was the first rocket whose design, structural fabrication, propellant and thermal protections were developed in the CTA. Several versions of this rocket were developed. The current version has a takeoff mass of about 370 kg, with an apogee of 50–100 km and capacity for 20–70 kg payload. In total, 61 SONDA II were launched, but the design was poorly documented, with no technical specifications of the materials constituting the system, probably because its reference configuration was based on the Canadian rocket BLACK BRANT III.[19]

Following the strategic line of increasing qualification, since 1971, SONDA III was developed, whose second stage was nothing more than a SONDA II rocket. SONDA III enabled CTA to familiarize itself with a more complex engineering work methodology, involving feasibility analyses and preliminary studies to define reference configuration, as well as the use of the program evaluation and review technique network, based on macro-events identified in a development plan. In 1971, during the development phase of SONDA III, the Federal Government decided to

---

[15] Comissión Nacional de Investigaciones Espaciales (CNIE).

[16] Beginning of the EXAMETNET rocket experiments involving NASA, Barreira do Inferno Launch Centre and INPE, in order to perform measurements of meteorological parameters between 30 and 60 km height. 88 operations had been performed between 1966 and 1978, totaling 207 entries, http://www3.inpe.br/50anos/english/timeline/66.html (accessed 23 May 2019).

[17] Gouveia, A., Esboço histórico da pesquisa espacial no Brasil, INPE-10467-RPQ/248, http://mtc-m12.sid.inpe.br/col/sid.inpe.br/jeferson/2003/07.02.08.39/doc/publicacao.pdf (accessed 23 May 2019).

[18] AEB, SONDA I, http://www.aeb.gov.br/programa-espacial-brasileiro/transporte-espacial/sonda-i/ (accessed 23 May 2019).

[19] AEB, SONDA II, http://www.aeb.gov.br/programa-espacial-brasileiro/transporte-espacial/sonda-ii/ (accessed 23 May 2019).

reorganize the space activities conducted in Brazil, giving it a systemic form.[20] To that end, on 20 January 2015, the Brazilian Space Activities Commission (COBAE)[21] was established under Decree-Law no. 68.099,[22] with the aim of advising the President of the Republic on the achievement of the National Policy for the Development of Space Activities (PNDAE).[23]

Also, within the framework of this major restructuring, the CNAE was abolished by Decree-Law no. 68.532,[24] giving way to the current National Institute of Space Research (INPE).[25] Likewise, in the same year, GETEPE was terminated by an ordinance act, passing its activities to the Institute for Space Activities, previously mentioned in Decree-Law no. 65.450[26] of 1969, which structured the Department of Research and Development, to which the CTA is subordinate.

In 1973, Brazil became the third country, after the USA and Canada, to have an operational station to receive satellite images, in the city of Cuiabá, Mato Grosso State.[27]

This was part of the strategy aimed at launching national satellites by means of a nationally developed launch vehicle from a Brazilian launch centre. Still motivated by the success of SONDA III, in the second half of 1976 the CTA/IAE began feasibility studies and technical specifications of what would become SONDA IV: an intermediate vehicle whose development would lead to the domain of critical technologies, without which it would not be possible to proceed consistently in an autochthonous space program.[28]

SONDA IV was also the first Brazilian space project to be divided into phases. It used the Work Breakdown Structure as a management tool, making it possible to control deadlines and costs, besides allowing the coding of responsibilities without ambiguities. As a reasonably sized propeller, it forced research on a new type of steel, a carbon-chromium-nickel-molybdenum class, with high silicon content, with treatment for the resistance level of 200 kg/mm^2.[29]

---

[20] AEB, SONDA III, http://www.aeb.gov.br/programa-espacial-brasileiro/transporte-espacial/sonda-iii/ (accessed 23 May 2019).

[21] Comissão Brasileira de Atividades Espaciais (COBAE).

[22] Brazil, Decree no. 68.099, 20 January 1971, https://www2.camara.leg.br/legin/fed/decret/1970-1979/decreto-68099-20-janeiro-1971-410111-publicacaooriginal-1-pe.html (accessed 23 May 2019).

[23] Política Nacional de Desenvolvimento das Atividades Espaciais (PNDAE).

[24] Brazil, Decree no. 68.532, 22 April 1971, https://www2.camara.leg.br/legin/fed/decret/1970-1979/decreto-68532-22-abril-1971-410268-publicacaooriginal-1-pe.html (accessed 23 May 2019).

[25] Instituto Nacional de Pesquisas Espaciais (INPE).

[26] Brazil, Decree no. 65.450 17 October 1969, https://www2.camara.leg.br/legin/fed/decret/1960-1969/decreto-65450-17-outubro-1969-406960-publicacaooriginal-1-pe.html (accessed 23 May 2019).

[27] AEB, 1973, http://www.aeb.gov.br/timelinr/1973/ (accessed 23 May 2019).

[28] AEB, SONDA IV, http://www.aeb.gov.br/programa-espacial-brasileiro/transporte-espacial/sonda-iv/ (accessed 23 May 2019).

[29] Ibid.

With the launch of four SONDA IV, the necessary bases for the beginning of the project of the Brazilian launch vehicle capabilities to place satellites in LEO were implemented. In 1978, at the request of the Brazilian government, the National Commission for Space Studies of France presented a proposal for the development of a launch vehicle and three satellites.[30] This proposal, after studies, was considered very expensive, not to mention that most of the development work would be carried out in French industries. Thus, in 1979, during the second Space Activities Seminar, held under the auspices of the Brazilian Space Activities Commission, INPE and the former CTA/IAE jointly presented an alternative proposal, which came to be known as the Brazilian Full Space Mission (MECB).[31,32]

In 1980, according to MECB, the development of the launch vehicle (which would become the VLS-1) and launch infrastructure was carried out by the CTA/IAE, while INPE was responsible for the development of two environmental data collection satellites and two others for remote sensing data.[33] In this context, the Alcântara Launch Centre (CLA),[34] in the state of Maranhão, was inaugurated in 1983,[35] with a structure adequate for the new scientific and technological undertaking. The VLS-1 can be understood as composed of four propulsion stages, a payload transport bay (satellite), sections (bays or modules) for housing instrumentation and various equipment, four functional electrical networks, and a set of 244 pyrotechnics, members of the so-called pyrotechnic network, although they do not constitute a network in the strict sense of that word.[36]

In 1993, the first Brazilian satellite, the Data Collection Satellite (SCD-1)[37] was launched, with the mission of collecting environmental data.[38] The following year, the Brazilian Space Agency (AEB)[39] was created to define, update and execute the National Policy for the Development of Space Activities (PNDAE),[40] to elaborate and update the National Space Activities Programs (PNAE's)[41] and the respective budget proposals, to analyze and propose agreements and conventions, in coordination with the Ministry of Foreign Affairs and Ministry of Science, Technology and Innovation,

---

[30]AEB, VLS-1, http://www.aeb.gov.br/programa-espacial-brasileiro/transporte-espacial/vls-1/ (accessed 23 May 2019).

[31]Missão Espacial Completa Brasileira (MECB).

[32]Orlando, V., Kuga, H., Os satélites SCD1 e SCD2 da Missão Espacial Completa Brasileira, http://www.cdcc.usp.br/cda/oba/aeb/a-conquista-do-espaco/Capitulo-5.pdf (accessed 23 May 2019).

[33]Ibid.

[34]Centro de Lançamento de Alcântara (CLA).

[35]G1, Centro de Lançamento de Alcântara completa 30 anos, http://glo.bo/ZhnBdo (accessed 23 May 2019).

[36]IAE, VLS-1, http://www.iae.cta.br/index.php/espaco/vls1 (accessed 23 May 2019).

[37]Satélite de Coleta de Dados-1 (SCD-1).

[38]Gunter's Space Page, SD 1, https://space.skyrocket.de/doc_sdat/scd-1.htm (accessed 23 May 2019).

[39]Agência Espacial Brasileira (AEB).

[40]Política Nacional de Desenvolvimento das Atividades Espaciais (PNDAE).

[41]Programa Nacional de Atividades Espaciais (PNAE).

and to advise the government on matters related to the aerospace area.[42] The agency was created as an organ of an expressly civil nature, with administrative and financial autonomy, and with its own staff and assets.

In 1995 the Brazilian government formally adhered to the Missile Technology Control Regime (MTCR),[43] an informal agreement on the non-proliferation of technologies considered sensitive in the aerospace area.[44] In effect, the MTCR prevented a series of agreements signed by AEB with foreign countries to acquire technologies for the development of the national VLS, forcing the country to develop such technological solutions autonomously.[45]

In the following years, AEB continued investing efforts in the development of VLS, a project under the responsibility of the CTA since the approval of the Brazilian Full Space Mission in 1980. The first VLS prototype was launched in November 1997, carrying the satellite data collector SCD-2A.[46] However, the launch was unsuccessful due to an ignition failure on one of the first stage rocket propellers. This failure forced AEB to launch another SCD-2 satellite by means of an US rocket the following year.[47]

In 1999, a new attempt to launch the VLS-1 was made, this time with the satellite SACI-2 (developed by INPE) on board, but resulted in a new failure, due to a failure in the second stage of the rocket.[48] The last attempt to launch the national VLS was scheduled on 25 August 2003, but three days before the date, a fire in the first stage of the rocket resulted in a tragedy that killed 21 technicians and scientists, as well as complete destruction of the launch pad at CLA.[49] The loss and the victims in the accident meant a major blow to the project, since some of the main advocates of PEB were among them.[50]

Figure 8.2 shows the Brazilian sounding rockets used for suborbital space exploration missions capable of launching payloads composed of scientific and technological experiments (with the exception of VS-43 which is still under project development). Brazil has operating vehicles in this class, which supply a good part of its needs with a successful history of launches.

---

[42]FGV, CPDOC, AEB, http://www.fgv.br/cpdoc/acervo/dicionarios/verbete-tematico/agencia-espacial-brasileira-aeb (accessed 23 May 2019).

[43]Missile Technology Control Regime (MTCR).

[44]Bowen, W., Report: Brazil's accession to the MTCR, https://www.nonproliferation.org/wp-content/uploads/npr/bowen33.pdf (accessed 23 May 2019).

[45]Ibid.

[46]Neto, R., Folha de S. Paulo, Falha de motor obriga controle da operação a explodir lançador no ar. 1997, https://www1.folha.uol.com.br/fsp/ciencia/fe031101.htm (accessed 23 May 2019).

[47]Ibid.

[48]Ottoboni, J., Folha de Londrina: INPE desiste do programa espacial Saci, https://www.folhadelondrina.com.br/geral/inpe-desiste-do-programa-espacial-saci-235771.html (accessed 23 May 2019).

[49]G1, Maior acidente do Programa Espacial Brasileiro completa 13 anos, http://glo.bo/2baWijW (accessed 23 May 2019).

[50]Ibid.

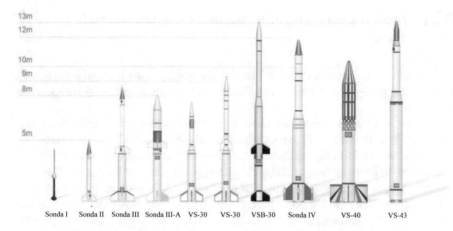

Fig. 8.2 Brazilian sounding rockets program. AEB, Transporte Espacial, http://www.aeb.gov.br/programa-espacial-brasileiro/transporte-espacial/ (accessed 25 September 2019)

Brazil has designed and produced a successful set of rockets for suborbital space flight, which has led to numerous scientific and technological experiments. The domain of sounding rockets' technology was the basis for developing the VLS, serving as R&D.[51]

A successful case concerns the VSB-30, which is a rocket aimed to conduct experiments in microgravitational environments, under the ESA Microgravity Programme. On 16 October 2009, the Ministry of Aeronautics and DCTA announced the VSB-30 Certification, qualifying it for mass production.[52] From 2004 to 2013, there were nine successful launches from Kiruna, Sweden and the CLA.

8.3 Economic Perspective of Space Program

Currently, AEB's budget is around R$194 million (USD 46 million) a year. Almost half a century since the beginning of space activities in Brazil and large budgetary variations throughout the period, there are great difficulties in enabling an aerospace industry in Brazil. In addition to a history of shifting financing, demand for products, services and professionals in the area also fluctuates year by year.

---

[51]AEB, VS-40, http://www.aeb.gov.br/programa-espacial-brasileiro/transporte-espacial/vs-40/ (accessed 25 September 2019).

[52]Folha de S. Paulo, Foguete nacional ganha selo para produção em série, https://www1.folha.uol.com.br/ciencia/2009/10/638857-foguete-nacional-ganha-selo-para-producao-em-serie.shtml (accessed 25 September 2019).

8.3.1 Institutional Space Budgets and Program Budget Trends

Based on the Transparency Portal of Brazil,[53] the budget of the last five years of AEB is described in Fig. 8.3. Agency expenditure can be made direct or by delegation of resources.

The current budget model for financing space activities in Brazil is linked to the Multi-Year Investment Plan (medium-term planning), which directs government actions, financed from the budget cycle of the Annual Budget Law (LOA).[54,55] In addition to the four-year and annual cycles, there is also the possibility, over the course of each year, of budget supplements and re-shifts through decrees or the sending of bills for approval by the National Congress, according to each publication of the Law of Budgetary Guidelines (LDO).[56] The budget authorized annually for the program has the form of a set of budgetary actions that make up the National Program of Space Activities. These actions are classified into activities or projects as the product or service is produced continuously (activity) or at a given time (project).

The negative impact on the program of this budgetary variation over time stems from the nature of the development cycle of space products and services. In fact, the construction of satellites, rockets and terrestrial infrastructure presents complexity and technological risks, high cost and long development cycles, usually between four and eight years. In this way, the management of projects and space activities becomes

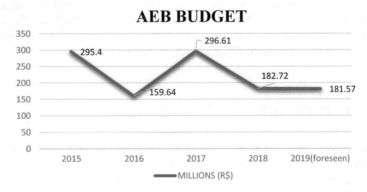

Fig. 8.3 AEB budget. Portal da Transparência, http://www.portaltransparencia.gov.br/ (accessed 23 May 2019)

---

[53]The Transparency Portal of the Federal Government is a channel through which citizens can monitor the use of federal resources collected with taxes in the provision of public services to the population, as well as inform themselves about other matters related to the Federal Public Administration.

[54]Lei Orçamentária Anual (LOA).

[55]Brazil, Transparency Portal, https://portaltransparencia.gov.br/orgaos/20402-agencia-espacial-brasileira-aeb (accessed 15 May 2018).

[56]Lei de Diretrizes Orçamentárias (LDO).

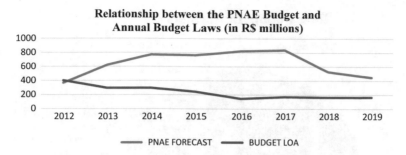

Fig. 8.4 Relationship between PNAE budget and LOA. AEB, Management Report, http://www.
aeb.gov.br/wp-content/uploads/2019/04/Relat.%C3%B3rio-de-Gest%C3%A3o-2018_AEB.pdf
(accessed 23 May 2019)

hostage to the long-term uncertainty about the financial support necessary to carry
out the tasks and contracts involved, which ends up generating constant problems for
continuity and exhaustive and continuous re-planning actions. In addition, periods of
shortage of resources translate into consecutive delays of schedule with consequent
obsolescence of infrastructure, technological backwardness, dissolution of international partnerships and dispersion or loss of specialized personnel. Among the main
losers of this situation is the national industrial sector, which is made up of small
and medium-sized enterprises, unable to cope with delays in contractual payments
or the lack of long-term contracts.

Historically, the difference between planned and realized investment reached R$
639 million in 2016 but this did not worsen due to the reduction in forecasts, given
the investment cycle, which followed the actual budget declines. The difference
between the envisaged budget and the approved budget has been reduced, not by
a possible increase in the approved budget, but by the reduction of the envisaged
budget (Fig. 8.4).

It is worth noting in the graph above that, although the projections of the PNAE
2012–2019 indicate an increase of resources until the year 2017, there was a
systematic reduction of resources for the space program.

8.3.2   Space Industry Market by Application/Sectors

The allocation of the 2018 AEB budget (R$ 140.9 million) to meet the demands of
the space program are divided into the following sectors: construction of launch vehicles (suborbital, VLM-I); Space, Education and Technology (E2T); maintenance of
the Vocational Space Technological Centre of Brazil (CVT); research and development in space technologies; maintenance and improvement of launch centres (CLA
and CLBI); development of satellites (CBERS-4A, Amazonia, scientific satellites
such as EQUARS , nanosatellites and others); maintenance of the Integration and

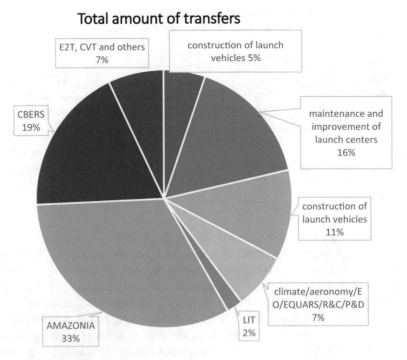

Fig. 8.5 AEB 2018 management report. AEB, Management Report, http://www.aeb.gov.br/wp-content/uploads/2019/04/Relat.%C3%B3rio-de-Gest%C3%A3o-2018_AEB.pdf (accessed 23 May 2019)

Testing Laboratory (LIT); stimulus to studies related to Climate, Aeronomy, EO (OBT), Tracking and Control (R&C), Research and Development (R&D) and others (Fig. 8.5).

Unlike the major space programs in the world, the Brazilian system does not privilege the participation of the national industry, according to the Brazilian Aerospace Industries Association.[57] The industry is considered only as a supplier of goods, components and equipment to space research bodies, such as INPE, which maintains the exclusive role in the area of development of satellite projects. The same occurs regarding the IAE, with regard to launchers. This logic makes it difficult to apply one of the main mechanisms adopted internationally to leverage high-cost, long-term and strategic programs: the use of state purchasing power to foster innovation and the competitiveness of local high-tech companies.

In order to capture more accurately the set of companies that provide goods and services to the Brazilian Space Program, INPE has shown that the main contributions to the space sector are provided by micro and small companies. It is worth pointing

---

[57]Rollemberg, R., Veloso, E., A política espacial brasileira, Câmara dos Deputados, Congresso Nacional, 2009, https://www2.camara.leg.br/a-camara/estruturaadm/altosestudos/arquivos/politica-espacial/a-politica-espacial-brasileira (accessed 15 May 2019).

out, from a comparative international perspective, that the main players globally in the space sector are large industrial conglomerates that, for the most part, have diversified business units and are not always linked only to space activity. Considering that Brazil is still at an early stage in the development of a space industry, the firms identified are concentrated around the manufacturing of subsystems and components of satellites, launchers, and supply of goods and services for the launching infrastructure and ground services.[58]

8.3.3 Space Development Economic Trends and Socio-economic Benefits

It is worth noting that the PEB irregular investments make it difficult to sustain its activities solely based on space activities, so it is possible that throughout their existence a large part of these identified firms has sought diversification of its activities. In this perspective, this skilled labor can be applied in other markets, not related to space. The profile of foreign trade probably is not only due to products and services related to space activity, although it is quite reasonable to suppose that the fact that almost half of the firms are importers is associated with the still existing dependence of Brazil on parts and components of systems that do not yet have complete endogenous development. In addition, regarding exports, it is quite plausible that the exported items are linked to other business units of the respective firms, such as aeronautics.

The competitiveness potential of the sector, in a medium- and long-term perspective, will, however, be directly linked to the strategic choices that these firms adopt in the coming years. Since they already have differentiated resources and competences in relation to the other companies, thus possessing internal skills that can be used, these choices will certainly be conditioned by the environmental circumstances that they will experience in that same period. By environmental circumstances, both the structure of the industry in which they are situated and the macro-environmental conditions, such as those related to politics, economy, socio-cultural, technological and legal changes, are understood. Among this group of factors, it is important to highlight that the most influential one is political, since government action to establish a flow of demand for space activity in Brazil is essential for the sustainability of the sector.

Regarding the participation of the private sector in space activities, the performance of private companies has been restricted to the supply of parts, components and subsystems commissioned by these national administration and organizations. In all the years of IPEA analysis conducted between 2000 and 2007, the number

---

[58]Ibid., p. 35.

of importing firms was larger than the national average.[59] Meanwhile, the average export values outweighed average import values in this period, so that although the sector's dependence on raw materials, hardware and components manufactured abroad is clear, the contribution of these firms to the national export of medium-high and high technological intensity products is evident.

Thus, the national space industry requires actions that contribute to the reduction of the growing gap between Brazil and other countries with space activities, since in these nations investments have been higher and more regular than those observed in the PEB. Even in a scenario in which the importance of private actors is growing, demand is still very much linked to the state's purchasing power, so there is no way to talk about the development of the local space industry without addressing the need to articulate government demand for space products and services, which requires a more coordinated and strategic approach to the issue than that currently in place, with regular and timely availability of resources.

In November 2011, an MoU was signed for a joint venture between Embraer and Telebras to operate in the Brazilian geostationary satellite project. In May 2012, the Ministry of Communications officially announced the signing of the shareholders' agreement for the effective incorporation of Visiona Tecnologia Espacial SA,[60] with 51% of the capital under Embraer control and the remaining 49% of Telebras, to operate in the PNAE. Visiona, which is headquartered at the Technological Park of São José dos Campos, is to operate the Brazilian geostationary satellite, which aims to meet the Federal Government's satellite communication needs, including the National Broadband Program and a broad spectrum of broadcasts strategies.

Government action in the sector was also manifested by the recent publication of Decree No. 7.769,[61] which provides for the management of the planning, construction and launch of the Geostationary Defense Satellite and Strategic Communications (SGDC).[62] In 2013, Visiona signed contracts with Thales Alenia Space (TAS) and Arianespace, who agreed to provide the satellite and its launch into orbit, respectively. The contract with the satellite supplier also provides the technology's transfer to Brazilian companies, through the coordination of AEB. The satellite was launched on 4 May 2017, by means of an ARIANE 5 ECA vehicle from Arianespace, launched from the Kourou Space Centre in French Guiana along with the KOREASAT 7 satellite. The SGDC-1 is equipped with 50 Ka-band transponders and five X-band transponders to provide broadband internet and communications to the Brazilian government and the military.

---

[59] Schmidt, F., Desafios e Oportunidades para uma Indústria Espacial Emergente: o caso do Brasil, p. 68, 2011, http://www.ipea.gov.br/portal/images/stories/PDFs/TDs/td_1667.pdf (accessed 5 May 2019).

[60] Estadão, CADE aprova joint venture de Embraer, Telebras e Visiona, https://economia.estadao.com.br/noticias/negocios,cade-aprova-joint-venture-de-embraer-telebras-e-visiona,133076e (accessed 15 May 2019).

[61] Brazil, Decree no. 7.769 28 June 2012, http://www.planalto.gov.br/ccivil_03/_ato2011-2014/2012/decreto/D7769.htm (accessed 15 May 2019).

[62] Satélite Geoestacionário de Defesa e Comunicações Estratégicas (SGDC).

Although this first satellite was acquired in the foreign market, due to the short term defined by the government for the launch, from this project on, the intention is to leverage the knowledge necessary for the second geostationary satellite to be launched in 2019, with partially domestic manufacturing. The success of this process of integrating large national projects with local industry will certainly boost the sector's development possibilities.

The participation of national companies in the Space Policy Thematic Program, as measured by the National Business Sector Participation Index in the Space Policy Thematic Program (IPSEN-2) indicator, also showed a trend: from 32.91% in 2017 to 33.89% in 2018. This means that, in relation to the previous year, the national supply of technological products and services for the projects and missions of the PNAE was greater.[63]

Regarding the development of space activities in Brazil in the coming years, changes and development are expected in three specific sectors.[64] First, is an effort to improve the governance of the Brazilian space sector. The challenge remains to strengthen the conditions for the integration of this new generation of professionals into the productive system of the chain of activities of the space sector, avoiding possible migrations to other branches of socioeconomic activity that do not need the specificity of the training of human resources of the space area, or even migrations of brains to other countries. It is also necessary to highlight an important advance in the governance of the sector, which occurred through the publication of Decree no. 9.279[65] that created the Development Committee of the Brazilian Space Program. This temporary Committee aims to establish guidelines and targets for the enhancement of the Brazilian Space Program and to supervise the implementation of the measures proposed for this purpose. At its first meeting, held on 1 March 2018, ten resolutions were approved, and nine inter-ministerial working groups were created.[66]

---

[63] AEB, Relatório de Gestão, 2018, http://www.aeb.gov.br/wp-content/uploads/2019/04/Relat.%C3%B3rio-de-Gest%C3%A3o-2018_AEB.pdf (accessed 15 May 2019).

[64] Brazil, Ministry of Defence, Brazilian Space Program, 2017, https://build.export.gov/build/groups/public/@eg_br/documents/webcontent/eg_br_122071.pdf (accessed 15 May 2019).

[65] Brazil, Decree no. 9.279, 6 February 2018, http://www.planalto.gov.br/ccivil_03/_ato2015-2018/2018/decreto/D9279.htm (accessed 15 May 2019).

[66] Andrade, I., O centro de Lançamento de Alcântara: abertura para o mercado internacional de satélites e salvaguardas para a soberania nacional, 2018, http://repositorio.ipea.gov.br/bitstream/11058/8897/1/td_2423.pdf (accessed 15 May 2019).

Second, is CLA modernization. In April 2019, the Parliamentary Front for Modernization of the Alcântara Launching Centre was established in the Chamber of Deputies.[67] The group, which has the participation of 200 deputies and two senators, defends the commercial use of the base. Third and finally, is the development of the Microsatellite Launch Vehicles (VLM).[68] In the area of satellite launch vehicles and sounding rockets, the country still does not have a company that assumes national leadership to act in the program as an industrial architect. This situation is due to the uncertainty and scarcity of the Brazilian budget for the space sector, specifically for the development and maintenance of satellite launch vehicles and sounding rockets. Currently, the focus is to develop the VLM. For this project, the nationally developed S-50 engine is in the test phase and is expected to be validated in 2019. AEB is preparing to launch the VLM in 2020 for orbiting satellite testing.[69]

The main results of the PEB were published in the AEB management report of 2018[70] (see Table 8.2).

In 2019, the process of technology transfer from the VSB-30 to the national industry is still ongoing. Another important step, which is the basis for the VLM, was the structural test of the S50 engine in 2018. In 2019, a flight test will be done with the VS50 vehicle. At the same time, the PSM project, which seeks to develop a payload for the VS-30 and VSB-30 sounding vehicles, which will serve as the basis for microgravity experiments, is in active status.

In relation to the CLA, with a view to make the centre commercially viable, the TSA between Brazil and US has passed both chambers of the National Congress on 19 November 2019 and entered into force on 16 December 2019.[71], In addition, in October 2019, the Ministry of Defense announced that the CLA reached the mark of 487 launches. On 20 December 2019, a China Long March 4B rocket launched the CBERS 4-A mission, together with FloripaSat-1, from Taiyuan Satellite Launch Center.

---

[67] Câmara dos Deputados, Lançada frente parlamentar pela modernização da base de Alcântara, 2019, https://www2.camara.leg.br/camaranoticias/noticias/ciencia-e-tecnologia/575418-lancada-frente-parlamentar-pela-modernizacao-da-base-de-alcantara.html (accessed 15 May 2019).

[68] AEB, Desenvolvimento do Projeto VLM é tema de reunião na AEB, 2014, http://portal-antigo.aeb.gov.br/desenvolvimento-do-projeto-vlm-e-tema-de-reuniao-na-aeb/ (accessed 15 May 2019).

[69] UFMS, Integra UFMS, Agência Espacial Brasileira se prepara para lançamento de foguete em 2020, https://integra.ufms.br/agencia-espacial-brasileira-se-prepara-para-lancamento-de-foguete-em-2020/ (accessed 25 September 2019).

[70] AEB, Relatório de Gestão AEB, 2018, http://www.aeb.gov.br/wp-content/uploads/2019/04/Relat.%C3%B3rio-de-Gest%C3%A3o-2018_AEB.pdf (accessed 15 May 2019).

[71] AEB, CBERS-4A e FloripaSat são lançados com sucesso, http://www.aeb.gov.br/satelites-cbers-4a-e-floripasat-ja-estao-em-orbita (accessed 25 February 2020).

Table 8.2 PEB main results

PEB main results		
1	Suborbital vehicle or survey	In December 2018, Operation MUTITI was carried out, whose main objective was the launching and screening of the survey vehicle VS-30/V14 manufactured by IAE, from CLA, using CLBI as Remote Station. This vehicle carried as a payload the Suborbital Reentry Platform (PSR-01), with five experiments from research institutions selected for technological development, including, among them, the qualification of a yo-yo system and a separation system to be used in PSM. Also, in 2018, the process of planning the technology transfer of the VSB-30 to the national industry began. The process is expected to be finalized in 2019
2	VLM-1 (Microsatellite Launch Vehicles)	The VLM-1[a] will have the capacity to place payloads of up to 150 kg in low orbit (LEO)—equatorial, polar or re-entry and is being developed in partnership with DLR. For the development of VLM-1 it is necessary to master new technologies in solid propulsion, the S50 engine, whose motor envelope consists of composite materials, instead of metallic envelopes, reducing the total weight. In 2018, the S50 motor envelope rupture tests and the Critical Design Review (CDR) of the VS-50/VLM-1 project occurred
3	PSM[b] (Suborbital Microgravity Platform)	The PSM project was delayed in its execution schedule. The company responsible for its development has encountered difficulties in importing items related to the system that will carry out measures of position, speed and attitude during the flight, and its parachute opening actuator. Both items were in the category of equipment with international restrictions for import, since they can be used with dual purposes. Although Brazil is a signatory of the MTCR and the project is not military in nature, China has vetoed exports. Faced with this impasse, the company Orbital Engenharia and IAE decided to internalize their technological developments. For the new schedule, the PSM qualification model will be finalized in 2019[c]

(continued)

Table 8.2 (continued)

PEB main results		
4	Launch Centres	Several preventive and corrective maintenance activities are being carried out annually, such as operational control and triggering, data processing, synchronization, tracking, telemetry, as well as the Mobile Tower Integration systems and equipment. The importance of these services is emphasized, since the launch centres have advanced technology systems, thus requiring specialized technical maintenance services and are difficult to repair if preventive maintenance is not performed. At CLA, several works are underway, such as the implementation of fire-fighting systems in the facilities of the Preparation and Release Sector. Electrical/electronic equipment (e.g. a portable weather station) was purchased to support the centre's operations staff. In 2018, four rocket launchings and a launch of a VS-30/V14 drill vehicle were successfully completed. At CLBI, maintenance services for the Adour and Bearn radars and the Telemedidas station are underway. In 2018 two launches of training rockets were carried out
5	Consolidation of the National Conformity Assessment System in the Space Area	In 2018, work began on revising the Technical Regulations on Space Safety and on the area of spatial certification. The participation of the Institute for Industrial Development and Coordination in the Interfaces Monitoring Meetings and in the Launch Interface Group is noteworthy. Regarding the activities of verification of conformity, highlight the process of monitoring the development and delivery of the S50 engine with the national industry and the participation of experts in the plenary of the International Organization for Standardization, making possible the defense of the normative aspects aerospace industries of the country's interest[d]

[a] Veículo Lançador de Microssatélite (VLM)
[b] Plataforma Suborbital de Microgravidade (PSM)
[c] AEB, Relatório de Gestão, 2018, p. 43, http://www.aeb.gov.br/wp-content/uploads/2019/04/Relat.%C3%B3rio-de-Gest%C3%A3o-2018_AEB.pdf (accessed 15 May 2019)
[d] Ibid.

8.4 Space Development Policies and Emerging Programs

The Brazilian space policy seeks to assist the country in its development, aimed at solving the problems of the Brazilian state for the welfare and improvement of the quality of life of society, in a sustainable way and with the participation of industry. During the military dictatorship (1964 to 1985) period, the PEB evolved as a natural extension of the strategy to transform Brazil into a medium-sized space power. Currently, research, missions and space projects in Brazil are indirectly linked to the government's foreign policy, which aims to project the country as a nation-continent with regional economic and geopolitical ambitions, although strictly committed to the use of technology for peaceful purposes, in line with the principles of international space law.[72]

With a limited budget,[73] international restrictions and operational difficulties in conducting the space program, the flagship of the PEB, which is the development of satellites, launchers and autonomous access to space, has not advanced much. There are, however, areas where Brazil has achieved results, such as satellite image processing, especially in the area of meteorology. Knowledge, however, is limited, since Brazil is dependent on foreign satellites.[74]

The dilemma about the direction of PEB was heightened after the accident that killed 21 technicians and scientists during the preparation for the launch of VLS-1 at CLA in August 2003.[75] The instability of resources and successive budgetary contingencies are pointed out as the main factors for the tragedy, demonstrating that, although it was a national strategic policy, the space program was not up to its mission because of governmental budgetary decisions. The accident evaluation report[76] pointed to three main causes: low investments in the area, shortage of qualified personnel, and problems in the organizational structure of the Brazilian Space Program, suggesting that the AEB should be subordinated directly to the Presidency of the Republic.

---

[72]Rollemberg, R., Veloso, E., A política espacial brasileira, Câmara dos Deputados, Congresso Nacional, 2009, https://www2.camara.leg.br/a-camara/estruturaadm/altosestudos/arquivos/politica-espacial/a-politica-espacial-brasileira (accessed 15 May 2019).

[73]R$181 Million (USD 44.15 millions) to 2019, http://www.portaltransparencia.gov.br/orgaos/20402-agencia-espacial-brasileira (accessed 15 May 2019).

[74]Brazil utilizes NOAA GOES series satellites from the US as "courtesy", subject to foreign operating decisions.

[75]New York Times, Explosion of Brazilian Rocket During a Test Kills at Least 16, 2003, https://www.nytimes.com/2003/08/23/world/explosion-of-brazilian-rocket-during-a-test-kills-at-least-16.html (accessed 8 May 2019).

[76]Brazil, Ministry of Defence, Relatório da Investigação do acidente ocorrido com o VLS-1 V03, em 22 de agosto de 2003, em Alcântara, Maranhão, 2004, http://www.aereo.jor.br/downloads/VLS-1_V03_Relatorio_Final.pdf (accessed 10 May 2019).

The PNDAE, instituted by Decree no. 1.332 of 1994,[77] seeks to consolidate and expand the Brazilian advance in this sector. In other words, it mandates the completion, maintenance and upgrading of the science and technology infrastructure in the sector, to increase and improve the human resources base dedicated to space activities, and to expand government participation and the national industrial park in the Brazilian Space Program.

In this context, PEB has developed in distinct phases of scientific and technological capacity building. Initially, with human resources coming from schools of recognized training capacity, such as the ITA, rockets were developed. An example of the success of this strategy is the VSB-30 sounding rocket.[78] This vehicle was approved by ESA to carry out flights in Europe carrying Texus and Maser scientific payloads from the European Microgravity Program, becoming the only PEB product to be commercialized internationally.[79]

The Brazilian Space Program's main source of investment is the resources from the Thematic Program—Space Policy, from the Government's Pluriannual Plan (PPA). The main executing budget units are the AEB and the Ministry of Science, Technology, Innovation and Communications (MCTIC).

The PNAE is a planning tool of the PEB that seeks to guide its actions for periods of ten years. It provides a set of guidelines and strategic guidelines as well as the main space missions to be developed in the period, under the National System for the Development of Space Activities (SINDAE). The current PNAE is in its fourth edition and covers the period 2012–2021, covering aspects related to: Missions, Means of Access to Space, Applications and Scientific, Technological and Educational Programs. The PNAE is based on three main strategic focuses: society, autonomy and industry.

The new PNAE 2012–2021 is being implemented in two phases, continuous and complementary. The first is consolidation. In it, projects already started in the past must be completed and others started, in order to expand and consolidate a set of actions aimed at raising industrial capacity, technological dominance, skills development and regulation of space activities, which will create better conditions to ensure greater sustainability of the program. The second phase is expansion. Here, new projects of greater technological complexity and high strategic value must be launched and developed, imposing unprecedented challenges on the program. In time, it should certainly have consolidated integrating companies, a more structured production chain, access to the conquered space, a broad technological domain and a much larger team of trained specialists. In order to fulfill all the proposals foreseen in this PNAE, it is necessary to have resources of the order of R$ 9.1 billion, of which 47% are destined to the projects of satellite missions, 17% for projects of access to

---

[77] Brazil, Decree no. 1.332, 8 December 1994, http://www.planalto.gov.br/ccivil_03/decreto/1990-1994/D1332.htm (accessed 10 May 2019).

[78] Lucca, E., The Brazilian Sounding Rocket VSB-30: meeting the Brazilian Space Program and COPUOS objectives, COPUOS 10 to 21 February 2014, http://www.unoosa.org/pdf/pres/stsc2014/tech-44E.pdf (accessed 10 May 2019).

[79] Maliq, T., Brazilian Rocket Launches Microgravity Experiment, 2005, https://www.space.com/1829-brazilian-rocket-launches-microgravity-experiment.html (accessed 10 May 2019).

space, 26% for space infrastructure and 10% for other special and complementary projects (see Table 8.3; Figs. 8.6 and 8.7).

In recent years, the focus of space activities has been the continued presence of national satellites in operational orbit, the closeness of completion of new satellite projects, the strong engagement of the academic community and university students in nanosatellite initiatives, recent preparation of a new generation of trained professionals abroad, and progress in the development of what will be the first national satellite launcher.[80]

In 2016, the Strategic Space Systems Program (PESE)[81] was approved, which establishes the strategy for the long-term implementation of the subprograms and space defense systems projects with dual use: military and civil. The PESE will allow the operations of the Armed Forces to have the necessary support of space applications in a coordinated and integrated way. The procedures and doctrine of command and control, as well as operational intelligence were improved using remotely piloted aircraft during the Olympic and Paralympic Games—Rio 2016. The transmission of videos in real time enabled the elevation of situational awareness and timely advice for decision-making.

In that sense, in 2016, a non-onerous lending agreement was signed, through which the Ministry of Defense received images of partner companies and developed the capacity to plan satellite passes for viewing and downloading images of areas of interest in the national territory. In order to increase information security, the Brazilian Air Force has issued a standard in which it establishes the central unit and the teams for handling and responding to security incidents in computer networks, especially those focused on air defense and air traffic control.[82]

The PEB planning, as a government public policy, continues the previous multiannual plans, and part of its recognition as an important initiative in favor of the national economic and social development is due to its high scientific-technological content and innovation, and by the apparent geopolitical and strategic position that it occupies in government policy, contributing to the sovereignty and autonomy of the country.

In this context, the main challenge of the Space Policy, established in the PPA 2016–2019 and in the PNDAE, continues to be to enable the country to develop and implement a set of space assets and applications for the solution of problems in Brazil

---

[80] Annual Report for the Evaluation of the Multi-Year Plan 2016–2017, https://www.camara.leg.br/proposicoesWeb/prop_mostrarintegra;jsessionid=7C018552539D316BED26248FF5221356.proposicoesWebExterno2?codteor=1564660&filename=MCN+8/2017+CN (accessed 10 May 2019).

[81] Ministry of Defence, Programa Estratégico de Sistemas Espaciais do Ministério da Defesa, https://www.defesa.gov.br/arquivos/ensino_e_pesquisa/defesa_academia/cadn/palestra_cadn_xi/xv_cadn/o_programa_estrategico_de_sistemas_espaciais_pese.pdf (accessed 10 May 2019).

[82] Annual Report for the Evaluation of the Multi-Year Plan 2016–2017, p. 360, https://www.camara.leg.br/proposicoesWeb/prop_mostrarintegra;jsessionid=7C018552539D316BED26248FF5221356.proposicoesWebExterno2?codteor=1564660&filename=MCN+8/2017+CN (accessed 10 May 2019).

Table 8.3 Schedule of investments PNAE 2012–2021

Schedule of investments (in R$ millions)

		2012	2013	2014	2015	2016	2017	2018	2019	2020	2021	Total
Space missions		81.1	100.2	183.6	273.9	248.6	184.9	45.6	36.8	0	0	1154.8
Consolidation phase	CBERS Satellites	45	34.7	53.7	24	15.3	6	6	0	0	0	184.6
	AMAZONIA Satellites (1 and 1B)	35.9	52.3	54.1	45	38.5	26	0	0	0	0	251.9
Expansion phase	AMAZONIA 2 Satellite	0	8.8	39.6	66	49.2	35.3	12.3	12.3	0	0	223.2
	LATTES Satellite	0	3.9	17.1	49.9	71	73.6	2.8	0	0	0	218.2
	SABIA-Mar Satellite	0.5	0.5	19.1	89	74.7	44.1	24.5	24.5	0	0	276.9
Access to space		94.2	112.4	179.6	206.7	252.2	294.2	180.2	139	110	9.2	1578.1
Consolidation phase	Suborbital Rockets	19.2	19.2	30.2	9.2	20.2	9.2	20.2	9.2	20.2	9.2	166
	Vehicle Launcher VLS-1	62.5	45.7	35.4	11.5	0	0	0	0	0	0	155.1
	Vehicle Launcher VLM-1	10	25	25	20	20	15	0	0	0	0	115
Expansion phase	Vehicle Launcher VLS ALPHA	2	19	33	98	130	120	40	0	0	0	442
	Vehicle Launcher VLS BETA	0.5	3.5	56	68	82	150	120	130	90	0	700
Infrastructure		156.9	339.3	319.9	150	181	211	158	141	122	123	1902.1
Infrastructure and Operations of Space Missions		17.2	31	60	60	61	61	38	41	42	43	454.2
Infrastructure and Access to Space		24.7	28.3	30	50	80	110	80	60	40	40	543
Specific Infrastructure of Alcântara Cyclone Space		15.6	206.7	127.3	0	0	0	0	0	0	0	349.6
General Infrastructure of the Alcântara Launch Centre		99.4	73.3	102.6	40	40	40	40	40	40	40	555.3
Critical Technologies and Skills Development		36	70.8	87.1	132.9	141.1	147	142.2	131	113	114	1114.9
Critical Technologies		22.5	47.5	52.5	57.5	62.5	67.5	72.5	77.5	82.5	87.5	630.5

(continued)

Table 8.3 (continued)

Schedule of investments (in R$ millions)

		2012	2013	2014	2015	2016	2017	2018	2019	2020	2021	Total
Small Satellites		5	10	10	10	10	10	10	10	10	10	95
Scientific and Technological Missions		0.3	0.3	9.6	50.4	53.6	54.5	44.7	28.5	5.9	1	248.5
Research in Space Science and Climate		5.2	10	10	10	10	10	10	10	10	10	95
Competency Development		3	3	5	5	5	5	5	5	5	5	46
Total		368.5	622.6	770.2	763.5	822.9	837.1	525.9	346	346	246	5749.8
Projects in Partnership (resources from other sources)		186	452.4	676	266.3	341.9	431.2	451	57.5	57.5	0	3343.8
Consolidation phase	Alcântara Cyclone Space (MCTI)	130	164.9	164.9	0	0	0	0	0	0	0	459.8
	Satellite SGDC-1 (Telebras/MD)	56	250	410	0	0	0	0	0	0	0	716
	Data Collection Satellite (ANA)	0	30	60	40	20	0	0	0	0	0	150
Expansion phase	GEOMET-1 Satellite	0	1	3	150	200	250	100	0	0	0	704
	SGDC-2 Satellite	0	0	0	0	0	56	250	0	0	0	716
	Radar Satellite	0	6.5	38.1	76.3	121.9	125.2	101	0	57.5	0	598
Total (with partnership projects)		554.5	1075	1446	1030	1165	1268	976.9	929	403	246	9093.6

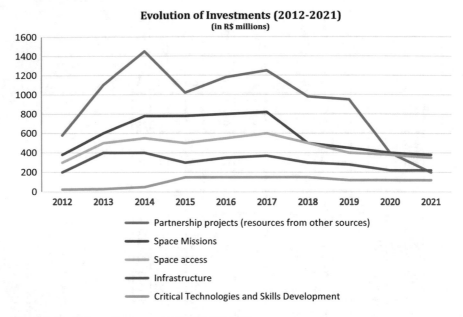

Fig. 8.6 Evolution of investments PNAE 2012-2021

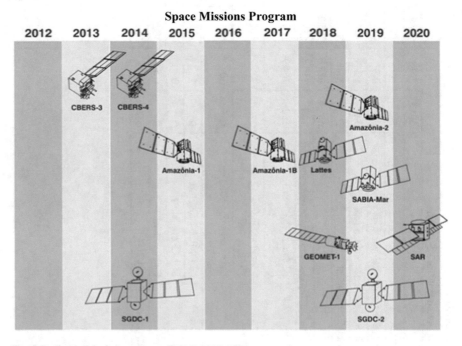

Fig. 8.7 Space missions program PNAE 2012-2021

and for the well-being and improvement of the quality of life of the populations and the society, in a sustained way and with the participation of industry.[83]

Space assets, understood as equipment that is or may be placed in space, such as a satellite or a launch vehicle, as well as ground-based equipment to support space activities such as satellite control stations and data reception, are of high political-strategic value for the development of nations. The training of the national industry to develop and produce complete space systems is of great importance and constitutes another important space policy challenge, which lies in the need to expand and organize its value chain and suppliers of goods and services space, still in nascent stage in Brazil.

Among the main results for the PPA 2014–2019 period, one indicator stands out: the IPSEN-2,[84] which expresses the percentage relation between the total amounts paid in contracts for the supply by Brazilian companies of technological products or services and the amount of total expenditures of the budgetary actions of a final nature of the Thematic Space Policy Program. The IPSEN-2 for the year 2016 was R$ 40.14 million, which reflects the increase in contracting technology transfer projects in the context of the acquisition of the SGDC, which generated contracts using economic subsidies to national companies.

In order to improve and expand the availability of images, data and services for the benefit of Brazilian society, through space missions using national satellites, it was observed that it was highly necessary and strategic to consolidate in Brazil the capacity to design, develop and manufacture indigenous EO satellites geared to applications of national interest in areas such as mineral, forestry and water resources, agriculture, environment, territorial surveillance, and monitoring of environmental disasters.

This objective is anchored in two goals and nine initiatives, all associated with the development of satellite missions that seek to meet the spectrum of applications considered a priority by PEB. To achieve this goal the aim is to launch the first AMAZONIA-1 EO Satellite, AMAZONIA-1, in 2020. The other goal is to launch the CBERS-4A Satellite, and have the year of 2016 dedicated to the elaboration of an agreement to support and to plan the necessary industrial contracts to secure Brazilian responsibility for the project. On 20 December 2019, a China Long March 4B rocket launched the CBERS-4A mission.[85]

These two goals will mean important progress in the country's capacity building in this area. The CBERS-4A Satellite will serve the users of today's products that use the CBERS-4 images, which are in orbit and operational, and will also feature data from a

---

[83] Brazil, Ministério do Planejamento, Orçamento e Gestão Secretaria de planejamento e investimento estratégico, PPA 2016-2019 Mensagem Presidencial, http://www.planejamento.gov.br/assuntos/planeja/plano-plurianual (accessed 10 July 2019).

[84] 1º Ciclo Estratégico da Agência Espacial Brasileira 2017-2019, Annex IX, p. 38, http://www.aeb.gov.br/wp-content/uploads/2018/10/Plano-Estrat.%C3%A9gico-v3111.pdf (accessed 10 May 2019).

[85] Ministério do Planejamento, Orçamento e Gestão Secretaria de planejamento e investimento estratégico, PPA 2016-2019 Mensagem Presidencial, http://www.planejamento.gov.br/assuntos/planeja/plano-plurianual (accessed 10 July 2019).

high-resolution camera. The Amazonia-1 Satellite will provide data like the CBERS Satellite, with the advantage of broadening the revisit rate and diversifying the data source. This will also be the first national remote sensing satellite, inaugurating a series of satellites that must meet varied national demands.[86] Among the missions indicated, there is a second satellite of the AMAZONIA series, SCD-Hidro, SABIA-Mar in partnership with Argentina, and the EQUARS mission. All these missions should be started during the current PPA period, but with reduced execution in relation to the priority missions, associated with the goals already mentioned.[87]

Regarding launch vehicles the knowledgeable technical-scientific community expects that market demand in the niche of launch vehicles of micro and mini-satellites will grow. In this context, taking advantage of all the learning, the concepts and technologies acquired with the design of the VLS-1 vehicle, IAE and AEB are working on the development and certification of a Microsatellite Launcher Vehicle (VLM-1), with lower cost and more current and commercially viable technologies.[88] The VLM-1 is a Satellite Launcher that will be able to place payloads (microsatellites and cubesats) up to 150 kg in LEO—equatorial, polar or re-entry. This vehicle is being developed in partnership with the German Aerospace Centre (DLR), aiming to meet the needs of both PEB and the German equivalent program and enter into a market niche not explored as there is a growing demand for specific vehicles for the launch of microsatellites due to the large number of new scientific and operational applications for satellites up to 100 kg.[89]

In its first mission, VLM-1 should launch a national technological payload, as stated in PPA 2016–2019. In this context, the most important event was the signing of a contract with the national industry in December 2016, to supply the rocket motors to be used in VLM-1, called S50.[90] Another important activity that occurred in 2016 was the Mission Definition Review, where the mission requirements of that vehicle were reviewed and approved. In 2017, the Preliminary Design Review, the Critical Design Review, and the delivery of the first S50 engines were carried out.[91]

In order to promote scientific and technological knowledge, human capital and the domain of critical technologies to strengthen the space sector, the aim is to

[86]Spaceflight now, China launches environmental satellite built in cooperation with Brazil, https:// spaceflightnow.com/2019/12/21/china-launches-environmental-satellite-built-in-cooperation-with-brazil/ (accessed 16 January 2020).

[87]Ministério do Planejamento, Orçamento e Gestão Secretaria de planejamento e investimento estratégico, PPA 2016–2019 Mensagem Presidencial, http://www.planejamento.gov.br/assuntos/ planeja/plano-plurianual (accessed 10 July 2019).

[88]IAE, http://www.iae.cta.br/index.php/espaco/vls1 (accessed 10 July 2019).

[89]Falcão, D., VLM-1, Sonho ou Realidade? Certamente Uma Necessidade, 2013, https:// brazilianspace.blogspot.com/2013/04/vlm-1-sonho-ou-realidade-certamente-uma.html (accessed 10 July 2019).

[90]IAE, IAE sedia a assinatura do contrato de produção dos motores S50, http://www.iae.cta. br/index.php/ultimas-noticias/406-iae-sedia-a-assinatura-do-contrato-de-producao-dos-motores-s50-3 (accessed 10 July 2019).

[91]IAE, Relatório de Atividades 2018, http://www.iae.cta.br/images/relatorios-atividades/Relatorio_ de_Atividades_2018_Final.pdf (accessed 10 July 2019).

broaden the domain of critical technologies, as well as to develop and consolidate competences and human capital to guarantee autonomy and sustainability for the development of space activities in Brazil. This objective has been achieved with the advancement of scientific and technological knowledge obtained in the execution of PNAE programs, projects and missions that, directly or indirectly, also support the training of new talents through the continuous training of specialists in the space area.

In the context of this objective, it is important to mention key developments in 2016 including the creation of the educational space within the E2T Program (Space, School and Technology), for the training of future talents in the space area, through actions of teacher scientific training for institutions of higher, middle and fundamental education, seeking to awaken in students the creativity and interest in science and technology and the space program, aiming at the development of new vocations for space.[92] The inauguration in 2017 of the Space Technological Vocational Centre (CVT-Espacial) of the Launch Centre of the Barrier of Hell (CLBI) was significant.[93] In the field of the development of small-scale satellite platforms and subsystems, university satellites of the SERPENS Program (Space System for Research Realization and Experiments with nanosatellites) will be launched. Also noteworthy is the training in France of a significant contingent of Brazilian technicians and engineers who are part of the Geostationary Satellite Defense and Strategic Communications Technology Absorption Plan (SGDC), who attended introductory and advanced theoretical courses. These professionals, after familiarizing themselves with the SGDC project and their respective mission requirements, followed specialized On-Job-Training in the different systems and subsystems of the satellite.[94]

It is important to recognize that the PEB has been able to advance in the greater goal of returning to Brazilian society, in products, services and technological innovation, the investments made in it. The presence of national satellites in orbit in operational condition, proximity to the completion of new satellite projects, advancement in the development of the first national satellite launcher, strong engagement of the academic community and university students in initiatives aimed at nanosatellites, the recent preparation of a new generation of professionals trained abroad, among others, are the results of projects initiated years ago, which continue to be supported in this PPA. The importance and necessity of the implementation of a set of governmental actions to boost Brazilian space policy should be emphasized. These include: the increase, stability and continuity of investments in the sector, due to the nature of space projects which have long-term developmental characteristics; continuous investment in the experts of the two main space research institutions in the country (INPE/MCTIC and DCTA/MD); revision of the legal framework for the sector, in

[92] AEB, E2T, http://www.aeb.gov.br/espaco-educacao-e-tecnologia/e2t/ (accessed 10 July 2019).

[93] AEB, Agência espacial inaugura primeiro Centro Vocacional Tecnológico espacial, http://portal-antigo.aeb.gov.br/agencia-espacial-inaugura-primeiro-centro-tecnologico-espacial-do-brasil/ (accessed 10 July 2019).

[94] AEB, Relatório de Gestão 2018, http://www.aeb.gov.br/wp-content/uploads/2019/04/Relat.%C3%B3rio-de-Gest%C3%A3o-2018_AEB.pdf (accessed 10 July 2019).

particular the legislation dealing with bids and government contracts, aimed at contracting projects involving technological developments; and the permanent prioritization of space policy among government actions, for its geopolitical and strategic importance to the country, in particular the improvement of the governance system established in the SINDAE.[95]

In relation to the governmental agenda in the area of defense, a national defense strategy was launched in 2008 through Decree no. 6.703,[96] which brought a new defense concept to Brazil based on three main axes of action: reorganization of the armed forces, restructuring of the defense materials industry; and re-composition of the armed forces. Within the scope of the reorganization of forces, the need to strengthen three sectors of strategic importance: space, cybernetics (information and communication technologies), and nuclear is emphasized. These sectors have in common their dual nature: technological civil applications—directly linked to the country's development project—and military applications—are essential to ensure flexibility, mobility and presence, as well as the defense of national critical infrastructures and of the interoperability between the three forces.

The priorities established for the space sector are thus detailed in the National Defense Strategy: (1) Design and manufacture satellite launch vehicles and develop remote guidance technologies, especially inertial systems and liquid propulsion technologies; (2) Design and manufacture satellites, especially geostationary, for telecommunications and those for high-resolution, multispectral remote sensing, and to develop attitude-control technologies for satellites. (3) Develop communication, command and control technologies from satellites, with ground, air and sea forces, including submarine forces, to enable them to operate in a network and to be guided by information received from them; (4) Develop technology for determining geographic coordinates from satellites.

In 2012, the Strategic Space Systems Program (PESE)[97] was created to meet the strategic needs of the armed forces and Brazilian society. Initially six low-orbit satellite fleets will be launched by 2022, and three geostationary orbit satellites—which include ground control, reception and data processing ground stations—will be launched to provide terrestrial observation, telecommunications, information mapping, positioning, space and a space systems operation centre.

In the military field, the space systems considered in the PESE must include the modernization of several systems in operation, such as the Brazilian Aerospace Defense System, the Aeronautical Digital Links System, the Military System, as well as others that are in the planning or implementation phase, such as the Blue Amazonia Management System and the Integrated Border Monitoring System. It is also planned to use these in support of civil initiatives, such as prevention and action in cases of major environmental catastrophes, in the Amazon Protection System, and in the National Broadband Program, among others. The main objective of the PESE

---

[95] Ibid., pp. 394–401.

[96] Brazil, Decree no. 6.703, 18 December 2008, National Defense Strategy, http://www.planalto.gov.br/ccivil_03/_ato2007-2010/2008/decreto/d6703.htm (accessed 18 May 2019).

[97] FAB, PESE, http://www2.fab.mil.br/ccise/index.php/o-que-e-o-pese (accessed 18 May 2019).

is to provide space infrastructure to be used strategically and in a positive way in the Blue Amazonia Management System, the Integrated Border Monitoring System, the Brazilian Aerospace Defense System, the Amazon Protection System and similar.

This infrastructure can also be used to support precision farming, prevention of environmental disasters, public safety and to increase the reach of the National Broadband Plan. Broadband service will reach everywhere, even in remote communities, where there is no possibility of setting up telephony service antennas. The systems proposed by the program have dual use—civil and military—with direct and indirect benefits for users of government services and Brazilian society. The PESE complements the PNAE by providing satellites with adequate capacity to support the missions of the armed forces and to enable concrete improvements in the lives of the population.

8.5 Current and Future Perspectives on Legal Framework Related to Space Activities

Article 21 of the Constitution of the Federative Republic of Brazil, adopted in 1988, states that "the Union shall have the power to: operate, either directly or through authorization, concession or permit air and aerospace navigation and airport infrastructure". Nowadays, Brazilian space law is composed of laws based on the creation of the AEB (Law no. 8.854, 10 February 1994)—the main legal framework for the development of space activities that set as a strategic goal the development of national space technology capabilities—and the National System of Space Activities—SINDAE (Law no. 1.953, 10 July 1996). To these can be added the numerous international and regional treaties and agreements on space activities.

The current policy instrument is the National Programme of Space Activities 2012–2021 (PNAE 2012–2021). It identifies priority actions, needed investments and international cooperation possibilities. It also foresees a calendar of future space missions and describes a set of specific projects.

Regarding registration, all space objects must have a responsible country. This service is provided by AEB and attests to the responsibility of Brazil State over the object and possible harmful consequences of its operation (collision with other objects or fall on the territory of other countries, for example) with the international community.

According to Ministerial Order no. 96, dated 30 November 2011,[98] published in the Official Journal of the Federal Government on 2 December 2011, AEB oversees the "implementation and operation of the National Registry of Objects space". The registration is mandatory, and this is an AEB service that complies with Decree

---

[98] Brazil, Ministerial Order no. 96, 30 November 2011, Regulation on the implementation and operation of the registry of space objects launched in outer space, http://www.aeb.gov.br/wp-content/uploads/2018/04/DOU-Registro-de-Objetos-Espaciais-Portaria-96-111130-1.pdf (accessed 9 April 2019).

Table 8.4 International space related treaties respected by Brazil[a]

International Space Treaties respected by Brazil

Treaty[b]	Adoption	Ratification
OST	1967	Ratification/Decree no. 64.362, 17 April 1969
ARRA	1968	Ratification/Decree no. 71.989, 26 March 1973
LIAB	1972	Ratification/Decree no. 71.981, 22 March 1973
REG	1975	Ratification/Decree no. 5.806, 19 June 2006
MOON	1979	Not signed
NTB	1963	Ratification/Decree no. 2.864, 7 December 1998
BRS	1974	Signature only
ITSO	1971	Ratification/Decree no. 74.130, 28 May 1974
IMSO	1976	Ratification/Decree no. 2.129, 17 January 1997
ITU	1992	Ratification/Decree no. 2.962, 23 February 1999

[a]Status of International Agreements relating to activities in outer space as at 1 January 2019, UNOOSA Doc, A/AC.105/C.2/2019/CRP.3, http://www.unoosa.org/documents/pdf/spacelaw/treatystatus/AC105_C2_2019_CRP03E.pdf (accessed 4 October 2019)
[b]Treaty on Principles Governing the Activities of States in the Exploration and Use of Outer Space, including the Moon and Other Celestial Bodies (OST); Agreement on the Rescue of Astronauts, the Return of Astronauts and the Return of Objects Launched into Outer Space (ARRA); Convention on International Liability for Damage Caused by Space Objects (LIAB); Convention on Registration of Objects Launched into Outer Space (REG); Agreement Governing the Activities of States on the Moon and Other Celestial Bodies (MOON); Treaty Banning Nuclear Weapon Tests in the Atmosphere, in Outer Space and under Water (NTB); Convention Relating to the Distribution of Programme-Carrying Signals Transmitted by Satellite (BRS); Agreement Relating to the International Telecommunications Satellite Organisation (ITSO); Convention on the International Mobile Satellite Organization (IMSO); International Telecommunication Constitution and Convention (ITU)

no. 9.094 of 17 July 2017. Legal entities can use this service and must send the following information to AEB: (a) Name of the launching state or states; (b) an appropriate designation of the space object or its registration number; (c) date and territory or place of launch; (d) basic orbital parameters, including: (i) Nodal period; (ii) Inclination; (iii) Apogee; and (iv) Perigee. It takes about 31–60 days for the provision of this service.

All Brazilian space objects launched into space must be registered with the Brazilian Space Agency. The registration of objects maintained by the AEB is subsequently sent for inclusion in the international registry of space objects maintained by the UN Secretary-General as provided for in the Treaty on Principles Governing the Activities of States in the Exploration and Use of Outer Space, including the Moon and Other Celestial Bodies (Outer Space Treaty). A list of space treaties respected by Brazil is Table 8.4.

On 14 June 2019, Decree no. 9.839[99] came into force, establishing the Committee for Development of the Brazilian Space Program. The Committee works as an advisory body for the President of the Republic to formulate proposals on: (i) necessary subsidies for the enhancement of the Brazilian Space Program; (ii) development and use of technologies applicable to the Brazilian Space Sector, following launch infrastructure, launch vehicles and orbital and suborbital artifacts; and (iii) the supervision of the execution of the necessary measures for the enhancement of the Brazilian Space Program. In addition, the Committee may establish technical groups to prepare studies on: (i) the development of the launching infrastructure and launching vehicles of orbital and suborbital artifacts; (ii) the development of projects aimed strengthening the national industry for the Brazilian space sector; (iii) the composition of the staff of science and technology careers for the Brazilian space sector; (iv) public policies, social actions and land issues related to areas of the national territory intended for launching centre facilities; and (v) proposals for establishing legal frameworks for the Brazilian space sector.[100]

8.6 Cooperation Programs

Brazil has relevant space cooperation agreements with countries all over the world. AEB, with the purpose of promoting the scientific and technological capacity of the Brazilian space sector to meet the needs of the country, began intense international cooperation in the second half of the 1980s. This cooperation is also an effective way to mitigate risks of conflict in space, since nations have shared goals and greater interest in preserving the peaceful uses of outer space.

Intergovernmental framework agreements on cooperation for the peaceful uses of outer space have already been signed with eleven countries. These agreements are, in principle, precursors to new international instruments and initiatives that lead to the bilateral and multilateral development of the space program and the acquisition of new technologies.

8.6.1 Regional Level (Intra-latin America)

Brazilian cooperation on the intra-Latin America level is mainly based on scientific, technical and technological cooperation. Argentina is highlighted with an ongoing cooperation program with Brazil on the development of a dual satellite joint EO mission.

---

[99]Brazil, Decree no. 9.839, http://www.planalto.gov.br/ccivil_03/_Ato2019-2022/2019/Decreto/D9839.htm (accessed 18 September 2019).
[100]Ibid.

8.6.1.1   Argentina

Argentina is a longtime Brazilian partner. The bilateral space partnership gained momentum on 24 August 1989 when the Joint Declaration on Bilateral Cooperation in the Peaceful Uses of Outer Space was signed by the presidents of both countries.[101]

In the 1990s, after the creation of the AEB, the relationship between the two countries became even closer. On 9 April 1996, Brazil and Argentina signed the Framework Agreement on Cooperation in Peaceful Applications of Space Science and Technologies.[102]

In 1998, the Cooperation Program between AEB and the National Commission for Space Activities of Argentina (CONAE) was launched for Suborbital Launch,[103] as well as the Cooperation Program for Compatibility of Procedures in the Solo Systems of Space Missions[104] and the AEB and CONAE agreement regarding the SAC-C Project to implement a cooperation program to carry out the environmental tests for the SAC-C Mission, to be executed by INPE.[105] On 11 November of the same year, the space agencies of the two countries signed a joint satellite development cooperation program, SABIA-3, whose objective was to monitor the environment, water resources and agricultural production in Brazil and in Argentina.[106]

In 2004, there was a preliminary agreement to carry out tests of the Argentine SAC-C and SAOCOM satellites at INPE's facilities and it was decided to exchange images for scientific purposes between the Brazilian CBERS-2 and the Argentine SAC-C.[107]

The Brazilian-Argentine satellite was redefined in February 2008 at a bilateral working group meeting in Buenos Aires.[108] The SABIA-Mar, as it has been called

---

[101] Brazil-Argentina Joint Declaration on Bilateral Cooperation in the Peaceful Uses of Outer Space, 24 August 1989, http://portal-antigo.aeb.gov.br/wp-content/uploads/2012/09/AcordoArgentina89. pdf (accessed 18 April 2019).

[102] Framework Agreement on Co-Operation in Applications Pacific Science and Space Technology among the Government of the Federative Republic of Brazil and the Government of the Argentine Republic, http://portal-antigo.aeb.gov.br/wp-content/uploads/2012/09/AcordoArgentina96.pdf (accessed on 18 April 2019).

[103] Program of Cooperation between the Brazilian Space Agency and the National Commission of Space Activities of the Argentine Republic Suborbital Release, 10 November 1988, http://www. aeb.gov.br/wp-content/uploads/2018/01/AcordoArgentina98c.pdf (accessed 18 April 2019).

[104] Document is not available on AEB's website, http://www.aeb.gov.br/programa-espacial-brasileiro/cooperacao-internacional/argentina/ (accessed 18 April 2019).

[105] Program of Cooperation between AEB and CONAE regarding the project SAC-C, 22 May 1988, http://portal-antigo.aeb.gov.br/wp-content/uploads/2012/09/AcordoArgentina98.pdf (accessed 18 April 2019).

[106] Program of Cooperation between AEB and CONAE concerning the SABIA-3 project, 10 November 1988, http://portal-antigo.aeb.gov.br/wp-content/uploads/2012/09/AcordoArgentina98b.pdf (accessed 18 April 2019).

[107] Estadão, Agência Estado, 2004, https://ciencia.estadao.com.br/noticias/geral,brasil-amplia-projetos-espaciais-com-a-argentina,20040807p2504 (accessed 18 April 2019).

[108] Document is not available on AEB's website, http://www.aeb.gov.br/programa-espacial-brasileiro/cooperacao-internacional/argentina/ (accessed 18 April 2019).

since then, will have as its objective the observation of the coast and ocean of South America. The natural partnership for the development of this type of satellite is with Argentina, with whom Brazil shares extensive terrestrial frontier and relative common interests in the Atlantic Ocean.

In 2014, the MoU between AEB and CONAE was signed for the joint development of the SABIA-Mar Space Mission,[109] which aims to provide fundamental ocean measurements for the understanding and study of its biosphere and the ways in which it is affected and reacts to anthropogenic activities.

8.6.1.2 Chile

In 1990, Brazil and Chile signed the Basic Agreement for Scientific, Technical and Technological Cooperation.[110] In 1993, a supplementary adjustment to the agreement was signed.[111] In 2002, the space agencies of the two countries signed an MoU[112] on the exploration and use of outer space for exclusively peaceful purposes, including study of natural phenomena of the earth, space technologies, as well as exchanges in the training of specialists in the area.

8.6.1.3 Colombia

In 1981, Brazil and Colombia signed the Agreement for Scientific and Technological Cooperation,[113] aimed at meetings to discuss aspects related to science and technology; exchanges of teachers, scientists, technicians, researchers and experts; exchange of scientific and technological information; joint or coordinated execution of programs and projects of scientific and technological research and technological development, application and improvement of existing technologies and development of new ones; and other mutually agreed forms of cooperation.

---

[109]Protocol of Intentions between AEB and CONAE for the joint development of the SABIA-Mar Space Mission, 25 November 2014, http://www.aeb.gov.br/wp-content/uploads/2018/01/AcordoArgentina2014.pdf (accessed 18 April 2019).

[110]Basic Agreement for Scientific, Technical and Technological Cooperation between Brazil and Chile, 26 July 1990, http://portal-antigo.aeb.gov.br/wp-content/uploads/2012/09/AcordoChile1990.pdf (accessed 18 April 2019).

[111]Complementary Agreement to the Basic Agreement for Scientific, Technical and Technological Cooperation between Brazil and Chile establishing a Bilateral Cooperation Program in the Space Area, 26 March 1996, http://portal-antigo.aeb.gov.br/wp-content/uploads/2012/09/AcordoChile93.pdf (accessed 18 April 2019).

[112]Memorandum of Understanding between AEB and the Chilean Space Agency (ACE) on Space Area Cooperation, 20 March 2002, http://www.aeb.gov.br/wp-content/uploads/2018/01/AcordoChile2002.pdf (accessed 18 April 2019).

[113]Agreement on Scientific and Technological Cooperation between Brazil and the Colombia, 12 March 1981, http://portal-antigo.aeb.gov.br/wp-content/uploads/2012/09/AcordoColombia1981.pdf (accessed 18 April 2019).

In 1988, an adjustment[114] was made to the agreement for the purpose of executing entities to jointly design joint projects and other forms of scientific and technological cooperation, especially in the areas of training and education in remote sensing, digital images and geographic information systems; development and joint research on topics of mutual interest; transfer of software developed by both parties for the fulfillment of activities of common interest, respecting their respective national laws, as well as the mutual provision of scientific advisory services.

In 2009, a new complementary adjustment[115] was signed to cover the areas of space science, space technology, assessment and monitoring of the environment and land resources through remote sensing and other space applications, the development of joint satellite missions for scientific, space access and launching, application and support services for the development of agriculture projects of precision and also active work on technologies related to GNSS.

8.6.1.4 Peru

On 2006, the countries signed the Framework Agreement on Cooperation in the Peaceful Uses of Outer Space.[116] This Agreement was approved by the Brazilian National Congress in September 2008 and is awaiting ratification by Peru.

8.6.1.5 Venezuela

On 27 June 2008, the Framework Agreement for Cooperation in Space Science and Technology[117] was signed between the two countries. This Agreement is currently under discussion in the Brazilian National Congress. Only after approval and subsequent publication, will it be valid throughout the Brazilian territory.

---

[114]Complementary Agreement to the Agreement on Scientific and Technological Cooperation between Brazil and Colombia in the Field of Space Activities, 9 February 1988, http://portal-antigo. aeb.gov.br/wp-content/uploads/2012/09/AcordoColombia1988.pdf (accessed 18 April 2019).

[115]Complementary Agreement for Cooperation in Peaceful Space Science and Technology Applications between Brazil and Colombia, 17 February 2019, http://www.aeb.gov.br/wp-content/uploads/2018/01/AcordoColombia2009.pdf (accessed 18 April 2019).

[116]Framework Agreement between the Brazil and Peru on Cooperation in the Peaceful Uses of Outer Space, 17 February 2006, http://www.aeb.gov.br/wp-content/uploads/2018/01/AcordoPeru2006.pdf (accessed 18 April 2019).

[117]Framework Agreement for Cooperation in Space Science and Technology between Brazil and Venezuela, 27 June 2008, http://www.aeb.gov.br/wp-content/uploads/2018/01/AcordoVenezuela2008.pdf http://www.aeb.gov.br/wp-content/uploads/2018/01/AcordoPeru2006.pdf (accessed 18 April 2019).

8.6.1.6   South American Space Agency Proposal

South America is home to two trade blocs: the Andean Community since 1969 and Mercosur since 1991. The two blocs, along with Chile, Guyana and Suriname, integrated an intergovernmental union in 2008: UNASUR, which prioritizes political dialogue, social policies, education, energy, infrastructure, financing and the environment, with a view to eliminating socioeconomic inequality in South America. The Defense Minister of Argentina on 30 August 2011 proposed the creation of the South American Space Agency,[118] following the successful model of the ESA, arguing that the initiative is closely linked to the defense of both countries, since in the future a greater part of the territory will be controlled and protected from space. This initiative was carried out by the Defense sector of Argentina, with military collaboration.

In November of the same year, Defense Ministers from Argentina, Brazil, Bolivia, Chile, Colombia, Ecuador, Paraguay, Peru, Suriname, Venezuela and the deputy minister of Uruguay agreed to create a South American Space Agency during a meeting of the UNASUR Defense Council.[119] The goals for this agency were to focus efforts by putting satellites into orbit using a regional launch vehicle to reduce costs and increase technological capabilities. Brazil was open to the proposal, but emphasized the costs involved in creating new structures. This initiative was to be the responsibility of the South American Defense Council and incorporated into its 2012 Action Plan, but there is no record of this.

In early 2019, the Brazilian government denounced the Constitutive Treaty of UNASUR,[120] formalizing its departure from the organization. The decision was officially communicated to the Government of Ecuador, the depositary country of the agreement, to take effect in six months. In this way, the implementation of SASA has become a distant dream.

8.6.2   European Level

Brazilian cooperation with ESA and EU is mainly based on satellite data open access, as well as receiving and distributing this data.

---

[118]G1, Argentina propõe ao Brasil criação de agência espacial sul-americana, 2011, http://glo.bo/rgo9aQ (accessed 19 April 2019).

[119]UNASUR vai criar agência espacial sul-americana, 2011, https://www.rtp.pt/noticias/mundo/unasur-vai-criar-agencia-espacial-sul-americana_n498648 (accessed 19 April 2019).

[120]Denúncia do Tratado Constitutivo da UNASUL, 2019, http://www.itamaraty.gov.br/pt-BR/notas-a-imprensa/20291-denuncia-do-tratado-constitutivo-da-uniao-de-nacoes-sul-americanas-unasul (accessed 19 April 2019).

8.6.2.1   European Space Agency

The main cooperation instrument between AEB and ESA refers to the use of tracking and telemetry facilities in Brazil to support the implementation of the ARIANE, VEGA and SOYUZ launcher programs from the Kourou Launch Centre in French Guiana.

The partnership began in 1977 and became effective on 30 October 1980. In 1994, the Brazilian Government and ESA signed a new agreement,[121] which became effective on 24 October 1996, and was extended until 23 October 2019, which guaranteed and authorized the use of the CLBI facilities in Natal and the adaptation of the necessary equipment for the tracking and telemetry of the ARIANE launches from French Guiana. In 2011, the agreement was extended to permit the use of telemetry and tracking facilities in Brazil during the remaining validity period of the 1996 agreement for the VEGA and SOYUZ launches. In 2015, an extension of the agreement between Brazil and ESA on Spatial Cooperation for Peaceful Purposes was signed.[122]

In March 2019, AEB, INPE and ESA agreed on the Copernicus Space Component Technical Operating Arrangement (TOA)[123] to coordinate technical implementation for the provision of free, full and open access to all Brazilian EO satellite data, and certain other satellite data acquired by INPE, including historical data sets, to the Copernicus program and its participating states. The relevant Brazilian satellites mission—CBERS and AMAZONIA-1—comprises a certain series of civilian land, ocean and atmospheric satellites.

8.6.2.2   European Union

In March 2018, in São Paulo, the European Commission agreed with Brazil, Chile and Colombia on a space cooperation agreement that allows the exchange of data on ground observation and surveillance within the COPERNICUS program.[124] This partnership allows Brazil to access the Sentinel satellites. In Latin America, INPE is responsible for receiving and distributing European mission data. The images should contribute to the monitoring of sensitive biomes, such as Tropical and Cerrado forest, coastal zone and inland water bodies, as well as urban development.

---

[121] Agreement between the Brazil and ESA for the establishment and use of tracking and telemetry facilities located in Brazil, http://portal-antigo.aeb.gov.br/wp-content/uploads/2012/09/AcordoESA.pdf (accessed 19 April 2019).

[122] AEB, Document is not available on AEB's website, http://www.aeb.gov.br/programa-espacial-brasileiro/cooperacao-internacional/agencia-espacial-europeia-esa/ (accessed 19 April 2019).

[123] COPERNICUS Space Component Technical Operating Arrangement (TOA), http://www.inpe.br/institucional/sobre_inpe/relacoes_internacionais/arquivos/acordoCopernicusSpaceComponent_Assinado.pdf (accessed 19 April 2019).

[124] EU, EEAS, Delegación de la Unión Europea en Colombia. Comisión Europea firma Acuerdos de Cooperación para observación espacial con Colombia, Chile y Brasil, https://eeas.europa.eu/delegations/colombia/40977/comisi%C3%B3n-europea-firma-acuerdos-de-cooperaci%C3%B3n-para-observaci%C3%B3n-espacial-con-colombia-chile-y_es (accessed 19 April 2019).

By receiving the facilities required to downlink data from the new Sentinel satellite constellation, INPE has greatly broadened the capacity of its data centre, as well as ensuring access to the next generation of EO technologies. This provides the latest generation of remote sensing products for the Latin American community.

8.6.3 Bi-Lateral Level

There are eight countries with which Brazil maintains some agreement to further develop your space program on a bilateral level. France, Germany and Ukraine are highlighted with big projects. France collaborated with technologies transferred for the geostationary satellite SGDC. Germany is still working with Brazil on the development of the VLM-1. Concerning Ukraine, although the partnership was unsuccessful and is already over, a binational company was created to develop a launch vehicle and launching platform.

8.6.3.1 Belgium

Brazil has a Cooperation Program with the Centre of Liège—CSL,[125] signed on 6 October 2009, during the visit of the President of the Republic of Brazil to Belgium. The preliminary framework for cooperation refers to the area of education—mainly in the field of space sciences and techniques; EO techniques; design of space instruments; instrument tests, payloads and satellites; nanosatellites with student participation; optical technologies and specific technologies linked to the space sector. With the signing, Brazil and Belgium committed themselves to seeking adequate technical and financial means to advance collaborative actions. It was also agreed that, as far as possible, exemption from customs duties and taxes, as well as taxes on imports and exports of the necessary equipment for cooperation activities, will be obtained.[126]

8.6.3.2 France

Space cooperation between Brazil and France began in the early years of Brazilian space activities in the 1960s.[127] Since then, AEB's partnership with the National Centre for Space Studies (CNES) has advanced significantly, particularly in the field of

---

[125] Brazil's Cooperation Program with CSL, http://www.aeb.gov.br/programa-espacial-brasileiro/cooperacao-internacional/belgica/ (accessed 19 April 2019).

[126] Ibid.

[127] In 1967, when INPE was still called the CNAE, a stratospheric balloon-probe was launched from the facilities of the Institute, in scientific cooperation with the University of Toulouse. The objective was to collect data related to the study of gamma rays in astrophysics, https://books.openedition.org/iheal/1761 (accessed 19 April 2019).

scientific research. In 2005, the two agencies signed a Cooperation Program[128] related to Brazil's participation in the project Convection, Rotation and Transit Planetary (Corot), which resulted in research in the field of asteroseismology and exoplanets.

In April 2017, the Geostationary Satellite of Defense and Strategic Communications (SGDC)[129] was launched, a project of relevance in the cooperation between the two countries and an important social inclusion tool for Brazil. The SGDC is the only geostationary satellite capable of providing the Internet to the interior of Brazil. Its data capacity, at 60 gigabits per second, is more than double that of all other geostationary satellites operating in the country together. With this satellite, Telebras will be able to take the Internet to all schools and hospitals in Brazil with the support of telecommunication operators. The SGDC is also a very important asset for national security, allowing for the first time the Armed Forces to use satellite communications operated by Brazilians and using a completely national satellite. The construction of the satellite also contributed to the advancement of the country's space industry. In the context of the SGDC project, AEB led the Technology Absorption Process, through which Brazilians were trained in space engineering in France, and the Technology Transfer Program, through which Brazilian space companies received technologies transferred from the French company responsible for the construction of the SGDC. The program has the potential to revolutionize the Brazilian space industry, placing it at the international level of quality.

In 1995, the presidents of AEB and CNES in Paris signed the Specific Understanding on Cooperation between AEB and CNES for mini-satellite propulsion systems.[130] The MoU establishes a Framework for Cooperation in Space Activities between AEB and CNES.[131] The following year, in 1996, during the visit of the President of Brazil to France, a new Specific Understanding of Cooperation between AEB and CNES was signed[132] for the development of a microsatellite. In 1997, the Framework Agreement[133] between the two countries was signed on cooperation in research and uses of outer space for peaceful purposes.

In 2004, with a view to complementing the agreement on technical and scientific cooperation, the agreement between AEB and CNES on stratospheric balloons was

---

[128] AEB, Cooperação Internacional, França, http://www.aeb.gov.br/programa-espacial-brasileiro/cooperacao-internacional/franca/ (accessed 19 April 2019).

[129] Satélite Geoestacionário de Defesa e Comunicações Estratégicas (SGDC).

[130] AEB, Specific Understanding of Cooperation between with CNES for propulsion systems in minisatellites, 16 June 1995, http://portal-antigo.aeb.gov.br/wp-content/uploads/2012/09/AcordoFranca1995-2.pdf (accessed 19 April 2019).

[131] Memorandum of Understanding establishing a Framework for Cooperation in Space Activities between AEB and CNES, 16 June 1995, http://portal-antigo.aeb.gov.br/wp-content/uploads/2012/09/AcordoFranca1995-1.pdf (accessed 19 April 2019).

[132] Specific Understanding of Cooperation between AEB and CNES for the development of a microsatellite, 28 May 1996, http://portal-antigo.aeb.gov.br/wp-content/uploads/2012/09/AcordoFranca1996.pdf (accessed 19 April 2019).

[133] Framework Agreement between Brazil and France on Cooperation in Research and Uses of Outer Space for Peaceful Purposes, 27 November 1997, http://portal-antigo.aeb.gov.br/wp-content/uploads/2012/09/AcordoFranca1997-2.pdf (accessed 19 April 2019).

signed.[134] The following year, in 2005, AEB and CNES established a Specific Program of Cooperation regarding Brazil's participation in the COROT Mission.[135] In 2007, agreement was reached on a specific program of cooperation between AEB and CNES regarding the execution of stratospheric balloon flights in Brazilian territory.[136] In 2008, several specific projects were established: based on Cooperation in the Climate area and the Global Precipitation Measurement (GPM) water cycle, on cooperation in space technology applied to multimedia platforms (PMM), and on cooperation in the SGB Project area.[137]

In 2011, the Protocol on the use of telemetry facilities located on Brazilian territory (CLBI) was established for the ARIANE, SOYUZ and VEGA launches between AEB, DCTA and CNES.[138] In 2013, with a view to the training of qualified human resources in space areas, as well as the exchange of specialists in projects of common interest, with fellowships from the Sciences without Frontiers Program, the Cooperation Program between AEB and CNES was signed for the training of specialists in areas of interest to Brazil.[139] In 2014, the Program for Cooperation in Space Education[140] between AEB, ASTRIUM, SAFRAN, Aeronautics and Space Institute of France, Aeronautical Technological Institute and University of Brasilia created teaching and tutorial missions in the "Astrium Launch Vehicle Project and Safran for Brazilian Students".[141]

8.6.3.3 Germany

Brazil has had a long partnership with Germany spanning more than four decades. In 1969, the General Agreement on Cooperation in the Sectors of Scientific Research and Technological Development between the two countries was signed in Bonn.[142]

In 1971, the governments of the two countries approved an agreement between the then Technical Aerospace Centre (now called the Department of Aerospace Science

---

[134] AEB, Agreement on Technical and Scientific Cooperation with CNES on Stratospheric Balloons, http://www.aeb.gov.br/programa-espacial-brasileiro/cooperacao-internacional/franca/ (accessed 19 April 2019).

[135] AEB, International Cooperation, France, http://www.aeb.gov.br/programa-espacial-brasileiro/cooperacao-internacional/franca/ (accessed 19 April 2019).

[136] Ibid.

[137] Ibid.

[138] Ibid.

[139] Ibid.

[140] Ibid.

[141] Ibid.

[142] General Agreement between Germany and Brazil on Cooperation in the Sectors of Scientific Research and Technological Development, 9 June 1969, http://portal-antigo.aeb.gov.br/wp-content/uploads/2012/09/AcordoAlemanha1969.pdf (accessed 19 April 2019).

and Technology—DCTA) and the German Aerospace Centre (DLR). In 1982, a similar international instrument extended to INPE the partnership with DLR.[143]

In partnership with DCTA and INPE, DLR has developed more than a dozen scientific projects. In 1996, the Framework Agreement[144] on Cooperation in Scientific Research and Technological Development was signed in Brasilia. In 2002, also in Brasilia, an Agreement between AEB and DLR on Cooperation for the exploration and use of outer space for peaceful purposes was signed.[145]

Since the mid-1970s, there has been successful cooperation between DCTA/IAE and the Mobile Rocket Base of DLR (MORABA) in the use and development of Brazilian rocket engines for high-atmosphere physics, microgravity research, technology development and other research areas. In 2001, it began the development of sounding rockets, in which all rocket microgravity flights were carried out in the national (DLR-TEXUS) and European (ESA-MASER) programs.[146] The rocket motor systems were provided by the Brazilian cooperation partners in the exchange of services. In return, DLR has been involved in the development of components, which the Brazilian partners could not carry out without the support of DLR.[147] The first flight took place on 23 October 2004 at CLA during Operation Cajuana.[148] The first launch on European land took place in 2005 with the VSB-30 V02 from the ESRANGE Launch Centre in Kiruna, Sweden.[149]

The VS rockets were developed to follow PNAE in the field of experiments in microgravity environment and, due to their technical quality, aroused the interest of DLR. With more than 20 successful releases, most being overseas, the VSB-30 is moving to its third version of the modernized model.[150]

In 2005, the work program for the exchange of services and equipment signed between AEB, DLR and CTA was launched.[151] In 2010, the Protocol of Intent was

---

[143] Special Agreement between the CTA and the German Institute for Research and Testing of Air and Space Navigation, 19 November 1971, http://portal-antigo.aeb.gov.br/wp-content/uploads/2012/09/AcordoAlemanha1971.pdf (accessed 19 April 2019).

[144] Framework Agreement between Brazil and Germany on Cooperation in Scientific Research and Technological Development, 20 March 1996, http://portal-antigo.aeb.gov.br/wp-content/uploads/2012/09/AcordoAlemanha96.pdf (accessed 19 April 2019).

[145] Agreement between AEB and DLR on Cooperation for the exploration and use of outer space for peaceful purposes, 14 February 2002, http://portal-antigo.aeb.gov.br/wp-content/uploads/2012/09/AcordoAlemanha2002.pdf (accessed 19 April 2019).

[146] IAE, VSB-30, http://www.iae.cta.br/index.php/espaco/vsb-30 (accessed 6 August 2019).

[147] Brazilian Air Force, Veículo suborbital VSB-30 terá modelo modernizado, 2017, http://www.fab.mil.br/noticias/mostra/29413 (accessed 19 April 2019).

[148] Harvey, B., Smid, H., Pirard T., Emerging Space Powers: The New Space Programs of Asia, the Middle East and South-America, Springer Science & Business Media, 2011, p. 344.

[149] IAE, VSB-30, http://www.iae.cta.br/index.php/espaco/vsb-30 (accessed 6 August 2019).

[150] Brazilian Air Force, Veículo suborbital VSB-30 terá modelo modernizado, 2017, http://www.fab.mil.br/noticias/mostra/29413 (accessed 19 April 2019).

[151] Program of work on the exchange of services and equipment, signed between AEB, DLR and CTA, 13 October 2005, http://portal-antigo.aeb.gov.br/wp-content/uploads/2012/09/AcordoAlemanha2005.pdf (accessed 19 April 2019).

MICROSATELLITE LAUNCHER VEHICLE
ANATOMY OF THE SATELLITE LAUNCHER

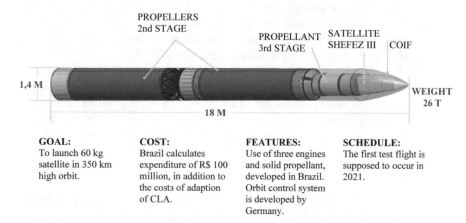

GOAL:
To launch 60 kg satellite in 350 km high orbit.

COST:
Brazil calculates expenditure of R$ 100 million, in addition to the costs of adaption of CLA.

FEATURES:
Use of three engines and solid propellant, developed in Brazil. Orbit control system is developed by Germany.

SCHEDULE:
The first test flight is supposed to occur in 2021.

Fig. 8.8 Microsatellite launcher vehicle (VLM). OGlobo, microsatellite launcher vehicle, https:// oglobo.globo.com/brasil/brasil-deve-lancar-foguete-no-espaco-em-2019-21516910 (accessed 19 April 2019)

signed between DLR, DCTA and AEB.[152] In 2011, DCTA/IAE and DLR agreed on a joint project for the development of a large and powerful sounding rocket (VS-50) and a microsatellite launcher (VLM-1). The two launch vehicles have a maximum of parts commonality and the goal is that the VS-50 will serve as an intermediate step towards the qualification of VLM-1. The launcher is based on solid propulsion and will target a launch price below €10 million. It is sought to be the baseline for a launcher family.

DLR's focus is on a research vehicle while Brazil's primary goal is access to space. When the project moved from the research phase to the development phase itself,[153] the tasks of developing and qualifying the subsystems for each participant were allocated, Brazil being responsible for the S50 and S44 propulsive systems, the reserve navigation system, infrastructure for launching and flight safety, and project documentation management. The development and qualification of the other VS-50 and VLM-1 vehicle systems is the responsibility of DLR (Fig. 8.8).

In 2013, a cooperation programme[154] was established for the Training of Qualified Human Resources in the Brazilian Space Area between AEB and DLR, based on grants from the Science Without Borders Program to train specialists in space areas

[152]AEB, Protocol of Intent between DLR and DCTA, http://www.aeb.gov.br/programa-espacial-brasileiro/cooperacao-internacional/alemanha/ (accessed 19 April 2019).

[153]IAE, Relatório de Atividades, 2018, http://www.iae.cta.br/images/relatorios-atividades/Relatorio_de_Atividades_2018.pdf (accessed 19 April 2019).

[154]Program of Cooperation for the Formation of Qualified Human Resources in the Space Area of Brazil between AEB and DLR, 24 September 2013, http://www.aeb.gov.br/wp-content/uploads/2018/01/AcordoAlemanha2013.pdf (accessed 19 April 2019).

useful to Brazil and to bring German teachers with the same goal. In 2014, as a result of an international scientific campaign, a Cooperation Program between AEB and DLR was signed for the implementation of ACRIDICON-Rain Project: Amazon Campaign which included ground equipment, with the objective of evaluating the impact of pollution on the life cycle of clouds, the formation of storm clouds, the radiation balance and the climate of the Amazon region.[155]

In addition to these projects, AEB and DLR intend to confirm and strengthen the cooperation agreed in the Protocol of Intent of 19 January 2011, which allows interested parties to share information and establish the development and testing of an injector head in combination with the combustion cabin for one L75 engine based on LOX-Ethanol, liquid fuel. In 2016, AEB and DLR signed an MoU with DCTA for the development of the L75 project—the Satellite Fire Monitoring Program, which aims to develop a liquid propellant rocket motor that will use liquid oxygen and kerosene.[156]

In March of the same year, AEB and DLR signed a letter of intent to develop the structure of a low-cost S50 motor envelope.[157] The S50 is the propulsive system of the VS-50 suborbital vehicle and the VLM-1, both developed under the Brazil-Germany partnership established by the protocol of intentions signed between the Brazilian and German space agencies and the DCTA in 2011.[158]

This partnership is important for the development of probes and launch vehicles in Brazil, as well as for the future provision of services in this area, in addition to training of national technical staff and industry. In addition, AEB, INPE and DLR share complementary interests in the advancement of remote sensing sciences, through observational research and development of satellite algorithms and climate models that improve understanding of vegetation, carbon and climate systems.[159]

---

[155]Cooperation Program between AEB and DLR for the implementation of ACRIDICON-Rain Project: Amazon Campaign, August 2014, http://www.aeb.gov.br/wp-content/uploads/2018/01/AcordoAlemanha2014.pdf (accessed 19 April 2019).

[156]Program of Cooperation between AEB and the DLR to further develop the satellite fire monitoring program (SAFIM), http://www.inpe.br/institucional/sobre_inpe/relacoes_internacionais/arquivos/programa_cooperacao_aebdlr_german_aerospace_research_establishment.pdf (accessed 17 July 2019).

[157]AEB, Letter of Intent to develop the structure of a low-cost S50 motor envelope with DLR, http://www.aeb.gov.br/programa-espacial-brasileiro/cooperacao-internacional/alemanha/ (accessed 19 April 2019).

[158]Revista da UNIFA, Universidade da Força Aérea, 1985, http://www2.fab.mil.br/unifa/images/revista/pdf/v29n1/ Revista.completa.1ed16.pdf (accessed 17 July 2019).

[159]Maltchiik, R., Brasil deve lançar foguete no espaço em 2019, https://oglobo.globo.com/brasil/brasil-deve-lancar-foguete-no-espaco-em-2019-21516910 (accessed 19 April 2019).

8.6.3.4   Italy

In 2008, the space agencies of both countries signed a Letter of Intent for Exploring Opportunities for Cooperation in the National and International Space in Technologies and Applications.[160] In 2013, a Program for Cooperation in Training of Qualified Human Resources in the Brazilian Space Area was established between the two agencies for the training of qualified human resources in the space area and the exchange of specialists in projects of common interest through the Sciences without Borders Program.[161]

8.6.3.5   Poland

The MoU between the AEB and the Polish Space Agency on Cooperation in the Exploration and Use of Outer Space for Peaceful Purposes was signed on 22 September 2015.[162] The document is traditional in offering legal and organizational support for the development and implementation of cooperation of agencies. The initiative envisages potential cooperation between the agencies in the exploration and use of outer space for peaceful purposes.

8.6.3.6   Portugal

In 2016, the Ministries of Science and Technology of the two countries signed a Joint Declaration to strengthen cooperation in the fields of scientific research and technology.[163]

8.6.3.7   Sweden

In 2014, based on the principles of equality and mutual benefit, AEB and the Swedish National Space Council (SNSB) signed a MoU on Cooperation on Spatial Activities

---

[160] AEB, Letter of Intent for Exploring Opportunities for Cooperation in the National and International Space in Technologies and Applications with ASI, http://www.aeb.gov.br/programa-espacial-brasileiro/cooperacao-internacional/italia/ (accessed 20 April 2019).

[161] Ibid.

[162] AEB, MoU on Cooperation in the Exploration and Use of Outer Space for Peaceful Purposes with POLSA, http://www.aeb.gov.br/programa-espacial-brasileiro/cooperacao-internacional/polonia/ (accessed 20 April 2019).

[163] Joint Declaration between the Ministry of Science, Technology, Innovation and Communications of Brazil and the Ministry of Science, Technology and Higher Education of Portugal for strengthening cooperation in the fields of scientific research and technology, 1 November 2016, http://www.aeb.gov.br/wp-content/uploads/2018/01/Declara%C3%A7%C3%A3oConjunta_2016.pdf (accessed 20 April 2019).

for Peaceful Purposes.[164] The memorandum provides a framework for collaborative activities between AEB and SNSB in programs and/or projects of common interest.

8.6.3.8 Ukraine

As Brazil does not dominate launching rocket technology, it has made a series of agreements with Ukraine, which dominates this field. Ukraine developed these space technologies along with the Soviet Union when it was part of the former USSR. However, despite having the technology, it does not have the geographical conditions to have its own launch centre. Both countries have found an alternative in cooperation since they are complementary: Ukraine was looking for a safer, more economical and more autonomous option, while Brazil was looking for its launch vehicle. The meeting of interests was combined when Brazil offered the infrastructural requirements while Ukraine develops the vehicle and launching platform. Thus, both countries agreed to create the binational Alcântara Cyclone Space (ACS), in a rented part of the CLA.[165] On 21 October 2003, the Treaty on long-term cooperation in the use of the CYCLONE-4 Launch Vehicle was signed in Brasilia.[166] This document led to the creation of the binational company ACS, which was responsible for promoting the launch of commercial launches.

Before that, both countries concluded some important agreements: "The Framework Agreement on Cooperation in Peaceful Uses of Outer Space in November 1999"[167] and "The Agreement on Technological Safeguards"[168] related to the participation of Ukraine in launches from the CLA in 2002.

In 2013, an MoU[169] on cooperation for the training of qualified professionals and students in the Brazilian space area between AEB and the State Space Agency of Ukraine was signed, aiming at cooperation in the training of qualified human resources in the space area and the exchange of specialists in projects of common interest through the Sciences Without Borders Program.

---

[164] AEB, MoU on Cooperation on Spatial Activities for Peaceful Purposes with Sweden, http://www.aeb.gov.br/programa-espacial-brasileiro/cooperacao-internacional/suecia/ (accessed 20 April 2019).

[165] AEB, Agreement with Ukraine to create the binational ACS, http://www.aeb.gov.br/programa-espacial-brasileiro/cooperacao-internacional/ucrania/ (accessed 20 April 2019).

[166] Treaty between Brazil and Ukraine on Long-term Cooperative Use of the Cyclone-4 Launch Vehicle at the Alcântara Launch Centre, 21 October 2003, http://www.aeb.gov.br/wp-content/uploads/2018/01/TratadoUcr%C3%A2nia2003.pdf (accessed 20 April 2019).

[167] Framework Agreement between Brazil and Ukraine on Cooperation in the Peaceful Uses of Outer Space, 18 November 1999, http://portal-antigo.aeb.gov.br/wp-content/uploads/2012/09/AcordoUcrania.pdf (accessed 20 April 2019).

[168] Agreement between Brazil and Ukraine on Technological Safeguards related to the participation of Ukraine in launches from the Alcântara Launch Centre, 16 January 2002, http://www.aeb.gov.br/wp-content/uploads/2018/01/AcordoUcr%C3%A2nia2002_a.pdf (accessed 20 April 2019).

[169] AEB, Cooperation for the training of qualified professionals and students between AEB and SSAU, http://www.aeb.gov.br/programa-espacial-brasileiro/cooperacao-internacional/ucrania/ (accessed 20 April 2019).

In 2015, the Treaty on Long-term Cooperation in the Use of the CYCLONE-4 Launch Vehicle, signed in 2003, was renounced by Brazil with the publication of Decree no. 8.494,[170] dated 24 July 2015, and ceased to be effective as of 16 July 2016. The justification was that, throughout the implementation of the Treaty, there was an imbalance in the technological-commercial equation that had justified the formation of the partnership between the two countries in the area of outer space.

8.6.4   Worldwide Level

On a worldwide level, China, Russia and the USA are the main Brazilian partners in space activities. The CBERS satellites, developed and launched in cooperation with China, are a successful case to be mentioned. Regarding Russia, represented by ROSCOSMOS, an agreement was signed to buy Russian materials and services for the Brazilian VLS. In addition, the Centenary Mission[171] carried the Brazilian astronaut Marcos Pontes onboard the SOYUZ TMA-8 spacecraft to the ISS, resulting in the first Brazilian astronaut in outer space. With respect to the USA, besides years of cooperation, the Technological Safeguards Agreement (TSA) is the main cooperation document that was signed with Brazil but is still being internally negotiated at the Brazilian legislative level.

8.6.4.1   Canada

AEB and the Canadian Space Agency (CSA) signed a cooperation program for the training of students and qualified professionals in the space area in 2014,[172] with the support of the space segment of the Science without Borders Program. The aim of the program is to further intensify the collaboration between AEB and CSA in the context of missions and projects in the space area to advance mutual knowledge in areas related to science, technology and innovation in space.

---

[170]Brazil, Decree no. 8.494, 24 July 2015, http://www.aeb.gov.br/wp-content/uploads/2018/01/Decreto-n%C2%BA-8494.pdf (accessed 20 April 2019).

[171]Colonel Pontes's space voyage is officially known as the Centennial Mission, a reference to Brazil's pioneering aviator Alberto Santos Dumont. Brazilians are taught that Santos Dumont, not the Wright Brothers, was the first man to fly, before cheering throngs in Paris in 1906.

[172]Program for Cooperation between AEB and the CSA on cooperation for the training of professionals and qualified students in space-related disciplines through the space division of the "Science Without Borders Program" to promote training of human resources in areas related to the space of interest of Brazil and Canada, 12 May 2014, http://www.aeb.gov.br/wp-content/uploads/2018/01/AcordoCanada2014.pdf (accessed 21 April 2019).

8.6.4.2 United States of America

Cooperation projects between Brazil and the USA are currently focused on the development of joint scientific research. Since the creation of the AEB, more than 15 cooperation instruments have been signed with NASA.[173] Among the programs underway, NASA-I, which supports the participation of Brazilian students in training programs and specialization in research centres of that agency, should be highlighted.

Between 1994 and 1996, several agreements were signed, including an MoU between NASA and COBAE for the rocket launching campaign DIP Equator or Guará,[174] an MoU between AEB and NASA for the smoke/sulphate, clouds and radiation experiment (SCAR-B),[175] the CCD Imager Instrument Experiment, the flight of the humidity sensor Brazil on the PM-1 spacecraft, the NASA EO System (EOS)[176] and the Brazil-US Framework Agreement on Peaceful Uses of Outer Space,[177] which is currently expired.

In 1997, during the visit of the President of the United States, Bill Clinton, to Brazil, the Complementary Agreement between the two countries was signed for the Project for the Development, Operation and Use of Flight Equipment and Useful Loads for the International Space Station Program.[178] This agreement is currently suspended. The following year, an Agreement between NASA and the AEB on training of an AEB Mission Specialist was signed[179] and in 1999, the Tuning for the Cooperation Program between AEB and NASA to monitor the orbiting of the Advanced X-ray Astrophysics Equipment was signed.[180]

---

[173] AEB, Cooperation Programs, http://www.aeb.gov.br/programa-espacial-brasileiro/cooperacao-internacional/estados-unidos/ (accessed 21 April 2019).

[174] Memorandum of Understanding between NASA and the CNAE for the Rocket Launch Campaign DIP Equator or Guará, 28 March 1994, http://portal-antigo.aeb.gov.br/wp-content/uploads/2012/09/AcordoEUA1994.pdf (accessed 21 April 2019).

[175] Memorandum of Understanding between AEB and the NASA for the Smoke/Sulphate, Clouds and Radiation (SCAR-B) experiment, 19 May 1995, http://portal-antigo.aeb.gov.br/wp-content/uploads/2012/09/AcordoEUA1995.pdf (accessed 21 April 2019).

[176] Memorandum of Understanding between AEB and NASA on Experiment with the CCD Imager Instrument (CIMEX), 5 December 1996, http://www.aeb.gov.br/wp-content/uploads/2018/01/AcordoEUA1996-3.pdf (accessed 21 April 2019).

[177] Framework Agreement between Brazil and the United States of America on Cooperation in the Peaceful Uses of Outer Space, 1st March 1996, http://portal-antigo.aeb.gov.br/wp-content/uploads/2012/09/AcordoEUA1996-1.pdf (accessed 21 April 2019).

[178] Complementary Agreement between Brazil and the United States of America for the Project for the Development, Operation and Use of Flight Equipment and Useful Loads for the International Space Station Program, 14 October 1997, http://portal-antigo.aeb.gov.br/wp-content/uploads/2012/09/AcordoEUA1997.pdf (accessed 21 April 2019).

[179] Agreement between NASA and AEB on training of an AEB Mission Specialist, 19 November 1998, http://portal-antigo.aeb.gov.br/wp-content/uploads/2012/09/AcordoEUA1998-2.pdf (accessed 21 April 2019).

[180] Adjustment for a Cooperation Program between AEB and NASA to monitor the placement in orbit of the Advanced X-ray Astrophysics Equipment, 3 May 1999, http://www.aeb.gov.br/wp-content/uploads/2018/01/AcordoEUA1999.pdf (accessed 21 April 2019).

In the 2000s, an Agreement on Technological Safeguards was signed between the two countries related to USA participation in launches from CLA.[181] Subsequently, the agreement was rejected by the National Congress.[182] In the same year, the Adjustment for a Cooperation Program was signed between the two agencies in Geodetic Space Research with emphasis on GPS.[183] In the first decade of the 2000s, only three agreements were signed. In 2001, they signed the Adjustment Agreement for a program of cooperation between the two agencies on spatial geodesy research with an emphasis on very long baseline interferometry.[184] They also renewed the Framework Agreement between the two countries on Outer Space Uses in 2006[185] and extended the Adjustment to a Cooperation Program between AEB and NASA in Geodetic Space Research with emphasis on GPS in 2010.[186]

In 2011, The Framework Agreement between Brazil and the USA on Peaceful Uses of Outer Space, the Complementary Cooperation Agreement between AEB and NASA for participation in the Global Precipitation Measurement Mission, and the Complementary Adjustment of Cooperation between AEB and NASA for participation in the Ozone Cooperation Mission were all successfully concluded.[187]

In 2015, internships for undergraduate and postgraduate Brazilian students were available in the NASA-I Program with a view to creating an international network of primary and secondary education researchers to study environmental issues (Reimbursement Agreement between the agencies for participating in the International Internship Program at NASA).[188] Also, that year, the two agencies signed the Agreement on Cooperation in Global Learning and Observations for the Benefit of

---

[181] Agreement between Brazil and the United States of America on Technological Safeguards related to the participation of the United States of America in launches from the Alcântara Launch Centre, 18 April 2000, http://portal-antigo.aeb.gov.br/wp-content/uploads/2012/09/AcordoEUA2000 2.pdf (accessed 21 April 2019).

[182] Brazil, Projeto de Decreto Legislativo n.° 1.446-A, 2001, https://www.camara.leg.br/proposicoesWeb/prop_mostrarintegra?codteor=1099427&filename=Avulso+-PDC+1446/2001 (accessed 21 April 2019).

[183] Adjustment for a Cooperation Program between AEB and NASA in Geodetic Space Research with emphasis on GPS, 11 April 2000, http://portal-antigo.aeb.gov.br/wp-content/uploads/2012/09/AcordoEUA1999.pdf (accessed 21 April 2019).

[184] AEB, Adjustment Agreement for cooperation between AEB and NASA on spatial geodesy research with an emphasis on very long baseline interferometry, http://www.aeb.gov.br/programa-espacial-brasileiro/cooperacao-internacional/estados-unidos/ (accessed 21 April 2019).

[185] AEB, Framework Agreement between Brazil and USA on Outer Space Uses, http://www.aeb.gov.br/programa-espacial-brasileiro/cooperacao-internacional/estados-unidos/ (accessed 21 April 2019).

[186] AEB, Extension of the Adjustment for the Cooperation Program between AEB and NASA in Geodetic Space Research with emphasis on GPS in 2010, http://www.aeb.gov.br/programa-espacial-brasileiro/cooperacao-internacional/estados-unidos/ (accessed 21 April 2019).

[187] AEB, International Cooperation, USA Agreements, http://www.aeb.gov.br/programa-espacial-brasileiro/cooperacao-internacional/estados-unidos/ (accessed 21 April 2019).

[188] Reimbursement Agreement between AEB and NASA for participating in the International Internship Program at NASA, 18 June 2015, http://www.aeb.gov.br/wp-content/uploads/2018/01/AcordoEUA2015_c.pdf (accessed 21 April 2019).

the Environment (GLOBE)[189] and the Complementary Cooperation Agreement on Heliophysics and Space Weather Research, data sharing involving heliophysics and research of climate and space.[190]

In 2016, the activities of the Globe Program in Brazil started. It is an international science and environmental education program that promotes the participation of students, teachers, scientists and citizens from all over the world in environmental data collection and scientific studies on the environment at local, regional and global scales. In the first year of the program, workshops were held in Brasilia to train 67 teachers from public and private schools in the Federal District and Goiás State, as well as AEB's own staff.

In May and June 2017, three other workshops were held with the objective of bringing the GLOBE program to other states of the federation.[191] These workshops also marked the beginning of the international campaign, promoted by the GLOBE and NASA, to collect data for scientific studies of the larva of the mosquito Aedes Aegypti, and the launch of the mobile application Mosquito Habitat Mapper. The GLOBE is underway in the cities of Brasilia, São José dos Campos, Rio de Janeiro and in the coastal region of Paraná. The program currently counts on the participation of 117 schools in Brazil.

In 2019, the TSA was re-discussed in the National Congress and, after 20 years of negotiation, Brazil and the USA concluded the terms of the agreement for the commercial use of CLA.[192] The AST has clauses that protect both the technology used by the USA and by Brazilians.[193] Under the agreement, the USA could launch satellites and rockets from the site.

Also, in 2019, on 18 March, in Washington D.C., the AEB President and Assistant Administrator of NASA signed an MoU[194] for development of a nanosatellite. It is a CubeSat called SPORT—Scintillation Prediction Observations Research Task. The SPORT mission aims to study ionosphere phenomena with an emphasis on an

---

[189] Agreement between AEB and NASA for Cooperation in GLOBE, 30 June 2015, http://www. aeb.gov.br/wp-content/uploads/2018/01/AcordoEUA2015_b.pdf (accessed 21 April 2019).

[190] Complementary Agreement between AEB and NASA on Heliophysics and Space Weather Research, 30 June 2015, http://www.aeb.gov.br/wp-content/uploads/2018/01/AcordoEUA2015_a. pdf (accessed 21 April 2019).

[191] AEB, GLOBE Program, http://www.aeb.gov.br/programa-espacial-brasileiro/cooperacao-internacional/estados-unidos/ (accessed 21 April 2019).

[192] Mello, P., Brasil e EUA assinam acordo que permite uso comercial de Alcântara, Folha de São Paulo, 2019, https://g1.globo.com/politica/noticia/2019/03/18/brasil-assina-acordo-que-permite-aos-eua-lancar-satelites-da-base-de-alcantara.ghtml (accessed 5 April 2019).

[193] AEB, Conhecendo o Acordo de Salvaguardas Tecnológicas, 2019, http://Www.Aeb.Gov.Br/Wp-Content/Uploads/2019/05/Folder_Ast2mai19-1.Pdf (accessed 21 April 2019).

[194] SPORT is a partnership between NASA and the AEB. Through AEB, two Brazilian institutes will contribute to the mission, INPE and the ITA. Through NASA, U.S. universities and the Aerospace Corporation will provide the scientific instruments. SPORT is managed by NASA's Marshall Space Centre in Huntsville, Alabama. The mission is part of NASA's Heliophysics Technology and Instrument Development for Science program. NASA's heliophysics directorate studies our star and how it influences the very nature of space—and, in turn, the atmospheres of planets and the technology that exists there, https://www.nasa.gov/mission_pages/sport/index.html (accessed 16 July 2019).

occurrence known as the magnetic anomaly of the southern hemisphere. The launch of CubeSat SPORT is scheduled for 2020, from the ISS.[195]

8.6.4.3 Russia

In 1988, Brazil signed a protocol of cooperation with the Russian Federation in the field of space research and the use of space for peaceful purposes.[196] Since 1992, contracts were signed between the then Technical Aerospace Centre (now the Department of Aerospace Science and Technology—DCTA) and research institutions and Russian companies to supply materials and services for the VLS.[197] The cooperation was fundamental for the maintenance and continuation of the work related to the development of the vehicle.

In 1996, DCTA entered into a contract with the International Centre for Advanced Studies (ICAS) of the Moscow Aviation Institute (MAI) to undertake a two-year postgraduate course in liquid propulsion.[198] The following year, the countries signed the Agreement on Cooperation in Research and Uses of Outer Space for Peaceful Purposes.[199]

In 2004, a contract was signed with State Rocket Centre Makayev for the revision of VLS-1, as well as an MoU between the Ministry of Science and Technology of Brazil and the Russian Space Agency regarding the Program for Cooperation on Activities.[200] In 2006, the General Command signed a contract with the Niichimash firm for the construction of a test bench for rocket propellant liquid.[201] In April 2016, AEB and ROSCOSMOS signed an agreement for the installation of a Russian space debris monitoring station in Itajubá, Minas Gerais, Brazil, at the premises of

---

[195] AEB, Brasil e Estados Unidos firmam acordo para desenvolver nanossatélite, http://www.aeb.gov.br/brasil-e-estados-unidos-firmam-acordo-para-desenvolver-nanossatelite/ (accessed 16 July 2019).

[196] Protocol between the Government of the Federative Republic of Brazil and the Government of the Union of Soviet Socialist Republics on Cooperation in the Field of Space Research and the Use of Space for Peaceful Purposes, 19 October 1988, http://portal-antigo.aeb.gov.br/wp-content/uploads/2012/09/AcordoRussia1988.pdf (accessed 21 April 2019).

[197] AEB, contracts between DCTA and Russian companies to supply materials and services for the VLS, http://www.aeb.gov.br/programa-espacial-brasileiro/cooperacao-internacional/russia/ (accessed 21 April 2019).

[198] Ibid.

[199] Agreement between the Brazil and the Russian Federation on Cooperation in Research and Uses of Outer Space for Peaceful Purposes, 21 November 1997, http://portal-antigo.aeb.gov.br/wp-content/uploads/2012/09/AcordoRussia1997.pdf (accessed 21 April 2019).

[200] Memorandum of Understanding between the Ministry of Science and Technology of Brazil and ROSCOSMOS regarding the Cooperation Program on Space Activities, 22 November 2004, http://www.aeb.gov.br/wp-content/uploads/2018/01/AcordoRussia2004.pdf (accessed 21 April 2019).

[201] AEB, Contract signed for the test bench for rocket propellant liquid, http://www.aeb.gov.br/programa-espacial-brasileiro/cooperacao-internacional/russia/ (accessed 21 April 2019).

the National Laboratory of Astrophysics.[202] The station will integrate the Panoramic Electro-Optical System for Detection of Spatial Debris, which aims to collaborate with international efforts to mitigate the risk of serious accidents from space debris, both in orbit (as well as the risk of possible collisions) and reentry of fragments and space objects in the Earth's atmosphere. The telescope was inaugurated in April 2017, at the Observatory of Pico dos Dias.[203] The installation of this station is an important contribution to Brazilian scientific and technological education due to its contribution to the knowledge and the development of research on data and applications.

In 2005, the two agencies signed the protocol on cooperation in modernization of the VLS-1 Launch Vehicle,[204] as well as a memorandum on the establishment of a joint working group[205] and the Joint Declaration on the Brazilian High-Level Cooperation Commission and the Brazilian-Russian Intergovernmental Commission on Economic, Commercial, Scientific and Technological Cooperation.[206]

In 2005, due to an agreement between Brazil and Russia, represented by their space agencies, the Centenary Mission[207] was carried out. In this mission, Marcos Pontes[208] became the first Brazilian to go to space. The vehicle used for the launch of the mission was the Soyuz TMA-8 spacecraft by ROSCOSMOS that was launched on 29 March 2006 at the Baikonur Launch Centre (Kazakhstan), with ISS as its destination.[209] The following year, they agreed on the Mutual Protection of Technology Associated with Cooperation in the Exploration and Use of Outer Space for Peaceful Purposes.[210]

In 2008, a cooperation program in the field of use and development of the Russian Global Navigation Satellite System (GLONASS) between AEB and ROSCOSMOS

---

[202] AEB, AEB participa de cerimônia de acordo entre Brasil e Rússia, 2016, http://portal-antigo. aeb.gov.br/aeb-participa-de-cerimonia-de-acordo-entre-brasil-e-russia/ (accessed 21 April 2019).

[203] AEB, LNA e ROSCOSMOS iniciam monitoramento de detritos espaciais no Brasil, 2017, http://portal-antigo.aeb.gov.br/em-parceria-com-agencia-russa-lna-inicia-monitoramento-de-detritos-espaciais/ (accessed 21 April 2019).

[204] AEB, ROSCOSMOS and AEB Protocol on cooperation for VLS-1 Launch Vehicle modernization, http://www.aeb.gov.br/programa-espacial-brasileiro/cooperacao-internacional/russia/ (accessed 21 April 2019).

[205] Ibid.

[206] Ibid.

[207] Colonel Pontes's space voyage is officially known as the Centennial Mission, a reference to Brazil's pioneering aviator Alberto Santos Dumont. Brazilians are taught that Santos Dumont, not the Wright Brothers, was the first man to fly, before cheering throngs in Paris in 1906.

[208] First South American and the first Lusophone to go into space when he docked onto the ISS aboard SOYUZ TMA-8 on 30 March 2006. He is the only Brazilian to have completed the NASA astronaut training program, although he switched to training in Russia after NASA's Space Shuttle program encountered problems. After Jair Bolsonaro's victory as President of Brazil, Pontes was officially nominated to be Minister of Science, Technology and Innovation, a position which he accepted days later.

[209] New York Times. Brazil's Man in Space: A Mere "Hitchhiker or a Hero?", 2006, https://www.nytimes.com/2006/04/08/world/americas/08rio.html (accessed 21 April 2019).

[210] Agreement between Brazil and the Russian Federation on the Mutual Protection of Technology Associated with Cooperation in the Exploration and Use of Outer Space for Peaceful Purposes, 14 December 2006, http://www.aeb.gov.br/wp-content/uploads/2018/01/AcordoRussia2006.pdf (accessed 21 April 2019).

was signed.[211] In 2012, the installation in Brazil of a correction ground station integrated to the GLONASS satellite localization system, with a view to improving the quality of the navigation system services in Brazil with the improvement of the precision of its signal in the hemisphere was announced.[212]

In 2014, a Letter of Intent was signed between the Open Joint-Stock Company Research-and-Production Corporation and the National Astrophysics Laboratory for a program entitled Panels.[213] The Joint-Stock Company Research-and-Production Corporation and the National Laboratory of Astrophysics declared their intention to explore the possibility and the conditions to install and operate, in Brazilian territory, an optical-electronic complex for detection of debris as part of a network of similar systems operated at various locations on Earth. In the following year, the agencies signed an MoU on the installation and use, in Brazilian territory, of the Optical-Electronic Complex.[214]

In the area of GNSS, due to the excellent results achieved in the collaboration between the University of Brasilia and GLONASS, contracts were signed for the installation of new correction stations at the Federal University of Santa Maria and at the Technological Institute of Pernambuco, which were inaugurated in 2016 and are fully operational.[215] The new stations create conditions to foster synergies and integrate the academic environment with the development of human, scientific and technological resources, in order to enable Brazil to become a provider of GNSS solutions.

8.6.4.4 China

In 1988, negotiations began with the Agreement on Research and Joint Production of the Sino-Brazilian Satellite (remote sensing between the two countries)[216] which

---

[211] Jornal de Brasília, AEB e agência russa promovem encontro empresarial em São Paulo, 2010, http://www.jornaldebrasilia.com.br/brasil/aeb-e-agencia-russia-promovem-encontro-empresarial-em-sao-paulo/ (accessed 21 April 2019).

[212] AEB, Primeira estação do GLONASS fora da Rússia é instalada na UnB, 2013, http://portal-antigo.aeb.gov.br/primeira-estacao-do-glonass-fora-da-russia-e-instalada-na-unb/ (accessed 21 April 2019).

[213] AEB, Letter of Intent between the Open Joint-Stock Company Research-and-Production Corporation and the National Astrophysics Laboratory for a program entitled Panels, http://www.aeb.gov.br/programa-espacial-brasileiro/cooperacao-internacional/russia/ (accessed 21 April 2019).
[214] Ibid.

[215] Russia Beyond, Brasil ganhará nova estação GLONASS em abril, 2016, https://br.rbth.com/ciencia/2016/03/30/glonass-na-america-latina_580395 (accessed 21 April 2019).

[216] Agreement for Exchange of Notes on Research and Joint Production of the Sino-Brazilian Remote Sensing Satellite between Brazil and China, 30 April 1988, http://www.aeb.gov.br/wp-content/uploads/2018/01/AcordoChina1988_b.pdf (accessed 22 April 2019).

led to signature, in July 1988, of the Protocol on the Approval of Research and Production of Satellite Earth Resources.[217]

In 1993, the Protocol on Cooperation in Peaceful Applications of Science and Technology of Outer Space[218] was signed between the Ministry of Science and Technology of Brazil and the China National Space Administration of China (CNSA). This protocol is about further development of the Sino-Brazilian satellites for monitoring terrestrial resources and related subjects, as well as on approval of research and satellite production of earth resources. The following year, the Framework Agreement on Cooperation in Peaceful Applications of Science and Technology for Outer Space[219] was signed. In 1995, on Technical Security Related to the Joint Development of Terrestrial Resources Satellites (the origin of the CBERS project)[220] was agreed In 1996, the two countries made a joint declaration on the peaceful applications of space science and technology.[221]

Space cooperation between Brazil and China has a horizontal character and allows the joint development of technologies and the training of human resources, meeting the strategic objectives of the Brazilian Space Program, notably regarding the expansion and consolidation of the space industry. The main program, the Sino-Brazilian Satellite Resources Program (CBERS),[222] has given rise to five satellites. The first, CBERS-1, was released on 14 October 1999.[223]

In the first decade of the 2000s, the two countries signed a Protocol on Space Technology Cooperation,[224] an MoU on Science and Technology Cooperation (2001),[225] and a Supplementary Protocol to the Framework Agreement on Co-operation in

---

[217]Protocol on the Approval of Research and Production of Earth Resources Satellite, between Brazil and China, 6 July 1988, http://portal-antigo.aeb.gov.br/wp-content/uploads/2012/09/AcordoChina1988.pdf (accessed 22 April 2019).

[218]Protocol between the Ministry of Science and Technology of Brazil and CNSA on Cooperation in Peaceful Applications of Science and Technology in Outer Space, 23 November 1993, http://www.aeb.gov.br/wp-content/uploads/2018/01/AcordoChina1993_a.pdf (accessed 22 April 2019).

[219]Framework Agreement on Cooperation in Peaceful Applications of Science and Technology in Outer Space between Brazil and China, 8 November 1994, http://portal-antigo.aeb.gov.br/wp-content/uploads/2012/09/AcordoChina1994.pdf (accessed 22 April 2019).

[220]Agreement between Brazil and China on Technical Security Related to the Joint Development of Terrestrial Resources Satellites—CBERS PROJECT, 13 December 1995, http://www.aeb.gov.br/wp-content/uploads/2018/01/AcordoChina1995.pdf (accessed 22 April 2019).

[221]AEB, China and Brazil jointly declared relation to the peaceful applications of space science and technology, http://www.aeb.gov.br/programa-espacial-brasileiro/cooperacao-internacional/china-2/ (accessed 22 April 2019).

[222]Soares, J., XVII SBSR—Simpósio Brasileiro de Sensoriamento Remoto, João Pessoa, PB, Brazil, 2015, https://www.b2match.eu/system/spaceweek-italy/files/Brazil_Soares.pdf?1445858501 (accessed 22 April 2019).

[223]INPE, 10 anos do lançamento do CBERS-1, 2009, http://www.cbers.inpe.br/noticias/noticia.php?Cod_Noticia=1978 (accessed 22 April 2019).

[224]Protocol on Space Technology Cooperation between Brazil and China, 21 September 2000, http://www.aeb.gov.br/wp-content/uploads/2018/01/AcordoChina2000.pdf (accessed 22 April 2019).

[225]AEB, MoU on Science and Technology Cooperation with China, 2001, http://www.aeb.gov.br/programa-espacial-brasileiro/cooperacao-internacional/china-2/ (accessed 22 April 2019).

Peaceful Applications of Space Science and Technology development of terrestrial satellites (2002).[226] In 2003, the CBERS-2 satellite was launched[227] with the same characteristics as CBERS-1. A further two MoUs were signed on the introduction of the ground system plan and the establishment of the intergovernmental coordination mechanism for space technology collaboration[228] on CBERS-3 and CBERS-4. In 2004, the Framework Agreement was signed together with the Protocol on Cooperation in Peaceful Applications of Science and Technology for Outer Space for Cooperation in the CBERS Application System and on Cooperation in Peaceful Applications of Science and Technology in Outer Space for Joint Development of Satellite CBERS 2-B.[229] Other MoUs[230] were also signed that year, aiming at Cooperation for the Development of an Application System for the Sino-Brazilian Satellite Program and on the establishment of the Sino-Brazilian High-Level Concertation and Cooperation Commission.

On 19 September 2007, the CBERS-2B satellite was launched[231] with CBERS-1 and CBERS-2 spare parts, with the IRS camera being replaced by a high-resolution (HR) camera. In 2009, two MoU were signed between departments of these countries, aiming at the reception and distribution of CBERS-3 data on Cooperation in EO and Digital Geoinformation Fields and the Protocol between AEB and China on Cooperation for the Continuity, Expansion and Applications of the CBERS Program.[232]

In 2010, another series of MoU were signed: to regulate Brazil-China cooperation around the International Space Climate Program of the Meridian Circle (ISWMCP); to define the CBERS Image Distribution Policy; on Direct Reception and Data Distribution CBERS-03 (together with the South African National Space Agency); to

---

[226] Supplementary Protocol to the Framework Agreement between Brazil and China on Cooperation in the Peaceful Application of Science and Technology in Outer Space for Continuing the Joint Development of Earth Resource Satellites, 27 November 2002, http://www.aeb.gov.br/wp-content/uploads/2018/01/AcordoChina2002.pdf (accessed 22 April 2019).

[227] World Meteorological Organization OSCAR, Satellite CBERS-2, https://www.wmo-sat.info/oscar/satellites/view/26 (accessed 22 April 2019).

[228] AEB, MoU on the introduction of the ground system plan and the establishment of the intergovernmental coordination mechanism for space technology collaboration with China, http://www.aeb.gov.br/programa-espacial-brasileiro/cooperacao-internacional/china-2/ (accessed 22 April 2019).

[229] AEB, Framework Agreement on Cooperation in Peaceful Applications of Science and Technology for Outer Space and Cooperation Agreement in Peaceful Applications of Science and Technology in Outer Space for Joint Development of Satellite CBERS 2-B, http://www.aeb.gov.br/programa-espacial-brasileiro/cooperacao-internacional/china-2/ (accessed 22 April 2019).

[230] Memorandum of Understanding between Brazil China on Cooperation for the Development of an Application System for the Sino-Brazilian Satellite Program, 24 May 2004, http://www.aeb.gov.br/wp-content/uploads/2018/01/AcordoChina2004_c.pdf (accessed 22 April 2019).

[231] World Meteorological Organization OSCAR, Satellite CBERS-2B, https://www.wmo-sat.info/oscar/satellites/view/27 (accessed 22 April 2019).

[232] AEB, Cooperation for the Continuity, Expansion and Applications of the CBERS Program, http://www.aeb.gov.br/programa-espacial-brasileiro/cooperacao-internacional/china-2/ (accessed 22 April 2019).

define the CBERS Image Distribution Data Policy; and also regarding Cooperation in the Area of Remote Sensing.[233]

In June 2012, the countries celebrated the 2012–2021 Sino-Brazilian Space Cooperation Plan,[234] which lists priority areas and programs in the framework of bilateral cooperation. Brazil and China also have a partnership in the field of education, which provides for the exchange for the training of Brazilian students and researchers in master's and PhD programs at Beihang University, in Beijing. Among the joint partnerships, support for the International Meridian Project, through the Sino-Brazilian Space Weather Laboratory, located at INPE is a highlight. The laboratory is responsible for jointly observing competing data from the meridian chain in the Western Hemisphere.

On 9 December 2013, the CBERS-3 satellite was launched[235] as an evolution of CBERS-1 and CBERS-2. Due to the failure of the launch vehicle, the satellite did not reach the target orbit.[236] The following year a new MoU was signed between the two space agencies on Remote Sensing Satellite Data Cooperation and its Applications.[237] Also, in 2014, the CBERS-4 satellite was launched[238] on 7 December, with the same characteristics as CBERS-3.

In 2015, they signed a Supplementary Protocol for the Joint Development of CBERS-4A on Cooperation in Peaceful Applications of Science and Technology in Outer Space,[239] as well as a Plan of Joint Action between the Government of the Federative Republic of Brazil and the Government of the People's Republic of China 2015–2021.[240]

In August 2016, the Brazilian National Congress approved the continuity of the program through the Supplementary Protocol for the Joint Development of the

---

[233] Ibid.

[234] Decennial Cooperation Plan between Brazil and China, 21 June 2012, http://www.aeb.gov.br/wp-content/uploads/2018/01/AcordoChina2012.pdf (accessed 22 April 2019).

[235] World Meteorological Organization OSCAR, Satellite CBERS-3, https://www.wmo-sat.info/oscar/satellites/view/28 (accessed 22 April 2019).

[236] NASA Space Flight, Brazil's CBERS-3 spacecraft lost following Chinese failure, 2013, https://www.nasaspaceflight.com/2013/12/chinese-long-march-4b-cbers-3/ (accessed 22 April 2019).

[237] Memorandum of Understanding between AEB and CNSA on Remote Sensing Satellite Data Cooperation and its Applications, June 2014, http://www.aeb.gov.br/wp-content/uploads/2018/01/AcordoChina2014.pdf (accessed 22 April 2019).

[238] World Meteorological Organization OSCAR, Satellite CBERS-4, https://www.wmo-sat.info/oscar/satellites/view/29 (accessed 22 April 2019).

[239] Supplementary Protocol for the Joint Development of CBERS-4A between the Brazilian Government and the Government of China on the "Framework Agreement between the Government of the Federative Republic of Brazil and the Government of the People's Republic of China on Cooperation in Peaceful Outer space", 19 May 2015, http://www.aeb.gov.br/wp-content/uploads/2018/01/AcordoChina2015_b.pdf (accessed 22 April 2019).

[240] Joint Action Plan between Brazil and China 2015–2021, http://www.aeb.gov.br/wp-content/uploads/2018/01/AcordoChina2015_a.pdf (accessed 22 April 2019).

CBERS-4A Satellite.[241] On 20 December 2019 the satellite was successfully launch from Long March 4B rocket.[242]

8.6.4.5 Japan

In September 1970, Brazil and Japan signed the Technical Cooperation Agreement, called the Basic Agreement.[243] As part of the activities foreseen in this instrument, a contract was signed between AEB and Japan Manned Space Corporation (JAMSS), linked to the Japan Aerospace Exploration Agency (JAXA), and two CubeSats (AESP-14 and SERPENS) were launched from the Kibo module of the ISS.[244] For Brazil, cooperation in the area of small satellites is an important contribution to the technical and scientific training of undergraduate and postgraduate students.

8.6.4.6 India

In 2002, AEB and the Indian Space Research Organization (ISRO) signed an MoU on outer space cooperation.[245] Two years later, in January 2004, the two countries signed the Framework Agreement on Cooperation in Peaceful Uses of Outer Space.[246] In 2007, the Complementary Adjustment for the Expansion of the Brazilian Remote Sensing Satellite Data Reception and Data Processing Station was signed with the purpose of stimulating commercial and industrial cooperation between the private

---

[241] Supplementary Protocol for the Joint Development of the CBERS-4A Satellite between the Brazilian Government and the Government of China on Cooperation in Peaceful Applications of Science and Technology in Outer Space, http://www.aeb.gov.br/wp-content/uploads/2018/01/AcordoChina2016.pdf (accessed 22 April 2019).

[242] NASA Space Flight, Long March 4B lofts CBERS-04A and Ethiopia's first satellite, https://www.nasaspaceflight.com/2019/12/long-march-3b-cbers-04a-ethiopias-first-satellite/ (accessed 25 February 2020).

[243] Technical Cooperation Agreement between Brazil and Japan, 22 September 1970, http://www.aeb.gov.br/wp-content/uploads/2018/01/AcordoJap%C3%A3o_1970.pdf (accessed 22 April 2019).

[244] Wakata, K., 25th UN/IAF Workshop on Space Technology for Socio-Economic Benefits "Integrated Space Technologies and Applications for a Better Society", 23 to 25 September 2016, https://www.aem.gob.mx/files/2016_UN_IAF_Workshop/25sep/1215%20UN-IAF%20WS%202016_KiboCUBE_set.pdf (accessed 22 April 2019).

[245] AEB, MoU on outer space cooperation with ISRO, http://www.aeb.gov.br/programa-espacial-brasileiro/cooperacao-internacional/india/ (accessed 22 April 2019).

[246] Framework Agreement between Brazil and India on Cooperation in Peaceful Uses of Outer Space, 25 January 2004, http://www.aeb.gov.br/wp-content/uploads/2018/01/AcordoIndia2004.pdf (accessed 22 April 2019).

sectors of both countries in the space field.[247] The Complementary Adjustment provides for the receipt of data from the RESOURCESAT-1 AwiFS and LISS-III payloads on its passage over the Cuiabá station, in addition to defining the terms and conditions for the expansion of the Brazilian terrestrial station to receive and process data from the satellites. In 2008, AEB, INPE and ISRO signed the Cooperative Program[248] for the direct reception and distribution of RESOURCESAT-1 data. Another Complementary Agreement[249] was signed in 2014, regarding the expansion of the Brazilian terrestrial station for receiving and processing data from Indian Remote Sensing Satellites (SRI), including RESOURCESAT-2.[250]

8.6.4.7 BRICS

The BRICS meetings on space cooperation began in 2015 with the signing of an MoU on Cooperation in Science, Technology and Innovation among participating states.[251] Two annual meetings were held in 2016 and 2017, respectively, to discuss the Coordination of Space Cooperation and Working Groups.[252]

The objectives of the agreement are: (a) To establish a strategic model for cooperation in science, technology and innovation among BRICS member countries; (b) Address regional and global socio-economic challenges in BRICS member countries through shared experiences and complementarities in science, technology and innovation; (c) Generate, in partnership, new knowledge and innovative products, services and processes in BRICS member countries using appropriate funds and investment instruments; (d) Promote, where appropriate, BRICS joint science, technology and innovation partnerships with other strategic actors in the developing world.

The areas of cooperation are: (a) Exchange of information on policies and programs and promotion of transfer of innovation and technology; (b) Food security

---

[247]Complementary Agreement between Brazil and India on Cooperation for the Expansion of the Terrestrial Data Reception and Data Processing Unit of the Remote Sensing Satellites of India, 4 July 2007, http://www.aeb.gov.br/wp-content/uploads/2018/01/Acordo%C3%8Dndia2007.pdf (accessed 22 April 2019).

[248]AEB, Cooperative Program between AEB, INPE and ISRO for the direct reception and distribution of RESOURCESAT-1 data, http://www.aeb.gov.br/programa-espacial-brasileiro/cooperacao-internacional/india/ (accessed 22 April 2019).

[249]Complementary Agreement between Brazil and India establishing Cooperation in the expansion of a Brazilian Terrestrial Station for Receiving and Processing of Data from Indian Remote Sensing Satellites, 16 July 2014, http://www.aeb.gov.br/wp-content/uploads/2018/01/AcordoIndia2014_a.pdf (accessed 22 April 2019).

[250]Cooperation Program between the Indian Space Research Organization and AEB for the Direct Reception and Distribution of RESOURCESAT-2, 16 July 2014, http://www.aeb.gov.br/wp-content/uploads/2018/01/AcordoIndia2014_b.pdf (accessed 22 April 2019).

[251]BRICS, Memorandum of Understanding on Cooperation in Science, Technology and Innovation between Brazil, Russian Federation, India, China and South Africa, 2015, http://brics.utoronto.ca/docs/BRICS%20STI%20MoU%20ENGLISH.pdf (accessed 22 April 2019).

[252]AEB, BRICS, Coordination of Space Cooperation and Working Groups, http://www.aeb.gov.br/programa-espacial-brasileiro/brics/ (accessed 22 April 2019).

and sustainable agriculture; (c) Natural disasters; (d) New and renewable energy, energy efficiency; (e) Nanotechnology; (f) High performance computing; (g) Basic research; (h) Space, aeronautics, astronomy and EO; (i) Medicine and biotechnology; (j) Biomedicine and life sciences (biomedical engineering, bioinformatics, biomaterials); (k) Water resources and pollution treatment; (l) High technology centres/technology parks and incubators; (m) Technology transfer; (n) Popularization of science; (o) Information and communication technologies; (p) Clean coal technologies; (q) Natural gas and unconventional gases; (r) Oceanic and polar sciences; (s) Geospatial technologies and their applications.[253]

The modalities of cooperation are: (a) Short-term exchange of scientists, researchers, technical specialists and students; (b) Training programs dedicated to supporting the development of human capital in science, technology and innovation; (c) Organization of science, technology and innovation workshops, seminars and conferences in areas of mutual interest; (d) Exchange of information on science, technology and innovation; (e) Formulation and implementation of collaborative research and development projects and programs; (f) Establishment of joint funding mechanisms to support BRICS research programs and large-scale research infrastructure projects; (g) Facilitating access to science and technology infrastructure among BRICS member countries; (h) Announcement of simultaneous calls in BRICS member countries; (i) Cooperation between the academies of science and engineering and national research agencies.[254]

There are two phases proposed for the BRICS Remote Sensing Satellite Constellation,[255] namely, Phase 1 comprising a virtual constellation of existing satellites, and Phase 2 to comprise a new satellite constellation to be discussed soon. The intention of the virtual constellation is to establish the remote sensing data sharing mechanism for, space solutions that meet the challenges faced by humanity, such as global climate change, major natural and technological disasters and environmental protection. The next steps will be to establish the legal framework, sign the Agreement and setup the mechanism.

8.7 Satellite Capacities of Brazil

Table 8.5 shows a list of spatial objects associated with Brazil, launched in partnership with other countries or under private initiative. According to Ordinance no. 96

---

[253]Ibid.

[254]Ibid.

[255]Hui, J., Progress of BRICS Remote Sensing Satellite Constellation. UN/UAE—High Level Forum: Space as a Driver for Socio-Economic Sustainable Development, Dubai, UAE, 6 to 9 November 2017, http://www.unoosa.org/documents/pdf/hlf/HLF2017/presentations/Day2/Session_7a/1._Progress_of_BRICS_Remote_Sensing_Satellite_Constellation-dubai.pdf (accessed 22 April 2019).

Table 8.5 Index of Brazilian Satellites launched into Outer Space[d]

Name	Launch date	Function	PE[a]	UNI[b]	PS[c]	Launch site	Launch vehicle	Launch provider	UN registered
BRASILSAT-A1	08-02-1985	Telecommunications service	X			Ko ELA-1	ARIANE-3	ARIANESPACE	Yes
BRASILSAT-A2	28-03-1986	Typical for geostationary satellites, Brasilsat-A2 was re-orbited on 6 March 2004 from a longitude of 63 degrees West and an inclination of 6.0 degrees. The Satellite is presently in a disposal orbit, at a perigee of 200 km above the geostationary orbit	X			Ko ELA-2	ARIANE-3	ARIANESPACE	Yes
DOVE (DO-17)	22-01-1990	Private amateur radio project			X	Ko ELA-2	ARIANE-40 H10	ARIANESPACE	No

(continued)

Table 8.5 (continued)

Name	Launch date	Function	PE$_a$	UNI$_b$	PS$_c$	Launch site	Launch vehicle	Launch provider	UN registered
SCD 1	09-02-1993	Brazilian Satellite designed for the collection of meteorological data relayed by data collection platforms spread throughout Brazilian territory	X			CC, B-52, RW15/33, USA	Pegasus	Orbital Sciences Corporation (Orbital ATK)	Yes
BRASILSAT-B1	10-08-1994	Telecommunications service	X			Ko ELA-2, French Guiana	ARIANE-44LP H10+	ARIANESPACE	Yes
BRASILSAT-B2	28-03-1995	Telecommunications service	X			Ko ELA-2, French Guiana	ARIANE-44LP H10+	ARIANESPACE	Yes
BRASILSAT-B3	04-02-1998	Telecommunications service	X			Ko ELA-2, French Guiana	ARIANE-44LP H10-3	ARIANESPACE	Yes
SCD 2	23-10-1998	Collecting data and transmitting to meteorological stations in the territory	X			CC, L-1011, RW13/31, USA	Pegasus-H	Orbital ATK	Yes

(continued)

Table 8.5 (continued)

Name	Launch date	Function	PE_a	UNI_b	PS_c	Launch site	Launch vehicle	Launch provider	UN registered
CBERS-1	10-10-1999	CBERS-1 is used for monitoring, through optic sensors, of terrestrial resources as well as for the promotion of the development and the use of remote sensing techniques in Brazil and in China	X			TY LC-7	CZ-4B	SAST (Shanghai Academy of Spaceflight Technology)	Yes
SACI 1	14-10-1999	SACI-1 is used to give support to the following scientific experiments: study of plasma blisters, photometer for study of aeroluminescence, observations of abnormal and solar cosmic rays in the magnetosphere, and geomagnetic experiments	X			TY LC-7	CZ-4B	SAST (Shanghai Academy of Spaceflight Technology)	Yes
BRASILSAT-B4	17-08-2000	Telecommunications Satellite	X			Ko ELA-2, French Guiana	ARIANE-44LP H10-3	ARIANESPACE	Yes

(continued)

Table 8.5 (continued)

Name	Launch date	Function	PE_a	UNI_b	PS_c	Launch site	Launch vehicle	Launch provider	UN registered
ESTRELA DO SUL	10-01-2004	Telecommunications Satellite			X	SL	Zenit-3SL (1)	SEA LAUNCH	Yes
AMAZONAS 1	05-08-2004	Telecommunications Satellite			X	Ba LC-200/39, Kazakhstan	Proton-M Briz-M (Ph.1)	International Launch Services	Yes
CBERS-2B	19-09-2007	Monitoring of terrestrial resources by optical sensors. Development of remote sensing techniques between Brazil and China	X			TY LC-7	CZ-4B	SAST (Shanghai Academy of Spaceflight Technology)	
STAR ONE C1	14-11-2007	Telecommunications Satellite			X	Ko ELA-3, French Guiana	ARIANE-5ECA	ARIANESPACE	Yes
STAR ONE C2	18-04-2008	Telecommunications Satellite			X	Ko ELA-3, French Guiana	ARIANE-5ECA	ARIANESPACE	Yes
AMAZONAS 2	01-10-2009	Telecommunications Satellite			X	Ko ELA-3, French Guiana	ARIANE-5ECA	ARIANESPACE	Yes
ESTRELA DO SUL 2	20-05-2011	Telecommunications Satellite			X				Yes
STAR ONE C3	10-11-2012	Telecommunications Satellite			X	Ko ELA-3, French Guiana	ARIANE-5ECA	ARIANESPACE	Yes
AMAZONAS 3	07-02-2013	Telecommunications Satellite			X	Ko ELA-3, French Guiana	ARIANE-5ECA	ARIANESPACE	Yes

(continued)

Table 8.5 (continued)

Name	Launch date	Function	PE_a	UNI_b	PS_c	Launch site	Launch vehicle	Launch provider	UN registered
NANOSATC-BR1	19-06-2014	First CubeSat project of Brazil; provides monitoring of Earth's magnetosphere by measuring the magnetic field over Brazil and to study the magnetic phenomena of the SAA (South Atlantic Anomaly) and the EEJ (Equatorial Electrojet). Note: This SAA is also referred to as SAMA (South Atlantic Magnetic Anomaly)	X	X		Do LC-370/13	Dnepr	ISC Kosmotras	Yes
CBERS-4	07-12-2014	Monitoring of terrestrial resources by optical sensors. Development of remote sensing techniques between Brazil and China	X			TY LC-9	CZ-4B	CASC (China Aerospace Science and Technology Corporation)	

(continued)

Table 8.5 (continued)

Name	Launch date	Function	PE$_a$	UNI$_b$	PS$_c$	Launch site	Launch vehicle	Launch provider	UN registered
AESP 14	05-02-2015	Investigate the generation mechanism of the equatorial plasma bubbles (also known as equatorial spread F) through the measurement of their occurrence and distribution characteristics on a global scale and as a function of local time, season and the ambient ionospheric conditions. Special attention will be given to the Brazilian longitude sector where their occurrence is known to be more frequent and intense as compared to other longitudes	X	X		CC SLC-40, USA	Falcon-9 v1.1	SPACE X	Yes
STAR ONE C4	15-07-2015	Telecommunications Satellite			X	Ko ELA-3, French Guiana	ARIANE-5ECA	Arianespace	Yes

(continued)

Table 8.5 (continued)

Name	Launch date	Function	PE$_a$	UNI$_b$	PS$_c$	Launch site	Launch vehicle	Launch provider	UN registered
SERPENS	17-09-2015	Consolidate new national aerospace courses; promote capacity-building; involve students in a hands-on project; serve as a laboratory for future missions		X		Ta YLP-2 (Tanegashima Space Centre, JAPAN)	H-2B-304	MITSUBISHI HEAVY INDUSTRIES	Yes
STAR ONE D1	21-12-2016	Telecommunications Satellite			X	Ko ELA-3, French Guiana	ARIANE-5ECA	ARIANESPACE	Yes
TANCREDO 1	20-01-2017	Integration, testing, assembly, coding, launching and operating the satellite, encouraging students to pursue careers in the science and technology areas, especially in the Space Engineering fields		X		Ta YLP-2 (Tanegashima Space Centre, JAPAN)	H-2B-304	MITSUBISHI HEAVY INDUSTRIES	Yes
INTELSAT 32E	14-02-2017	Telecommunication Satellite			X	Ko ELA-3, French Guiana	ARIANE-5ECA	ARIANESPACE	Yes
SGDC	04-05-2017	Telecommunication Satellite	X			Ko ELA-3, French Guiana	ARIANE-5ECA	ARIANESPACE	Yes

(continued)

Table 8.5 (continued)

Name	Launch date	Function	PE[a]	UNI[b]	PS[c]	Launch site	Launch vehicle	Launch provider	UN registered
SES-14	25-01-2018	Telecommunication Satellite	Equipments occupy Brazilian orbital position but are not Brazilian Satellites.			Ko ELA-3, French Guiana	ARIANE-5ECA	ARIANESPACE	Yes
AI YAH 3	25-01-2018	Telecommunication and broadband Internet services				Ko ELA-3, French Guiana	ARIANE-5ECA	ARIANESPACE	Yes
ITASAT	03-12-2018	Develop an engineering model that remains in soil to perform tests and simulations of the training of station operators		X		Vandenberg AFB Space Launch Complex 4; USA	Falcon-9 v1.2	SPACE X	Yes
CBERS 4A	20-12-2019	Monitoring of terrestrial resources by optical sensors. development of remote sensing techniques between Brazil and China	X			TY LC-9	CZ-4B	CASC (China Aerospace Science and Technology Corporation)	Yes
FLORIPASAT	20-12-2019	Technological development		X		TY LC-9	CZ-4B	CASC (China Aerospace Science and Technology Corporation)	Yes

[a]PE: Launched Satellites funded by Public Entities
[b]UNI: Launched Satellites funded by Universities/Students
[c]PS: Launched Satellites funded by Private Sector
[d]UNOOSA, Online Index of Objects Launched into Outer Space, http://www.unoosa.org/oosa/osoindex/search-ng.jspx?lf_id=   (accessed 25 February 2020)

of 30 November 2011,[256] AEB oversees the "implementation and operation of the National Register of Spatial Objects". Such registration is updated each time Brazil is the "launching state" as defined in the Convention on the Registration of Objects Launched into Outer Space. Internally, the record keeping is under the responsibility of the Satellite, Applications and Development Board[257] and the International Cooperation Office,[258] both being bodies of AEB.

8.8 Launch Capability and Sites

Brazil has two launch centres located in the northeast region. The first—Barreira do Inferno Launch Centre—was established in 1965 and designed to execute and support activities of launching and tracing aerospace devices, as well as collecting and processing data from their payloads, performing tests, experiments, basic or applied research and other technological development activities of aeronautics interest, related to the Aeronautics Policy for PNDAE. The second and most important—the Alcântara Launch Center (CLA)—located at latitude 2° 18′ South (below the equator), was created in 1983 and has the mission to carry out the launching and tracking activities of aerospace engagements and the collection and processing of data of its payloads, as well as the execution of tests and experiments of interest of the Aeronautical Command, related to the PNDAE.

Since the first launch in 1989 at CLA, when 15 SBAT-70 rockets and two SBAT-152 rockets were launched, numerous operations have been carried out. In March 2018, CLA carried out its 100th operation, reaching a total of 479 launched rockets (operations are the set of planned activities whose objective is the launching of space vehicles for scientific experiments or for the placement of satellites in orbit).[259] In addition, the CLA benefits from a privileged location for the placement of satellites in equatorial orbits, because of the fuel economy for the launch of the rocket and the increase of capacity for satellization. These factors guarantee the CLA attributes such as security, economy and availability, giving it important competitive advantages that can make it one of the best space centres in the world. Also, the region has a low population density and favourable climatic conditions with a well-defined rainfall regime and small temperature variations.

There is a current project of expansion of the CLA based on the use of 12,000 ha, besides the 8,700 ha of today. After the expansion, it would be possible to meet the demands of launching vehicles of medium orbit, which would expand the market

---

[256] AEB, Ordinance no. 96, http://www.aeb.gov.br/wp-content/uploads/2018/04/DOU-Registro-de-Objetos-Espaciais-Portaria-96-111130-1.pdf (accessed 22 September 2019).

[257] Diretoria de Satélites, Aplicações e Desenvolvimento, AEB.

[258] Assessoria de Cooperação Internacional.

[259] IPEA, O Centro De Lançamento De Alcântara: abertura para o mercado internacional de satélites e salvaguardas para a soberania nacional, 2018, http://repositorio.ipea.gov.br/bitstream/11058/8897/1/td_2423.pdf (accessed 22 April 2019).

to be served. With the development of new platforms and launch sites, CLA could serve up to six countries, in addition to enabling further development of proprietary technologies that could be used both for the construction of national satellites and for future versions of launch vehicles.

Despite having two launch centres, Brazil has a deficit in relation to satellite launch technology. One of the goals of the MECB concerns the development of the VLS. During the development of the technology, three launch attempts failed. The first one in 1997, with VLS-1 V01, did not achieve complete success after one of the powerplants failed at the time of take-off. The second attempt to launch the VLS occurred in 1999, when the VLS-1 V02 was successfully launched, but it failed in its pyrotechnic system, causing its explosion. On the third and final attempt, in 2003, a major accident occurred days before the scheduled launch date. There was an accidental ignition of the first stage of the rocket, causing a great fire in the integration platform and the death of 21 Brazilian technicians, as well as the destruction of the rocket, the two satellites on board—SATEC and UNOSAT—and structures of the launch pad.

One year after the accident, in 2004, Brazil and Ukraine signed a Treaty on Long-term Cooperation in the use of the CYCLONE-4 Launch Vehicle, which created the binational company Alcântara Cyclone Space (ACS), to build and operate safe and competitive commercial launches involving the use of the CYCLONE-4 rocket from the Alcântara Centre. In 2010, ACS began construction of its space complex within the CLA area. With this project, Brazil would have had the opportunity to launch its own medium-sized satellites for different purposes. In 2015, after almost twelve years of delays, the government cancelled the bilateral agreement for the launching of Ukrainian rockets with commercial satellites from Alcântara. The two governments spent approximately R$1 billion (USD 250 millions). The claimed reason was the high cost of the CYCLONE-4 satellite launcher, in a period of fiscal contraction. Ukraine has not fulfilled its agreed financial contributions and has even requested funding from the Brazilian Development Bank (BNDES) of Brazil to do so.

In May 2017, the FAB Commander stated in a public hearing in the Congress that there is Brazilian interest in the commercialization of CLA, emphasizing that "safeguard agreements are indispensable", since "no country will place its equipment's in Brazil without having the guarantee of safeguarding this equipment". The approval and implementation of the CLA expansion project has faced difficulties due to land obstacles. Part of the area foreseen in the original project is challenged, and is under judicial consideration, due to conflicts of interests with traditional and ethnic communities (it is noteworthy that of the 3,500 native Quilombola communities in Brazil, about 155 are in the region of Alcântara).

Aiming at economic benefits and, above all, the development of proprietary technologies, the opening of CLA to other countries' launches is a viable option that could benefit PEB if it occurs from balanced negotiations and offers benefits to counterparts, to the positioning strategy and to national interests. The PNAE does not consist of a government policy, but rather a long-term state project, as well as a proper and stable allocation of resources for the sector.

8.9 Space Technology Development

Over the decades, Brazil has sought to develop its own technology to launch a satellite. Several attempts were made to construct the VLS, which was the top priority. Over the last years, the country has abandoned its goal with a view to building a microsatellite launch vehicle in partnership with DLR. In tandem with this target, the Brazilian strategy is to expand partnerships with other countries mainly to foster qualification of needed specialists and stimulate the private sector to become involved with space activities.

8.9.1 Primary Space Needs of Brazil

Considering that Brazil is still at an initial stage in the development of a space industry, Brazilian companies that are dedicated to the sector are basically concentrated around the manufacture of subsystems and components of satellites, launchers and supply of goods and services for the infrastructure of launch and ground services. The country is capable of handling the basic technology of satellite construction, although the satellite launch sector is more critical. The research and construction of satellite launch vehicles focuses on the Institute of Aeronautics and Space (IAE), which develops technology for launching several smaller spacecrafts, but not satellite launchers. However, there is no Brazilian company that dominates the entire process of building satellites, not even launching vehicles.

The activities of private companies are restricted to the supply of parts, components and subsystems commissioned by INPE and the IAE, since the structure of the National System of Activities (SINDAE) defines that these institutions are responsible for projects, assembly, integration of systems and tests of satellites and launch vehicles, respectively.

8.9.2 Priority List

The priorities established for the space sector are detailed in the Ministry of Defence's National Defence Strategy (END),[260] which lists them as: "Designing and manufacturing satellite launch vehicles and developing remote guidance technologies, especially inertial systems and liquid propulsion technologies; Design and manufacture satellites, especially geostationary, for telecommunications and those for high resolution, multispectral remote sensing, and to develop satellite attitude control technologies; Develop communication, command and control technologies from satellites, with ground, air and sea forces, including submarine forces, to enable them

---

[260]Brazil, Estratégia Nacional de Defesa/Ministério da Defesa, https://www.defesa.gov.br/arquiv os/2012/mes07/end.pdf (accessed 22 April 2019).

to operate in a network and to be guided by information received from them; Develop technology for determining geographic coordinates from satellites".[261]

In addition to the END, the domain of the space cycle is included in the "Plan Brazil 2022"[262]—a long-term planning document, under various headings: "to increase the supply of meteorological information (Agriculture); master the technologies of manufacturing satellites and launch vehicles (Science and Technology); eradicate illiteracy, universalize school attendance from 4 to 17 years and reach the mark of 10 million university students (Education); promote the digital and technological inclusion of youth (Youth); ensure full access to broadband at a speed of 100 megabits to all Brazilians and have in orbit two Brazilian geostationary satellites (Communications); and ensure full monitoring of land borders and jurisdictional waters and launch the first satellite launch vehicle built in Brazil".[263]

8.9.3 Space Applications for Social Benefits

The current PNAE 2012–2021[264] clearly stated its peaceful character and focused on environmental control, aiming at the use of space applications for social benefits, maintaining the strategic importance of the investment in aerospace technology to obtain autonomy in space exploration, observation, monitoring, communications, navigation, air traffic, health and education. The strategic guidelines presented in the PNAE[265] are as Table 8.6.

Although the PNAE 2012–2021 schedules some launch vehicles for the next few years it still does not have the minimum autonomy to monitor Brazil's territory. The developed satellites are useful only for the most basic observations. The technical-political autonomy so desired has not yet been achieved.

---

[261] Ibid., p. 79.

[262] Brazil, Brasil 2022: Trabalhos Preparatórios/SEA, 2010, http://www.mda.gov.br/sitemda/sites/sitemda/files/user_arquivos_64/Brasil%202022%20-%20Trabalhos%20Preparat.%C3%B3rios.pdf (accessed 22 April 2019).

[263] Ibid.

[264] Brazil. PNAE 2012–2021, http://www.aeb.gov.br/wp-content/uploads/2013/01/PNAE-Portugues.pdf (accessed 24 April 2019).

[265] Ibid., p. 8.

Table 8.6 PNAE 2012–2021[a]

	PNAE Strategic Guidelines
1	To consolidate the Brazilian space industry, increasing its competitiveness and increasing its capacity for innovation, including using state purchasing power and partnerships with other countries
2	Develop an intense program of critical technologies, encouraging training in the sector, with greater participation of academia, government S&T institutions and industry
3	Expand partnerships with other countries, prioritizing the joint development of technological and industrial projects of mutual interest
4	Stimulate the financing of programs based on public and/or private partnerships
5	Promote greater integration of the system of governance of space activities in the country, by increasing the synergy and effectiveness of actions among its main actors and the creation of a National Council on Space Policy, conducted directly by the Presidency of the Republic
6	Improve legislation to streamline space activities, favouring and facilitating government procurement, increasing resources for the Space Sector Fund, and the release of industry
7	Foster the training and qualification of specialists needed by the Brazilian space sector, both in Brazil and abroad
8	Promote public awareness of the relevance of the study, use and development of the Brazilian space industry

[a]Ibid.

8.9.4 National Space Programs: Accomplishments

Despite the aspirations, some significant achievements must be highlighted. The launch of NASA's ERSTS-I satellite, operated between 1972 and 1978, motivated the country's first attempts at capacity-building in the promising remote sensing area. A Brazilian team was trained in the USA to decode the images of this satellite, whose applications were distributed in many areas of knowledge, such as agriculture, geology, geomorphology, geodesics, vegetation studies and oceanography among others.[266] Subsequently, this team participated in the "Radam Brasil project",[267] which mapped the country's natural resources through an airborne mission equipped with a radar system that transmitted the images through the satellite LANDSAT I.

Currently, there are countless projects related to the monitoring of the country's natural resources, developed with the use of space technology, such as the PRODES project (monitoring and quantifying Amazon deforestation) and Projeto Deter, a project of INPE/MCTIC with the support of Ministry of the Environment and IBAMA—Brazilian Institute of Environment, which is also part of the Federal Government's Plan to Combat Deforestation in the Amazon Region, PROARCO

---

[266]KRUG, T., Tecnologias espaciais como suporte à gestão dos recursos naturais, http://seer.cgee. org.br/index.php/parcerias_estrategicas/article/view/90/83 (accessed 24 April 2019).

[267]Escobar, I. and others, XII Simpósio Brasileiro de Sensoriamento Remoto, Goiânia, Brasil, 16–21 April 2005, pp. 4395–4397, http://marte.sid.inpe.br/col/ltid.inpe.br/sbsr/2004/11.18.10.17/doc/4395.pdf (accessed 24 April 2019).

(project to control fires and forest fire prevention), and the Economic Ecological Zoning, for studies related to biomass burning.[268] Although these projects go back to past decades, they remain active and INPE and its Weather Prevision Centre and Climate Studies (CPTEC) remain the safest institutes for obtaining environmental data in the country.

8.9.5 Main Space Actors in Brazil

Historically, the Brazilian Space Program has had two major actors, located in the city of São Paulo: INPE, which is responsible for the development of Brazilian satellites and IAE, which is a subordinate of DCTA, responsible for the development of national launch vehicles. This agglomeration in the city of São Paulo is due to the opening of the Aerospace Technological Centre (CTA) in the 1940s, with its Aerospace Technological Institute (ITA) training centre in São José dos Campos, constituting strong technical expertise promoted by government institutions, teaching and research units, thus providing fertile ground for the development of the aerospace industry in the locality.

The development of the space sector and the development of scientific capabilities go hand in hand. Thus, the hypothesis that companies associated with the Brazilian Space Program will be more engaged in innovation activities, which could initially be supported by the percentage of engineers, researchers and scientific professionals, can be further confirmed by the intensity of the connections of these firms with groups of research for the National Council for Scientific and Technological Development[269] and their participation in projects of the National Fund for Scientific and Technological Development.[270]

Both as an area of scientific and technological production and as a sector of the economy, space research in Brazil is in a peculiar and perhaps dislocated situation. Currently, there is a community of scientists, researchers or engineers and technicians with diverse specializations dedicated to space activities.

---

[268]KRUG, T., Tecnologias espaciais como suporte à gestão dos recursos naturais, http://seer.cgee. org.br/index.php/parcerias_estrategicas/article/view/90/83 (accessed 24 April 2019).

[269]Brazil, Law no. 1.310, http://www.planalto.gov.br/ccivil_03/LEIS/1950-1969/L1310.htm (accessed 24 April 2019).

[270]Brazil, Law no. 11.540, http://www.planalto.gov.br/ccivil_03/_Ato2007-2010/2007/Lei/ L11540.htm (accessed 24 April 2019).

8.9.6   Academic and Other Research for Space in Brazil

Recently, ITA, through undergraduate and graduate students, designed and manufactured the AESP-14[271] nanosatellite (standard cubesat satellite with 10 cm of edge), which arrived at the ISS aboard the DRAGON capsule along with more than 250 scientific experiments and other supplies. The project received financial support from AEB and CNPq, as well as institutional support from ITA and INPE. The initiative intends to teach systems engineering to undergraduate and graduate students using a real space mission project as a case study.

Brazil is able to demonstrate a vigorous scientific community, researchers dedicated to applications in remote sensing and meteorology, and a base in space technology and engineering. Just the IAE in the R&D area has 100 laboratories, and 54 research and development projects. In the Human Resources area, 822 civilians (71%), 176 military personnel (15%), 107 employees (14%) and 57 people in general services have participated, totalling 1162 persons. The infrastructure has 2,086,443 m^2 of total area, 54,078 m^2 comprising 256 buildings, and a green area of 975,150 m^2. It is a heritage—intellectual and material—that can enable Brazil to design commercial systems.[272]

8.10   Capacity Building and Outreach

The AEB is making efforts to bring services and information to people through public policies that include programs and ventures whose efforts are aimed at promoting the autonomy of the space sector. These initiatives aim to promote the well-being of Brazilian society through services derived from space products such as communication satellites, EO, exploration of the universe, meteorology, and the development of launch vehicles.

In 2003, the "AEB Escola"[273] program was created for schools all over Brazil. The program develops activities focused on themes such as satellites and space platforms, spacecraft, astronomy and space applications. Through the integration of the school community and Brazilian actions in the space field, AEB strengthens a culture of knowledge that enables the understanding of current technological reality to stimulate students through practical activities that encourage creativity, fueling the spirit of research and maintaining a close relationship with the daily life of Brazilian science.

---

[271] ITA, First Cubesat Developed in Brazil, http://www.aer.ita.br/~aesp14/ (accessed 24 April 2019).

[272] Rocha, M., Programa Espacial Brasileiro: desafios e perspetivas, 30 de outubro de 2013 (accessed 24 April 2019).

[273] Idealized as an instrument for the dissemination of the PNAE in secondary and elementary schools. http://aebescola.aeb.gov.br/ (accessed 24 April 2019).

Within the AEB Escola program, there is the "Espaço Educação Program",[274] a collaborative network for the dissemination and continuous updating of basic knowledge in the space area. This network consists of public and private institutions, researchers, university professors, students and technicians interested in the popularization of space sciences in the school environment and in Brazilian society. The program starts from the premise that the spatial theme permeates all areas of knowledge, and its applications are present in students' daily life, which facilitates the contextualization of the knowledge to be constructed.

The Program's instruments of action are: the Space Day (students who obtained the best results in astronautics issues in the test of the Brazilian Astronomy and Astronautics Olympiad are invited to participate, together with their professors, in a week of activities focused on the theme of space sciences); Brazilian Youth Society for the Advancement of Science (SBPC) (activities that include the construction of rockets out of plastic bottles, a small exhibition on robotics, astronomy, astronautics and related science sessions in the Digital Planetarium and other workshops that demonstrate aspects of the spatial activity); Courses for Professors (a basic program with activities on astronomy, rockets, satellites and applications); and a basis of teaching materials to be used in the Education Space (books produced by AEB in partnership with the Ministry of Education and collaborators on Astronomy, Astronautics, Climate Change, Meteorology, Remote Sensing, Didactic Projects, Space Vehicles, Satellites etc.).

In 2015, the Brazilian government officialized its entry into the NASA GLOBE program, through an agreement signed between AEB and NASA.[275] As Brazil is a key player in environmental studies, and has great environmental diversity and extensive territory, its participation in the GLOBE program contributes to research and development related to climate change around the world. In June 2016, the First GLOBE Workshop was held in Brasilia, where teachers from the public and private education networks of the Federal District were trained. The program seeks to expand the number of participating schools to reach several regions of Brazil. In March 2018, the city of Natal received the first GLOBE workshop held at the Space Technological Vocational Centre (CVT-E), located at CLBI.

This CVT-E,[276] inaugurated in 2017, is an area where it is possible to gather cutting-edge technological and scientific knowledge, where the student can learn not only fundamental principles from many sciences, but mainly apply this knowledge; a space where the multidisciplinary experience is encouraged from the proposal of missions; where science and technology are associated with the exercise of fellowship, team spirit and competition; where, through the application of well-defined goals, knowledge is learned efficiently, not only as information but as a tool for the

---

[274] AEB, Espaço Educação, http://www.aeb.gov.br/espaco-educacao-e-tecnologia/espaco-educacao/ (accessed 24 April 2019).

[275] NASA signs Scientific and Education Agreements with Brazil, 30 June 2015, https://www.nasa.gov/press-release/nasa-signs-scientific-and-education-agreements-with-brazil (accessed 27 April 2019).

[276] AEB, AEB inaugurates first technological vocational centre, http://www.aeb.gov.br/espaco-educacao-e-tecnologia/centro-vocacional-tecnologico/ (accessed 27 April 2019).

rest of life. It is assumed that students should move from their common learning space into a new environment, fully integrated with the concept of mission, where they will have the opportunity to exercise collective behavior, that is, to learn by dependence on their teammates in achieving a pre-established goal.

As this goal involves the manipulation of palpable technology, and not just the visualization of tasks on a computer screen or on sheets of paper, learning is accomplished through reinforcement involving all the senses. This is the principle in which one learns by doing, covering a wide range of natural sciences focused on the simulation of a space mission.

All the programs described above were unified in the design of the Space, Education and Technology Platform (E2T),[277] with the objective of reorganizing the fundamental programs that gave rise to the actions of development of technologies and competences of the AEB Directorate of Satellites, Applications and Development by aggregating other initiatives in the form of an efficient interrelated and operating system. The E2T is foreseen in the PPA 2016–2019 (Multi-Year Plan),[278] which establishes as an initiative "Promoting scientific and technological knowledge, human capital and the domain of critical technologies to strengthen the space sector". The program runs through elementary, high school, universities, institutes and industry.[279]

Appendix

See Table 8.7.

---

[277] AEB, E2T, http://www.aeb.gov.br/espaco-educacao-e-tecnologia/e2t/ (accessed 27 April 2019).

[278] Brazil, Law no. 13.249, January 2016, http://www.planalto.gov.br/ccivil_03/_ato2015-2018/2016/lei/l13249.htm (accessed 27 April 2019).

[279] AEB, E2T, http://www.aeb.gov.br/espaco-educacao-e-tecnologia/e2t/ (accessed 29 July 2019).

Table 8.7 Brazil institutional contact information overview

Institution	Details
Brazilian Space Agency (AEB)	SPO—Setor Policial, Área 5, Quadra 3, Bloco A, Brasília/DF—Brazil, Postal Code 70610-200, Phone: +55 (61) 2033-4000 Website: http://www.aeb.gov.br/ e-mail: presidente@aeb.gov.br
Ministry of Foreign Affairs (Itamaraty)	Zona Cívico-Administrativa BL H, Brasília/DF—Brazil, Postal Code 70.170-900, Phone: +55 (61) 2030-6199 Website: http://www.itamaraty.gov.br/en/ e-mail: http://www.itamaraty.gov.br/en/ contact-us (directly from the website)
Ministry of Science, Technology, Innovations and Communications (MCTIC)	Esplanada dos Ministérios, Bloco E, Brasília/DF—Brazil, Postal Code 70.067-900, Phone: +55 (61) 2033-7500 Website: http://www.mctic.gov.br e-mail: ouvidoria@mctic.gov.br
National Institute for Space Research (INPE)	Av. dos Astronautas, 1758—Jardim da Granja, São José dos Campos/SP—Brazil, Postal Code 12.227-010, Phone: +55 (12) 3208-6000 Website: http://www.inpe.br e-mail: http://www.inpe.br/contato/ (directly from the website)
Alcântara Launch Centre (CLA)	Av. dos Libaneses, 29—Aeroporto Tirirical, São Luís/MA, Brazil, Postal Code 65.055-040, Phone: +55 (98) 3311-9202 Website: http://www.cla.aer.mil.br/ e-mail: dir@cla.aer.mil.br
Barreira do Inferno Launch Centre (CLBI)	Rodovia RN 063—Km 11—Caixa Postal 54, Parnamirim/RN, Brazil, Postal Code 59.140-970, Phone: +55 (84) 3216 1200 Website: http://www.clbi.cta.br e-mail: comunicaosocial@clbi.cta.br
Departamento de Ciência e Tecnologia Aeroespacial (DCTA)	Av. Brig. Faria Lima, 1941—Jardim da Granja, São José dos Campos/SP, Brazil, Postal Code 59.140-970, Phone: +55 (12) 3947 6600 Website: http://www.cta.intraer/ e-mail: protocolo.dcta@fab.mil.br

Chapter 9
Chile

Contents

Abstract This chapter presents a general analysis of Chile's space activities. As with other countries of the region, Chile started its space activities at the beginning of the space race and it has recently adopted a national space program to be implemented in the coming years. A general country review and the history of Chile's space activities are presented in the first sections. The next sections outline the national space policy and space program with a description of the main functions and responsibilities of Chile's space actors. Next, there is a brief description of Chile's legal framework on space matters and the space cooperation programs adopted by Chile and other regional and international partners. Then, there are two sections concerning Chile's satellite capacities and the development of space technology in the country. The last section concerns Chile's capacity building on space issues, giving a detailed description of current space education programs.

© Springer Nature Switzerland AG 2020

A. Froehlich et al., *Space Supporting Latin America*, Studies in Space Policy 25,
https://doi.org/10.1007/978-3-030-38520-0_9

9.1   General Country Overview

Chile is located in South America and is the longest and one of the narrowest countries in the world, bordering Peru in the north, Bolivia and Argentina in the east, Antarctica in the south, and the Pacific Ocean in the west. It has 2,006,096 km[2] of territory, without taking into account its territorial sea and Exclusive Economic Zone. The country has an estimated population of 17,574,003 (2017), of which 51.1% are women and 48.9% are men. Chile's main language is Spanish, and its currency is the Chilean Peso. The population is mestiza, a mix of European and indigenous people.[1]

Chile is divided in three geographic zones (continental Chile, insular Chile, and the Chilean Antarctic Territory) and five natural regions (Norte Grande, Norte Chico, Zona Central, Zona Sur and Zona Austral (Fig. 9.1). Concerning Chile's topography, there are three main entities: the Andes, the Coast Mountains, and the Intermediate Depression. Chile's climate is diverse: the north is extremely arid, the centre zone is mostly warm, and the south has a lot of cold rains. Easter Island and the Juan Fernandez archipelago located in the Pacific Ocean have a subtropical climate.[2]

Fig. 9.1 Chile's natural regions

---

[1] Nuestro país, Chile's government, https://www.gob.cl/nuestro-pais/ (accessed 30 May 2019).

[2] Chile nuestro país, Biblioteca del Congreso Nacional de Chile, https://www.bcn.cl/siit/nuestropais/index_html (accessed 30 May 2019).

9.1.1 Political System

Chile possesses a republican, democratic and representative political system, and it has a presidential government. The state is divided in three independent powers or branches: executive, legislative and judicial. The executive is chaired by the president, whose term lasts four years without possibility of re-election. The government and the administration correspond to the president, who is the head of the state. The legislative branch is based in the National Congress and it is composed of two chambers: the Senate (38 members) and the Chamber of Deputies (120 members). The judicial power is an independent and autonomous organ, and at the head there is the Supreme Court.[3]

Chile is a unitary state and its administration is functional and territorially decentralized. Chile's capital is Santiago and the country is divided in regions (Fig. 9.2), regions are divided in provinces, and provinces in communes. The administration of each region lies in the regional government, which is constituted by a regional governor and a regional council. Regional councils are normative, decision-making and auditor bodies. Every region has a regional presidential delegation, which is responsible for the exercise of the functions and competences of the president in the region.[4]

Sebastian Piñera, leader of Chile Vamos, the centre right coalition party, won the 2017 presidential elections after defeating Alejandro Guillier, representative of the centre left official association Nueva Mayoría. Considered as the third richest person in Chile, Piñera had already been president of the country from 2010 to 2014. In its last presidential campaign, he promised to double economic growth, as well as to improve citizen security, health, education, transport and quality of life.[5]

---

[3]Political System, Chile en el Exterior, 3 August 2017, https://chile.gob.cl/chile/sistema-politico (accessed 20 May 2019).

[4]Government and Regional Administration, articles 110–115 bis of the Political Constitution of the Republic of Chile, 24 October 1980, last reform, 16 June 2018, https://www.leychile.cl/Navegar?idNorma=242302 (accessed 30 May 2019).

[5]Elecciones en Chile: el expresidente Sebastián Piñera gana la segunda vuelta y gobernará por los próximos cuatro años, BBC Mundo, 17 December 2017, https://www.bbc.com/mundo/noticias-america-latina-42382186 (accessed 31 May 2019).

Fig. 9.2 Chile's political division

9.1.2 Economic Situation

Chile's economy has rapidly grown in the last decades thanks to a solid macroeconomic framework, which has helped to reduce the population living in poverty from 30% in 2000 to 6.4% in 2017. The economy grew 4% in 2018 compared to 1.3% in 2017 due to greater confidence by the private sector, low interest rates, and higher copper prices as its main export good. Besides, mining, wholesale trade, commercial services, and manufacture increased. The current account deficit increased from 2.2% of GDP in 2017 to 3.1% in 2018 due to an increase in capital goods imports and net foreign assets. This deficit was mainly financed by foreign investment, which enabled the stabilisation of international reserves. The central government's debt went down from 2.7% of GDP in 2017 to 1.7% in 2018, although public debt increased from 24 to 26% of GDP in the same period.[6]

However, there are no signs of capital redistribution in Chile, keeping high levels of inequality. Bad distribution of incomes in the country is associated with low remunerations in the private sector.[7] Self-employment and involuntary part-time employment have increased, which undermines incomes and the financing of social protection. This situation is exacerbated by low levels of activity and employment rates for women, youth, low-skilled and indigenous groups.[8]

9.1.3 Social Development

Chile is transforming into a country with a high standard of living. However, there are many challenges to accomplish this transition, such as the large proportion of low-skilled workers, low workforce engagement, and high levels of inequality. In terms of job creation, Chile surpasses the OECD average concerning employment and unemployment across almost all levels of educational attainment. For example, the employment rate of male university graduates in Chile is the second highest in the OECD, at 93.3%. Chilean women in the same graduate level enjoy an 80.7% employment rate, which is also over the OECD average. The unemployment rate is roughly 6% below the OECD average.[9]

---

[6]Chile Overview, World Bank, 9 April 2019, https://www.bancomundial.org/es/country/chile/overview (accessed 31 May 2019).

[7]Velásquez Francisco, Economía chilena: la más próspera y desigual de Latinoamérica, diarioUchile, 10 October 2018, https://radio.uchile.cl/2018/10/10/economia-chilena-la-mas-prospera-y-desigual-de-latinoamerica/ (accessed 31 May 2019).

[8]OECD Economic Surveys, Chile, February 2018, p. 7.

[9]Reviews of National Policies for Education: Education in Chile, OECD, 2017, pp. 37–38.

Nevertheless, according to the OECD, Chile is the most unequal country among its members. Moreover, economic growth and prosperity vary in Chile's regions. For example, inhabitants of the Santiago Metropolitan region are 50% richer than the median Chilean. In addition, inequality persists concerning gender and ethnicities. In this sense, women's participation in the labour market is among the lowest in the OECD. Chile has important problems concerning income inequality, which is generally associated with lower literacy and educational attainment.[10]

9.2 History of Space Activities

Chile's space history goes back to 1959 when Chile agreed with the USA to track NASA's missions. During the following decades, the development of the space sector in the country was slow, although there were some important breakthroughs such as the membership of Chile in UNCOPUOS and the establishment of a Space Affairs Committee within Chile's Air Force. The launching of the first Chilean satellite in 1995 marked significant progress in Chile's satellite sector, followed by two additional satellite launches in 1998 and 2011. Nowadays, the adoption of a National Space Policy (NSP) in 2014 and the 2019 cooperation agreement between Chile's Air Force and the University of Chile aim to strengthen the space sector in the coming years (Table 9.1).

Table 9.1 Chile's space history

Year	Chile's space history[a]
1958	First collaboration between the academy and the National Defense of Chile on space issues through the Minitrack-Network program
1959	Agreement between the Government of the United States of America and Chile for the installation of a Control and Monitoring Ground Station for NASA's missions
1959	Creation of the Centre of Space Studies[b] of the University of Chile
1968	Establishment of the "LONGOVILO" Satellite Communications Station
1973	Chile joins UNCOPUOS
1980	Establishment of the Space Affairs Committee[c] of Chile's Air Force
1984	Adoption of the "Mataveri Agreement" between NASA and Chile to study the weakening of the ozone layer in the Antarctic
1984	First direct satellite connection between the Chilean university computer network and the Inter American university network
1993	Chile hosted II Space Conference of the Americas

(continued)

[10]Ibid., p. 39.

Table 9.1 (continued)

Year	Chile's space history[a]
1995	Launching of satellite FASAT-ALFA
1996	Chile integrates the Satellite System COSPAS-SARSAT[d]
1998	Launching of satellite FASAT-BRAVO
1999	"CHINITAS" space experiment
2000	Chile presents its first astronaut candidate
2001	Abolition of the Space Affairs Committee and creation of the Presidential Advisory Commission known as the Chilean Space Agency[e]
2002	Chile hosted the IV Space Conference of the Americas
2008	The Center of Space Studies of the University of Chile sells its assets to the Swedish Space Corporation
2011	Launching of FASat-Charlie, a Chilean EO satellite
2011	The Chilean Space Agency ends operations
2014	Adoption of the NSP
2014	Creation of the Council of Ministers for Space Development[f]
2015	Installation of the Reception Satellite Centre in the University of Conception
2017	Launching of the nano-satellite CubeSat SUCHAI 1 of the University of Chile from India
2018	Agreement between the Telecommunications Under Ministry (SUBTEL)[g] of Chile and the Copernicus Program of the European Commission
2018	Agreement between SUBTEL and the University of Chile
2019	Cooperation agreement between Chile's Air Force and the University of Chile for the development of the national space program

[a]See General background of space technologies, Fuerza Aérea de Chile, Grupo de Operaciones Espaciales http://www.ssot.cl/antecedentes.php, and Gutiérrez Héctor, Informe Espacial para la Comisión de Ciencia y Tecnología de la Cámara de Diputados, Asociación Chilena del Espacio, 21 November 2018, https://www.camara.cl/pdf.aspx?prmID=157010&prmTIPO=DOCUMENTOCOMISION (accessed 31 May 2019)
[b]Centro de Estudios Espaciales de la Universidad de Chile
[c]Comité de Asuntos Espaciales de la Fuerza Aérea de Chile
[d]COSPAT- SARSAT (Cosmicheskaya Sistyema Poiska Avariynich Sudov (COSPAS) - Search and Rescue Satellite-Aided Tracking (SARSAT)) is a satellite system that provides disaster and position alert data to assist search and rescue operations (SAR), see Servicio de Búsqueda y Salvamento Aéreo, Fuerza Aérea de Chile, https://www.fach.mil.cl/sar/Sistema%20CS.htm (accessed 10 September 2019)
[e]Comisión Asesora Presidencial/Agencia Chilena del Espacio
[f]Consejo de Ministros para el Desarrollo Espacial
[g]Subsecretaría de Telecomunicaciones de Chile (SUBTEL)

9.3 Space Development Policies and Emerging Programs

In the last decade, there have been important changes at the institutional and political level concerning Chile's space activities. On one hand, the adoption of an NSP and a space program will contribute to develop Chile's space activities for military and civil purposes, and position the country in the international space arena. On the other hand, the strengthening and restructuration of space actors have the purpose of improving the application of the space policy and program.

9.3.1 Space Policies and Main Priorities

For a long time, Chile did not possess an NSP, even when its space activities began in the early years of the space race. It was not until 2013 when the Chilean government decided to adopt a multiannual NSP for the development of its space and satellite sectors through a series of axes. The continuation of the space policy for the next years has already been planned. Moreover, the recent adoption of the national space program between Chile's Air Force and the University of Chile intends to give impulse to the space sector by taking advantage of the strengths and experiences of both institutions.

9.3.1.1 Chile's National Space Policy

Chile's first NSP was adopted by SUBTEL in 2013 for the period 2014–2020.[11] The NSP is the result of an inter-ministerial and multidisciplinary collaboration carried out by a work-table headed by SUBTEL with representatives of the Ministry General-Secretary of the Presidency, the Ministry of Foreign Affairs, the Ministry of National Defence, Chile's Air Force, the Ministry of National Assets, the National Commission for Scientific and Technological Research (CONICYT)[12] and the Data Centre of Natural Resources (CIREN).[13]

The NSP presents general guidelines to promote the development of space activities in Chile, mainly the satellite policy, bearing in mind that the diverse applications to human activities derived from the exploration and use of outer space may generate social and economic benefits for the country. NSP's vision aims to transform Chile in 2020 into a country that takes advantage of the economic and social benefits of space, with greater opportunities for the development of knowledge, innovation and entrepreneurship in space science and technology, all of this in a favourable environment for the citizens, the productive sector and the administration of the state. NSP's mission is to implement institutional changes, promote efficient coordination of state's resources, and foster continuous progress in the fields of space science, technology and innovation, with the aim of strengthening the human and productive capital of Chile.[14]

To delineate its space strategic objectives, Chile's NSP enumerates a series of benefits that can be exploited by space activities and technologies. For example, EO or remote sensing can contribute to manage Chile's territory and its natural resources, namely in the sectors of agriculture, mining, fishing, forestry, aquaculture, geology, and energy. Satellite data can be used for urban planning, cartography, oceanography,

---

[11] Política Nacional Espacial 2014–2020, Subsecretaría de Telecomunicaciones/2013, Santiago, Ed. Ograma Impresores, February 2014.

[12] Comisión Nacional de Investigación Científica y Tecnológica (CONICYT).

[13] Centro de Información de Recursos Naturales (CIREN).

[14] Política Nacional Espacial 2014-2020, op. cit., pp. 23–24.

and habitat protection. Remote sensing is also useful for public health applications, national defence and security, as well as to prevent natural disasters.

In the sector of satellite telecommunications, digital technology services such as the Internet, data communication, multimedia and videoconference applications, radio and TV signal broadcasting, satellite telephony, among others, can improve the development of the country and the connection between its citizens. Health and education are other important areas where satellite technology can be useful. Moreover, satellite telecommunications can provide benefits to the Chilean government in terms of coverage, defence and security.

Satellite navigation can be used in civil and military applications. Air, maritime and land navigation reduces operation costs and navigation mistakes. They are also useful for environmental studies and natural phenomena monitoring, such as movements of glaciers and icebergs, tectonic plates, species migration, and species displacements. In the military sector, global positioning systems improve the location of fixed and mobile points of interest.

Understanding the universe and astronomical sciences is another benefit directly related to space technology. Chile's territory has excellent locations to observe the sky in order to expand human knowledge. Besides, the use of space technology to track space debris and other space objects could be useful for the security of the national territory.

The NSP also presents the current situation of Chile's space sector in the form of potentials and weaknesses, which are presented in Tables 9.2 and 9.3, respectively.

The aforementioned potentials and weaknesses have been taken into account in the development of the NSP, which aims to solve these shortcomings as a necessary step to incorporate and develop concrete initiatives to carry out the NSP's objectives. In this sense, the NSP is a long term project to foster Chile's space sector and to improve the current space infrastructure through practical strategic axes. Each strategic axis possesses specific action lines, objectives and initiatives (Table 9.4).

Strategic Axe: Environment for Space Development

The environment for space development implies the implementation of institutional capacities in the interior of the state's administration to improve the development of satellite science and technology. The objective is to avoid duplication of functions, align incentives, and improve the quality of goods and services. This axis has three action lines: institutional space capacities, promotion and access to information, and space infrastructure and applications.

Initiatives for the improvement of institutional space capacities include organic legal reform, provisional institutionalization, a space development plan, and a technical cooperation program. The promotion and access to information foresees a program for the promotion of space activity to spread space activities information to state organs and the community. Meanwhile, initiatives for infrastructure and space applications consist of the development of space data infrastructure, provisional institutionalization for space data infrastructure to integrate and homogenize available

Table 9.2 Potentials of Chile's space sector

	Potentials of Chile's space sector
1	National EO satellite "FASat-Charlie"; the developed capacity of the Space Operations Group for its operation and self-maintenance; the elaboration of value-added products obtained from images and remote sensors from third countries and organisations through a process executed by the Aerial Survey Service of Chile's Air Force
2	Chile's geographical location and environmental characteristics for astronomical observations (north region) and other potential areas for satellites tracking (south region), control of navigation systems, and a test laboratory for space vehicles and exploration instruments
3	Chile's positioning in hosting and supporting complex and expensive space related projects, such as the installation of big observatories and pilot space vehicles experiments
4	Experience and specialization in satellite planning, implementation and operation to carry out future space projects based on the capacity building of the Ministry of National Defence to develop an Earth Observation Satellite System project
5	The existence of institutionalisation for the coordination and management of public geospace data, centralized in the National System of Territorial Information
6	Acquired knowledge in the application of space technology in the agriculture and livestock sector
7	An international legal space framework adopted by Chile through the ratification of UN Space Treaties
8	Chile's integration in the international space cooperation framework through the adoption of bilateral agreements on space matters, as well as Chile's participation in different international bodies and commissions

Table 9.3 Weaknesses of Chile's space sector

	Weaknesses of Chile's space sector
1	Lack of a long-term public space policy and the integration of competences for its design, implementation and control into the state's administration
2	Weak institutions for the coordination and management of public geospace data due to the lack of legal capacity of the National System of Territorial Information
3	Data asymmetry in the public and private sectors on the benefits and potentialities of space activities and the development of space technology
4	Limited specialized human resources in science and technology related to the exploration and use of space, as well as lack of transversal capacities for the use of space goods and services
5	Lack of useful orbital slots that enable the development of GEO satellite's programs

public geospace data, land registry and state demand projection in satellite telecommunications and EO matters, satellite goods and services, a continuity program of the FASat-Charlie satellite, as well as a solution program of satellite telecommunications capacity for emergencies and connectivity of isolated areas.[15]

---

[15]Política Nacional Espacial 2014–2020, Eje Estratégico 1. Entorno para el desarrollo espacial, pp. 31–38.

Table 9.4 Strategic objectives of the National Space Policy	Strategic objectives of Chile's national space policy	
	1	Environment for space development
	2	Innovation and entrepreneurship
	3	Human capital

Strategic Axis: Innovation and Entrepreneurship

The NSP foresees the sustainable growth of Chile's economy, productivity and competitiveness through innovation and entrepreneurship. In this regard, Chile aims to create the necessary conditions so that the private sector and society develop their creative and productive potential. Actions lines of this axis include research and development (R&D), space industry, and geographic position and climate and environmental characteristics.

R&D initiatives comprise the creation of a space technology R&D plan, the elaboration of a promotion program of space activities, and the adoption of technical cooperation agreements. Space industry initiatives include a promotion program of space activities, a synergies study, the adoption of international cooperation agreements, a study of development capacities on space infrastructure, and a study on public-private cooperation potentialities. Concerning geographic position and climate and environmental characteristics, the creation of a registry of geographic zones, a regulatory framework study, and the adoption of international cooperation agreements in the sector are planned.[16]

Strategic Axis: Human Capital

This axis requires the formation of specialists in the space sector, to deliver the necessary skills for the use of space applications, and to maximize the development of scientific and technological talent. Foreign talent attraction is also important to enhance local human capital. The three lines of action of this axis are the formation of specialized human capital, users training, and space science and research.

The formation of specialized human capital includes a study of offer and demand for space related specialists, a registry of capacity-building and education institutions, the promotion of state scholarships, and dissemination of foreign talent attraction programs. Users training initiatives consist of training programs, dissemination seminars, and public-private alliances. Similarly, the science and research line of action includes the creation of a registry of research centres, the dissemination of promotion programs, and the adoption of international cooperation agreements.[17]

---

[16]Política Nacional Espacial 2014–2020, Eje Estratégico 2. Innovación y emprendimiento, pp. 39–43.

[17]Política Nacional Espacial 2014–2020, Eje Estratégico 3. Capital humano, pp. 44–47.

9.3.1.2   The New National Space Policy

Since the current NSP was created for the period 2014 to 2020, an update of the NSP has been planned to cover at least the following ten years (up to 2030). The revision should take into account the results of several expert meetings held between 2016 and 2017, and the analysis of subjects not included in the current policy, such as the sustainable goals of the 2030 Agenda, a governance framework and a satellite management model, and a permanent public institutional framework, and should include the draft of the new NSP.[18] Moreover, it has been stated that the revised NSP should be delivered to the new Ministry of Science, Technology, Knowledge and Innovation to integrate it into the National Policy of Science, Technology, Knowledge and Innovation.[19] The official announcement of the creation of the new NSP was given by Chile's President Sebastian Piñera in June 2019. The new NSP will include the acquisition and operation of a modern satellite that will replace the current Chile's satellite FASat-Charlie.[20]

9.3.1.3   Space Program for Chile

In 2018, the Faculty on Physical Sciences and Mathematics of the University of Chile, at request of the Commission of Future Challenges, Science and Technology of the Senate, presented a proposal for the creation of a Space Program for Chile (SPC).[21] According to the initiative, the strategic importance of data for disaster prevention and management, and for the development of agriculture and smart mining, have forced Chile to invest in this sector. The investment will cover research and innovation, human resources training, and incentives for the creation of enterprises aimed at developing goods and services from the available data. The benefits of this program will consist of independence and sovereignty concerning data and knowledge, the development of civil applications, the opening of new research and services sectors, as well as international collaboration on space matters. Social benefits such as the monitoring of environment, territorial management, public security, national defence,

---

[18]Gutiérrez Héctor, Proceso de actualización de la Política Nacional Espacial de Chile, Simposio y Taller UIT Sistemas de Regulación y Comunicación de Satélites Pequeños, 7–9 November 2016, Consejo de Ministros para el Desarrollo https://www.itu.int/en/ITU-R/space/workshops/2016-small-sat/Documents/H-GUTIERREZ-MTT-SUBTEL.pdf (accessed 4 June 2019).

[19]Gutiérrez Héctor, Informe Espacial para la Comisión de Ciencia y Tecnología de la Cámara de Diputados, Asociación Chilena del Espacio, 21 November 2018, https://www.camara.cl/pdf.aspx?prmID=157010&prmTIPO=DOCUMENTOCOMISION (accessed 4 June 2019).

[20]Basoalto Héctor, Un nuevo programa satelital: presidente anuncia futura puesta en marcha del reemplazante del FASat-Charlie, La Tercera, 1 June 2019, https://www.latercera.com/nacional/noticia/nuevo-programa-satelital-presidente-anuncia-futura-puesta-marcha-del-reemplazante-del-fasat-charlie/680937/ (accessed 7 June 2019).

[21]Programa Espacial para Chile.

access to information for health and education, among others, are also envisaged. In addition, economic benefits could reach US 1.537 billion in 2020.[22]

The proposal included the development of a constellation of nine to twelve satellites with sensors for the observation of the Earth and data transmission of atmospheric variables, the integration of sensors networks, and the development of a centre for the integration and analysis of geospace data.[23]

On 18 March 2019, the University of Chile and Chile's Air Force[24] signed an agreement to develop a national space program. The Framework Cooperation Agreement on Space Issues,[25] which took into account the 2018 proposal of the University of Chile, aims to strengthen the national legal framework on space activities, to consolidate national space institutions, and to facilitate multi sectoral integration. In particular, the agreement foresees the improvement of professional training, research and technology development in the aerospace sector; the improvement of aerospace applications for economic, social, environmental, and defense development; the definition of a specific, regular and complementary public budget for the bodies linked to space activities, and an increase in public access to geospace data. Thirteen national institutions joined the agreement, which will use nanosatellites for monitoring forest fires and light pollution, in a first stage, integrate land networks, and develop a Center for Geospace Data Integration and Analysis.[26]

9.3.2 Space Actors

After the abolition of Chile's Space Agency in 2011,[27] Chile's national space activities fell to the competence of three ministries: the Ministry of Transportation and Telecommunications; the Ministry of National Defence, and the Ministry of Science, Technology, Knowledge and Innovation. Moreover, the Council of Ministers for Space Development also has a vital role in Chile's space activities. Finally, the University of Chile has a significant role in the development of the space sector in the country.

---

[22]Programa Espacial para Chile, Facultad de Ciencias Físicas y Matemáticas, Universidad de Chile, 6 November 2018, http://ingenieria.uchile.cl/noticias/148745/programa-espacial-para-chile (accessed 1 July 2019).

[23]Presupuesto, Facultad de Ciencias Físicas y Matemáticas, Universidad de Chile, 6 November 2018, http://ingenieria.uchile.cl/noticias/148735/presupuesto (accessed 1 July 2019).

[24]Fuerza Aérea de Chile (FACH).

[25]Acuerdo Marco de colaboración en materias espaciales.

[26]U. de Chile y FACH acuerdan desarrollar programa espacial nacional, diarioUchile, 19 March 2019, https://radio.uchile.cl/2019/03/19/u-de-chile-y-fach-acuerdan-desarrollar-programa-espacial-nacional/ (accessed 10 June 2019).

[27]Campos Felipe, Política espacial chilena 2014-2020, Cosmo noticias, 28 July 2014, https://www.cosmonoticias.org/politica-espacial-chilena-2014-2020/ (accessed 10 June 2019).

9.3.2.1   Ministry of Transportation and Telecommunications

In general, the Ministry of Transportation and Telecommunications[28] is responsible for the proposal, implementation, management and control of national policies in the sectors of transportation and telecommunications. More specifically, SUBTEL is in charge of the coordination, promotion, encouragement, and development of Chile's telecommunication sector. It proposes and implements national policies on telecommunications according to government guidelines, and supervises public and private companies that work in the field.[29] Moreover, the Department of Management of the Radio Spectrum and Numbers is responsible for the administration of the efficient use of the radio spectrum assigned by SUBTEL.[30] The Department of International Relations acts on behalf SUBTEL in international fora related to the telecommunications sector, such as the International Telecommunication Union (ITU) and the Inter-American Telecommunications Commission (CITEL),[31] as well as trade organizations and fora such as the Asia-Pacific Economic Cooperation (APEC) and the World Trade Organization (WTO). It also participates in bilateral negotiations in the framework of free trade agreements and memorandums of understanding.[32]

9.3.2.2   Ministry of National Defense

The Ministry of National Defence supports Chile's President in the political conduct of national defence. It is through Chile's Air Force that the Ministry exercises its functions in the space sector. The Space Operations Group (GOE), that works under the Combat Command of Chile's Air Force, is responsible for the operation of the FASat-Charlie satellite (see Sect. 9.7) and other future satellites and space objects.[33] It is also responsible for the general development of national space capacities.[34]

---

[28] Institutional review, Ministerio de Transportes y Telecomunicaciones, Gobierno de Chile, https://www.mtt.gob.cl/resenainstitucional (accessed 4 June 2019).

[29] Quiénes somos, Subsecretaría de Telecomunicaciones, https://www.subtel.gob.cl/quienes-somos/ (accessed 4 June 2019).

[30] Departamento Administración del Espectro Radioeléctrico y Números, Subsecretaría de Telecomunicaciones, https://www.subtel.gob.cl/departamento-administracion-del-espectro-radioelectrico-y-numeros/ (accessed 8 June 2019).

[31] Comisión Interamericana de Telecomunicaciones.

[32] Relaciones Internacionales, Subsecretaría de Telecomunicaciones, https://www.subtel.gob.cl/quienes-somos/relaciones-internacionales/ (accessed 8 June 2019).

[33] Grupo de Operaciones Espaciales, Fuerza Aérea de Chile, http://www.ssot.cl/grupo_operaciones.php (accessed 4 June 2019).

[34] During the development and launching of FASat-Charlie, the organization's name was Satellite Operation Centre (Centro de Operaciones Satelitales). Then, during the initial operations phase, it was called Satellite Operations Group. Ibid.

9.3.2.3 Ministry of Science, Technology, Knowledge and Innovation

The recently created Ministry of Science, Technology, Knowledge and Innovation collaborates with and advises Chile's President in the design, elaboration, coordination, implementation, and assessment of policies, plans and programmes intended for the promotion and strengthening of science, technology and innovation derived from scientific and technological research. Among its functions, the Ministry is responsible for the development and promotion of space activities in the country. In this respect, the Ministry must coordinate its responsibilities with the Ministry of National Defence, and the Ministry of Transportation and Telecommunications. In addition, the National Agency of Research and Development, which is linked to Chile's President through the Ministry of Science, Technology, Knowledge and Innovation, manages and implements the programs and instruments destined for the promotion, fostering and development of scientific and technological research.[35]

9.3.2.4 Council of Ministers for Space Development

Created in 2014, the Council of Ministers for Space Development (the Council) is the body responsible for advising Chile's President on the elaboration of public policies, plans, programs and actions that contribute to the promotion of space activities and use of space applications and technology. It is composed of the Secretary-General of the Presidency, and the Ministers of Transportation and Telecommunications (who chairs the Council), Interior, Foreign Affairs, National Defence, Treasury, Social Development, Education, Agriculture, National Assets, and Economy, Development and Tourism.[36] The Executive Secretary is the Under-Minister of Telecommunications.

In addition, the Council proposes to Chile's President the NSP, functions as the appropriate coordination interface with public bodies that have space related activities compctences, identifies and proposes the elaboration of studies for the determination of national satellite needs and the model of use of satellite resources, and proposes the adoption of national and international cooperation agreements related to space science and technology.[37]

---

[35]Guía legal sobre el Ministerio de Ciencia, Tecnología, Conocimiento e Innovación, Biblioteca del Congreso Nacional de Chile, 17 August 2018, https://www.bcn.cl/leyfacil/recurso/ministerio-de-ciencia,-tecnologia,-conocimiento-e-innovacion (accessed 8 June 2019).

[36]Consejo de Ministros para el Desarrollo Espacial, Subsecretaría de Telecomunicaciones, 9 September 2015, http://espacial.subtel.gob.cl/consejo-de-ministros/ (accessed 8 June 2019). It is likely that the Minister of Science, Technology, Knowledge and Innovation will integrate the Council.

[37]Decree N° 181 that establishes the Presidential Advisory Commission entitled the Council of Ministers for Space Development, 28 October 2015, https://www.leychile.cl/Navegar?idNorma=1087964 (accessed 20 May 2019).

9.3.2.5   University of Chile

The role of the University of Chile has been extremely important in the development of Chilean space activities. From 1959 to 2008, the University of Chile, through its Centre for Space Studies of the Faculty of Physics Sciences and Mathematics, had many tasks concerning the monitoring of satellites of the US space program, the study of agricultural lands, volcanoes and sea pollution, as well as telemetry, tracking and remote control of more than 370 space missions of different countries.[38] In 2019, the University of Chile, together with Chile's Air Force, has been carrying out the new space program for Chile, which aims to develop the national space sector for the following years (see Sect. 9.3.1.3).

9.4   Current and Future Perspectives on the Legal Framework Related to Space Activities

Chile's space activities are governed, on the one hand, by a series of national regulations that generally stipulate the purposes and functions of Chile's space and telecommunications actors. On the other hand, Chile's international obligations on space matters are based on a significant number of space treaties and other related instruments ratified or adopted by it.

9.4.1   National Space Legislation and Instruments

Chile's national legislation on space matters consists of laws and decrees concerning the general regulation of the telecommunications sector and the creation of national bodies or mechanisms responsible for the management and operation of Chile's space and telecommunications activities (see Sect. 9.3.2). These regulations define the purposes, functions, structure and procedures of these bodies.

Chile's space related legislation includes Decree N° 181 of 2016 that creates the Presidential Advisory Commission entitled the Council of Ministers for Space Development, and Law N° 21.105 of 2018 that establishes the Ministry of Science, Technology, Knowledge and Innovation.

Concerning the telecommunications sector, some of the most relevant instruments are Decree N° 1.762 which created the Under-Ministry of Telecommunications in 1977; Supreme Decree N° 423 of 1978, which adopted the National Policy of Telecommunications; Law N° 18.168 of 1982, known as the General Law of Telecommunications, which deregulated the telecommunications sector, allowing

---

[38] Centro de Estudios Espaciales: el inicio, Facultad de Ciencias Físicas y Matemáticas, Universidad de Chile, 27 March 2018, http://ingenieria.uchile.cl/noticias/142034/centro-de-estudios-espaciales-el-inicio (accessed 3 July 2019).

private companies to participate in the sector,[39] and Decree N° 240 of 2011 concerning the use of the radio spectrum.[40] Other decrees include the promulgation of international space and telecommunications treaties and international cooperation agreements.[41]

9.4.2 Chile and International Space Instruments

As a member of UNCOPUOS, Chile supports the codification and progressive development of international space law. In this regard, Chile has adopted all UN Space Treaties and other space and telecommunications related instruments. Moreover, Chile's international obligation concerning the creation of a national registry of its space objects launched into outer space has been fulfilled. In addition, Chile supports several international mechanisms on space debris mitigation.

9.4.2.1 Space Treaties and Other Instruments

Chile is one of the few states in the world that have ratified the five UN Space Treaties.[42] Chile has also ratified other international agreements related to space or telecommunications such as the Treaty Banning Nuclear Weapon Tests in the Atmosphere, in Outer Space and under Water (NTB), the Convention on the International Mobile Satellite Organization (IMSO) and the ITU Convention and Constitution (Table 9.5).[43]

---

[39] History of the Under-ministry of Telecommunications, Subsecretaría de Telecomunicaciones, https://www.subtel.gob.cl/quienes-somos/historia-2/ (accessed 8 June 2019).

[40] Decreto N° 240 que modifica Decreto Supremo N° 127, de 2006, que aprobó el Plan General del Uso del Espectro Radioeléctrico, Ministerio de Transportes y Telecomunicaciones, Biblioteca del Congreso Nacional de Chile, 25 August 2010, https://www.leychile.cl/Navegar?idNorma=1022440 (accessed 8 June 2019).

[41] For example, the Agreement on the Rescue of Astronauts, the Return of Astronauts and Return of Objects Launched into Outer Space, 22 January 1982, Biblioteca del Congreso Nacional de Chile, https://www.leychile.cl/Navegar?idNorma=8778 (accessed 8 June 2019).

[42] Only 14 States worldwide have ratified the five Space Treaties: Australia, Austria, Belgium, Chile, Kazakhstan, Kuwait, Lebanon, Mexico, Morocco, the Netherlands, Pakistan, Peru, Turkey and Uruguay. COPUOS, Status of International Agreements relating to activities in outer space as at 1 January 2019, A/AC.105/C.2/2019/CRP.3, 1 April 2019, http://www.unoosa.org/documents/pdf/spacelaw/treatystatus/AC105_C2_2019_CRP03E.pdf.

[43] Ibid.

Table 9.5 International space related treaties ratified by Chile

International Space Treaties ratified by Chile		
Treaty	Adoption	Chile's ratification
OST	1967	R
ARRA	1968	R
LIAB	1972	R
REG	1975	R
MOON	1979	R
NTB	1963	R
BRS[a]	1974	R
ITSO[b]	1971	R
IMSO[c]	1976	R
ITU	1992	R

[a]Convention Relating to the Distribution of Programme-Carrying Signals Transmitted by Satellite, 21 May 1974
[b]Agreement Relating to the International Telecommunications Satellite Organization, 20 August 1971
[c]Convention on the International Mobile Satellite Organization, 3 September 1976

9.4.2.2   National Registry

According to Article II of the Registration Convention, ratified by Chile in 1981, states parties shall notify the UN Secretary-General of the creation of a national registry to enter their space objects launched to outer space. Chile notified UNOOSA on 18 January 2016 of the creation of its register, which is kept by the Directorate for International and Human Security of the Ministry of Foreign Affairs.[44] In this regard, Chile has submitted several notifications to UNOOSA on its space objects launched into outer space from 1998 to 2018.[45]

9.4.2.3   Space Debris Mitigation Instruments

Although Chile has not yet adopted any national mechanisms on space debris mitigation, it fully adheres to the non-binding UNCOPUOS Space Debris Mitigation Guidelines, and it supports the IADC Space Debris Mitigation Guidelines, ISO

---

[44]Dirección de Seguridad International y Humana del Ministerio de Relaciones Exteriores, Información proporcionada de conformidad con el Convenio sobre el Registro de Objetos Lanzados al Espacio Ultraterrestre, UN Secretary-General, ST/SHG/SER.E/INF/33, 2 March 2016, https://cms.unov.org/dcpms2/api/finaldocuments?Language=es&Symbol=ST/SG/SER.E/INF/33 (accessed 4 June 2019).

[45]See Online Index of Objects Launched into Outer Space, UNOOSA, http://www.unoosa.org/oosa/osoindex/search-ng.jspx?lf_id = (accessed 4 June 2019).

Space Systems-Space Debris Mitigation Requirements (ISO 24113:2011) and ITU Recommendation ITU-R S.1003.[46]

9.5 Cooperation Programs

Since the beginning of its space activities, Chile has adopted several space related cooperation agreements with national and international actors.[47] While several of these agreements have already expired,[48] there are some agreements that are still in force.

At the national level, the University of Chile and Chile's Air Force adopted on 18 March 2019 the Framework Cooperation Agreement on space issues[49] with the aim of developing a new NSP (see 9.3.1.3).

At the international level, one of the most important space cooperation agreements is the agreement adopted on 6 April 2001 in Tokyo, Japan, between the Government of Chile and the representatives of Europe, Japan and North America to construct and operate together the Atacama Large Millimetre/submillimetre Array (ALMA), a radio telescope comprised of sixty-four transportable 12-meter diameter antennas distributed over an area 14 km^2 in extent, in the Atacama region, in northern Chile.[50]

This cooperation agreement has benefited Chile in different ways. For example, the relationship between ALMA and Chile's Government is constantly being strengthened thanks to the collaboration and communication with many ministers, such as the Ministries of Foreign Affairs, Public Lands, and Education. Also, the construction and operation of ALMA has contributed to the development of astronomy in the country through the ALMA-CONICYT Fund,[51] which promotes human

---

[46]Compendium of space debris mitigation standards adopted by States and international organizations, Chile, UNCOPUOS A/AC.105/C.2/2014/CRP.15, 18 March 2014, p. 18, http://www.unoosa.org/pdf/limited/c2/AC105_C2_2014_CRP15E.pdf (accessed 26 April 2019).

[47]For example, in 1959 Chile and NASA signed a cooperation agreement to create the Center for Space Studies of the University of Chile and the installation of the first Satellite Tracking Station in Chile; China and Chile adopted a space cooperation agreement in 1997; Chile and Ecuador adopted a cooperation agreement in 2006; and Brazil and Chile subscribed to a complementary protocol in 1994. These instruments can be found at Biblioteca del Congreso Nacional de Chile, Búsqueda de Tratados, https://www.leychile.cl/Consulta/buscador_tratados.

[48]For example, the Agreement between the Republic of Chile and the Government of the United States of America on the use of the Mataveri Airport, Easter Island, as an emergency landing place for rescue of space shuttles, adopted in 1985 and promulgated by Chile's Decree N° 917, Ministry of Foreign Affairs, Library of the National Congress of Chile, 6 November 1985.

[49]Acuerdo Marco de colaboración en materias espaciales.

[50]Europa, Norteamérica y Japón resuelven construir en conjunto ALMA en Chile, Atacama Large Millimeter/submillimeter Array, 6 April 2001, https://www.almaobservatory.org/es/comunicados-de-prensa/europa-norteamerica-y-japon-resuelven-construir-en-conjunto-alma-en-chile/ (accessed 29 June 2019).

[51]The National Commission on Scientific and Technological Research (Comisión Nacional de Investigación Científica y Tecnológica (CONICYT)). The Commission became the Ministry of Science, Technology, Knowledge and Innovation.

resources, instrumentation, teaching, outreach and fellowships. ALMA also encourages the training of specialized human resources, and fosters the development of engineering and software, namely for astronomical instrumentation and data processing. As to the telecommunications sector, ALMA has contributed to the creation of a fiber optic link between the San Pedro Atacama and Catama communities. In this sense, ALMA contributes to the ALMA Region II Fund, which promotes productive and social-economic development in the San Pedro de Atacama community. In terms of employment, 80% of ALMA's staff is locally hired, and Chilean astronomers have access to 10% of the observing time with ALMA.[52]

The cooperation agreement on EO between the European Commission and Chile (through SUBTEL) adopted on 8 March 2018 is another relevant space agreement. The agreement enables Chile to get information from the COPERNICUS program provided by SENTINEL satellites through the use of broadband connections between data centres. In addition, Chile facilitates European Commission access to in situ data from its regional observation networks, including geophysical and meteorological networks. Chile also provides technical support to COPERNICUS for the calibration and assessment of collected data by SENTINEL satellites for Latin America. The agreement is also intended to promote the joint development of products and services.[53]

Other bilateral cooperation frameworks include an agreement between Chile and France for the reinforcement of their mutual common defense relations. In this regard, the two countries agreed to articulate a renewed strategic dialogue on emergent threats, and to improve cooperation concerning defense industries, in particular in the naval, military and space sectors.[54] A Memorandum of Understanding was adopted between Chile and South Korea, specifically between CONICYT and the Korea Astronomy and Space Science Institute (KASI), on the preparation of a joint call for research projects in astronomy, the implementation of postdoc mobility opportunities, the promotion of student training, and the participation of academics in cooperation activities of the Korea University of Science and Technology undergraduate program.[55] Finally, in 2008 Chile and Argentina adopted a space cooperation framework agreement to promote their mutual interests in the peaceful exploitation and utilization of outer space, and to work together to reach their respective space

---

[52]Beneficios para la Comunidad, Atacama Large Millimeter/submillimeter Array, https://www.almaobservatory.org/es/sobre-alma/alma-en-chile/beneficios-para-la-comunidad/ (Accessed 30 June 2019).

[53]Comisión Europea firma Acuerdos de Cooperación para observación espacial con Colombia, Chile y Brasil, Delegación de la Unión Europea en Colombia, 7 March 2018, https://eeas.europa.eu/delegations/colombia/40977/comisi%C3%B3n-europea-firma-acuerdos-de-cooperaci%C3%B3n-para-observaci%C3%B3n-espacial-con-colombia-chile-y_es (accessed 10 June 2019).

[54]Chile, Declaración Presidencial sobre una hoja de ruta bilateral, France Diplomatie, 9 October 2018, https://www.diplomatie.gouv.fr/es/fichas-de-paises/chile/eventos/article/chile-declaracion-presidencial-sobre-una-hoja-de-ruta-bilateral-09-10-18 (accessed 10 Juin 2019).

[55]Nuevas oportunidades para la colaboración Chile-Corea del Sur en Ciencia y Tecnología, Programa de Cooperación Internacional, CONICYT, Ministry of Education, 27 April 2015, https://www.conicyt.cl/pci/2015/04/nuevas-oportunidades-para-la-colaboracion-chile-corea-del-sur-en-ciencia-y-tecnologia/ (accessed 10 June 2019).

program's goals.[56] This agreement also mentions that both parties recognize the need to establish a Regional Space Agency.[57]

9.6 Satellite Capacities

National satellite capacities are currently concentrated in the FASat-Charlie satellite and the corresponding ground segment. This satellite, launched into LEO in 2011, is an EO satellite for civilian and defence purposes integrated in the Earth Observation Satellite System (SSOT) Project.[58]

SSOT Project's emerged in 2007 from a joint initiative between academics from the University of Concepción and Chile's Air Force.[59] That year, the Ministry of National Defence stipulated the acquisition of an optical EO satellite system. In 2008, a contract was signed with the EADS-ASTRIUM company for the development of the project, which included the design and assembly of the satellite in France, a ground segment (control station, bi-band antenna station, and image processing station), complete launch services (launcher, launch campaign, and launch insurance), training in the operation of the satellite, and technology transfer. Thus, the SSOT System is composed of a space segment consisting of the FASat-Charlie satellite,[60] and a ground segment located in the Air Force's airbase "El Bosque" (see Fig. 9.3).[61] GOE is responsible for the operation and maintenance in orbit of FASat-Charlie, and the Aerophotogrammetric Service of Chile's Air Force organises the commercialization

---

[56]Firma de acuerdos bilaterales en Chile, Casa Rosada, Presidencia de la Nación, 5 December 2008, https://www.casarosada.gob.ar/informacion/archivo/20313-blank-49596941 (accessed 10 June 2019).

[57]Acuerdo Marco de Cooperación en el Campo de las Actividades Espaciales entre la República Argentina y la República de Chile, Punta Arenas, Chile, 4 December 2008, https://tratados. cancilleria.gob.ar/tratado_archivo.php?tratados_id = kp + ilZc = &tipo = kg ==&id = 9332& caso = pdf (accessed 28 September 2019).

[58]SSOT Project, Space Operations Group, Chile's Air Force, http://www.ssot.cl/proyectossot.php (accessed 7 June 2019).

[59]The National Defense Book stipulates participation in the creation of own space capacities. In this regard, capacity building must be orientated in the following areas: space elements (creation of technology); space infrastructure (laboratories, control centers); information systems, and space activities oriented towards the creation of a critical mass), Ibid.

[60]FASat-Charlie is based in an AstroSat 100 platform developed by the Centre national d'études spatiales (CNES) of France, and EADS Astrium. It weighs 130 kg, measures 1.4 m high and 0,95 wide. Its 620 km polar orbit enables the satellite to obtain images of a point on Earth every three days on average, and to communicate with the Satellite Operations Group four times a day, see SSOT System, Space Operations Group, Chile's Air Force, http://www.ssot.cl/sistemassot.php (accessed 7 June 2019).

[61]Ibid.

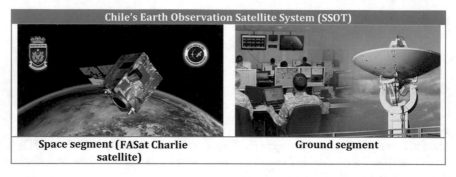

Fig. 9.3 Chile's Earth Observation Satellite System (SSOT). *Images source* SSOT System, Space Operations Group, Chile's Air Force. http://www.ssot.cl/sistemassot.php (accessed 7 June 2019)

and distribution of the images generated by the satellite, which can be acquired by public and private entities.[62]

SSOT's applications include precision farming, flood prevention and control, volcanic eruption control, forest fire control, cartography, urban growth control, defence, and national development.[63]

Due to the fact that FASat-Charlie satellite has exceeded its useful life, Chile's President Sebastian Piñera has recently announced the acquisition of a new satellite that will replace FASat-Charlie. The new satellite will function within a network of other similar satellites, and it will serve Chile's Air Force, the state and civil society.[64]

Apart from FASat-Charlie, undergraduate students, engineers and professors from the University of Chile designed and developed a CubeSat called SUCHAI in 2017 ("Satellite of the University of Chile for Aerospace Investigation"), the first satellite totally built in Chile.[65] The satellite was launched from the territory of India, and it made studies on space physics, technical tests with structures, electronic components, materials and flight software.[66] Two new nanosatellites, SUCHAI II and SUCHAI III, are currently in development (Table 9.6).[67]

---

[62]Exploitation models and access policies on FASat-Charlie images, Space Operations Group, Chile's Air Force, http://www.ssot.cl/modelo.php (accessed 7 June 2019).

[63]SSOT applications, Space Operations Group, Chile's Air Force, http://www.ssot.cl/aplicaciones.php (accessed 7 June 2019).

[64]Saez Javier, Chile tendrá nuevo programa espacial y satélite FASat-Charlie será reemplazado, Tele 13, 1 June 2019, https://www.t13.cl/noticia/politica/chile-tendra-nuevo-programa-espacial-y-satelite-fasat-charlie-sera-reemplazado (accessed 7 June 2019).

[65]Suchai, Space and Planetary Exploration Laboratory, http://spel.ing.uchile.cl/suchai.html (accessed 7 June 2019).

[66]Lazcano Patricio, Los récords que dejó el Suchai 1, el primer satélite construido en Chile, Qué pasa, La tercera, 8 January 2019, https://www.latercera.com/que-pasa/noticia/los-records-dejo-suchai-1-primer-satelite-construido-chile/476111/ (accessed 7 June 2019).

[67]Suchai ha dado más de 5 mil vueltas a la Tierra en su primer año de operación, Facultad de Ciencias y Matemáticas, Universidad de Chile, 22 June 2018, http://ingenieria.uchile.cl/noticias/144476/suchai-ha-dado-mas-de-5-mil-vueltas-a-la-tierra-en-su-primer-ano (accessed 7 June 2019).

Table 9.6 Chile's Satellites launched from 1995 to 2017

Chile's Satellites launched to Outer Space[a]

Satellite	Launched	Orbit	Features	Focus
FASat-A (Alfa)	1995	No parameters	• Orbital position: launch into orbit was unsuccessful and the satellite is currently coupled with the Ukrainian satellite SICH-1 • Launch vehicle: TSIKLON-3 • Launch site: Plesetsk Cosmodrome (Russian Federation)	• Owner: Chile's Air Force • Status: inactive • General functions: ozone layer monitoring experiment; data transfer experiment; Earth imaging system; GPS receiver
FASat-B (Bravo)	1998	Low-Earth orbit (LEO). Sun-synchronous near-circular orbit	• Launch vehicle: ZENIT-2 • Launch site: Baikonur Cosmodrome (Kazakhstan)	• Owner: Chile's Air Force • Status: inactive (currently in orbit) • Expected life: six years • General functions: ozone layer monitoring experiment; data transfer experiment; Earth imaging system; GPS receiver
FASat-C (Charlie)	2011	Low-Earth orbit (LEO). Sun-synchronous circular orbit	• Launch vehicle: SOYUX 2 • Launch site: Guiana Space Centre, Kourou, French Guiana	• Owner: Chile's Air Force • Status: active • Expected life: six years, (it has already exceeded its expected life) • General functions: EO
SUCHAI-I	2017	Low-Earth orbit (LEO)	• Launch vehicle: PSLV (Polar Satellite Launch Vehicle) rocket • Launch site: Satish Dhawan Space Centre, India	• Owner: University of Chile • Status: active • General functions: Educational and scientific study of the atmosphere

[a]See Chile's verbal notes of 13 December 2012 and 9 January 2018 addressed to the Secretary-General, Information furnished in conformity with the Convention on Registration of Objects Launched into Outer Space, UNOOSA, Doc. ST/SG/SER.E/660, 26 February 2013, http://www.unoosa.org/documents/pdf/ser660E.pdf, and Doc. ST/SG/SER.E/832, 22 March 2018, respectively

9.7   Space Technology Development

Space technology development and research are mostly concentrated in the academia.[68] The Faculty of Physical Sciences and Mathematics of the University of Chile has different research laboratories related to space matters, such as the laboratories of Millimetric and Submillimetric Waves, the Mini Radio Telescope, the GOTO Telescope, and a Mechanical Workshop in the Department of Astronomy, and the Laboratory of Space and Planetary Exploration in the Department of Electrical Engineering.[69] The Faculty also has the Centre of Astrophysics and Related Technologies (CATA),[70] which is an internationally renowned entity dedicated to research and technology development related to astronomy.[71] Also, the Astrophysics Institute of the Faculty of Physics of the Pontifical Catholic University of Chile does research in astrophysics, both theoretical and observational.[72]

It is also appropriate to mention the observatories built in Chile, which have largely contributed to the understanding of the universe and to the development of technology in astronomy and other disciplines. The international observatories installed in Chile collaborate to the development of the national space sector. These observatories are the Cerro Tololo, Pachón, Las Campanas, La Silla, the VLT of Paranal in the Cordillera of la Costa, in the region of Antofagasta, and ALMA, in San Pedro de Atacama.[73]

9.8   Capacity Building and Outreach

Education and training of professionals in the space sector in Chile is conducted in different ways. On the one hand, space education in Chile goes from undergraduate and postgraduate programs in Astronomy, Aeronautics and even Law, to workshops

---

[68] According to a Chilean expert, "Chile has been declining in its timid leadership in the use of space technology in South America since 2001 to date, in which Chile it occupied the third place after Brazil and Argentina, down to sixth place after Venezuela, Bolivia and Peru, taking into account the investment in space infrastructure and in the decision to implement a space agency by the law" (unofficial translation), En medio de una nula política satelital como país, la Universidad de Chile culminó con éxito su primera mission especial, Nodal Tec, 15 January 2019, https://www.nodal.am/2019/01/en-medio-de-nula-politica-satelital-como-pais-universidad-de-chile-culmino-con-exito-su-primera-mision-espacial/ (accessed 3 July 2019).

[69] Laboratorios de Investigación, Facultad de Ciencias Físicas y Matemáticas, Universidad de Chile, http://ingenieria.uchile.cl/investigacion/presentacion/89805/laboratorios (accessed 1 July 2019).

[70] Centro de Astrofísica y Tecnologías Afines (CATA).

[71] Desarrollo académico de la astronomía, memoriachilena - Biblioteca Nacional de Chile, http://www.memoriachilena.gob.cl/602/w3-article-92384.html (accessed 1 July 2019).

[72] Instituto de Astrofísica, Pontificia Universidad Católica de Chile, http://astro.uc.cl/instituto/sobre-el-instituto (accessed 3 July 2019).

[73] Observatorios, memoriachilena - Biblioteca Nacional de Chile, http://www.memoriachilena.gob.cl/602/w3-article-92388.html (accessed 1 July 2019).

and conferences lectured by the Space Operations Group of Chile's Air Force. On the other hand, several Chilean NGOs, such as the Chilean Space Association, promote space education by carrying out dissemination of information activities.

9.8.1 Space Education

Chile's space education programs focus mainly on aeronautics, astronomy and telecommunications. Half of the programs are taught at the undergraduate level, while the other half are taught at the postgraduate level, both in master's and PhD degrees. The University of Concepción and the University of Chile account for most of these programs (see Table 9.7).[74]

Apart from the university programs, GOE of Chile's Air Force gives lectures, workshops and conferences on diverse space matters, namely issues concerning the operation of the FASat-Charlie satellite and the processing of its images.[75] There are also many amateur or professional groups for the diffusion of astronomy and other space disciplines, such as the Engineers Professionals Association, which organizes lectures on space issues, and the Aerospace Students Society of the University of Concepción.[76] The National Aerospace Museum, located in the capital Santiago, also displays exhibitions in aeronautics and space for educational purposes.[77]

---

[74]Education Opportunities in Space Law: a Directory, UNCOPUOS, op. cit., pp. 6–7. See also Plan de Estudios + Contenidos Mínimos, Universidad de Belgrano, Buenos Aires, Licenciatura en Relaciones Internacionales, Plan 2015, http://www.ub.edu.ar/sites/default/files/contenidos_minimos_Relaciones_Internacionales.pdf; Ingeniería Civil Aeroespacial, Departamento de Ingeniería Mecánica, Facultad de Ingeniería de la Universidad de Concepción, http://www.dim.udec.cl/pregrado/ingenieria-civil-aeroespacial/; Licenciatura en Ciencias con mención en Astronomía, Magíster en Ciencias con mención en Astronomía, and Doctorado en Ciencias con mención en Astronomía, Facultad de Ciencias Físicas y Matemáticas de la Universidad de Chile, http://ingenieria.uchile.cl/estudiarenlafcfm; Doctorado en Astrofísica, Instituto de Astrofísica de la Pontificia Universidad Católica de Chile, http://astro.uc.cl/item-4-menu-izquierdo#programa-de-estudio, (accessed 3 July 2019).

[75]Grupo de Operaciones Espaciales, Fuerza Aérea de Chile, http://www.ssot.cl/noticias18.php (accessed 1 July 2019).

[76]Space Generation Advisory Council, https://spacegeneration.org/regions/south-america/chile (accessed 2 July 2019).

[77]Museo Nacional Aeronáutico y del Espacio, official website: http://www.museoaeronautico.gob.cl/home/servicios.

Table 9.7 Space programs in Chile in 2019

Space Programs in Chile (2019)			
University	Program	Education level	Characteristics
Pontificia Universidad Católica de Chile	Astrophysics (Ph.D.)	Postgraduate	Apart from research work, the program consists of mandatory and optional courses. The mandatory or minimum courses comprise physical process in astrophysics, daily meetings for discussions, and bi-annual research courses. Optional courses include courses on physics and astrophysics (in particular extragalactic and stellar astrophysics)
Universidad de Chile	Space Law Course (Bachelor in Law)	Undergraduate	Three terms of eleven weeks each: The topics "Public International Law & Human Rights", and "International Organizations & International Environmental Law" which comprise 50% of the syllabus, deal with space law issues. There is also a moot court open to space law cases
Universidad de Chile	Specialization in Astronomy (Bachelor in Sciences)	Undergraduate	Eight semesters The first two years are part of the Common Plan of Engineering and Sciences of the Faculty of Physical Sciences and Mathematics. Then, it can be chosen as a specialization in Astronomy
Universidad de Chile	Specialization in Astronomy (Master in Sciences)	Postgraduate	Three to four semesters The cycle comprises courses and research workshops. It also includes the drafting of a dissertation in a period not of less than six months duration

(continued)

Table 9.7 (continued)

Space Programs in Chile (2019)			
University	Program	Education level	Characteristics
Universidad de Chile	Ph.D. in Astronomy	Postgraduate	Six to eight semesters The program is aimed at Physics or Astronomy graduates and those who want to complete their professional training in Astrophysics. The program has an exchange agreement with Yale University
Universidad de Concepción	Aerospace Civil Engineering	Undergraduate	Six years program. Its main development fields of expertise are: aerodynamics, flight mechanics, and astronautic; electronics and control of aerospace devices; propulsion; design of aircraft and spacecraft; economy and management; manufacturing processes, and complementary humanist studies Three cycles of two years each: 1. Basic cycle: Science courses (Maths, Physics, Chemistry), and basic principles of Engineering 2. Engineering Sciences cycle—scientific and technologic specialization 3. Professional cycle: application and technology specialisation courses, and dissertation
Universidad de Concepción	Civil Engineering in Telecommunications (Bachelor degree)	Undergraduate	A six years program that prepares professionals in the telecommunications sector. It is composed of three cycles (the same as the Aerospace Civil Engineering program)

(continued)

Table 9.7 (continued)

Space Programs in Chile (2019)			
University	Program	Education level	Characteristics
Universidad de Concepción	Astronomy (Bachelor degree)	Undergraduate	The program of two years trains graduates in the field of astrophysics. There are three research areas: 1. Stellar and galactic astrophysics 2. Extra-galactic astrophysics 3. Theoretical and instrumental astrophysics
Universidad de Concepción	Astronomy (Master degree)	Postgraduate	The program of four years trains professionals in basic sciences (mathematics and physics), as well as in the fundamental subjects of astronomy. The program is adapted to the labour demand in Chile, specifically at a technical and operational level in observatories and research centres

9.8.2 Chile and Non-governmental Space Institutions

Perhaps the main Chile's NGO on space matters is the Chilean Space Association A.G. (ACHIDE),[78] which promotes and disseminates information on space and satellite related activities. The association has three commissions: the statutes commission, which revises and updates the association's internal regulations; the education commission, which is responsible for communication with universities, institutes, technical training centres, and other education institutions; and the FIDAE Commission, which prepares the activities of the association for the International Air and Space Fair (FIDAE).[79]

ACHIDE, the Technical University Federico Santa Maria, and the College of Engineers of Chile A.G., are currently developing a program called "Vamos@Marte" (Go to Mars), which aims to celebrate the 50th anniversary of the arrival of man on the Moon and to invite young people to propose solutions for arrival and colonization of Mars.[80]

---

[78] Asociación Chilena del Espacio A.G, official website: http://achide.org/.

[79] Feria Internacional del Aire y del Espacio.

[80] Vamos@Marte, Desarrollando Ingeniería para Colonizar Marte, http://www.vamosamarte.cl/ (accessed 15 June 2019).

Appendix

See (Table 9.8).

Table 9.8 Institutional Contact Information in Chile (2019)

Institution	Details
Ministry of Transportation and Telecommunications	139 Amunátegui, Santiago, Chile Website: https://www.mtt.gob.cl/
Ministry of National Defense	45 Zenteno 4th Floor, Santiago, Chile Website: https://www.defensa.cl/
Chile's Air Force	5000 Avenue Pedro Aguirre Cerda, Cerrillos, Chile Website: https://www.fach.mil.cl/
National Commission on Scientific and Technological Research (CONICYT)	1375 Moneda, Santiago, Chile Website: https://www.conicyt.cl/
University of Chile	Avenue Libertador Bernardo O'Higgins 1058, Santiago, Chile Website: http://www.uchile.cl/

Chapter 10
Colombia

Contents

Abstract This chapter presents some of the principal characteristics of Colombia's space activities. A general overview is described in Section One and Colombia's space history is explained in Section Two. Section three focuses on Colombia's space actors, Section Four analyses the country's space legislation and Colombia's ratification of international legal space instruments, Section Five analyses the cooperation programs adopted by Colombia and other international partners, and Section Six explains Colombia's satellite capacities. Finally, Section Seven mentions some of the space related programs offered by Colombia's universities.

10.1 General Country Overview

With a territory of 1,141,748 km^2, the Republic of Colombia (Colombia) is located in northeast South America. It borders the Antilles to the north, Venezuela and Brazil to the east, the Pacific Ocean to the west, and Peru and Ecuador to the south. It has an estimated population of 50,400,000 inhabitants (2019) of which 6,673,000 live

© Springer Nature Switzerland AG 2020   305
A. Froehlich et al., *Space Supporting Latin America*, Studies in Space Policy 25,
https://doi.org/10.1007/978-3-030-38520-0_10

in the capital, Bogota. Spanish is the official language, although there are several co-official indigenous languages in different territories. Colombia's currency is the Peso.[1]

Geographically, Colombia is divided into six regions: the Caribbean Region which comprises the Colombian Caribbean's coasts plains and the mountainous groups of Montes de Maria and Sierra Nevada de Santa Marta; the Pacific Region, a humid region with a great array of bays and coves; the Andean Region, a populated region occupied by the Andes; the Orinoquia and Amazonia Regions, both covered by huge plains, and the Insular Region, which comprises all of the Colombia's islands beyond the continental shelf.[2]

10.1.1 *Political System*

Colombia is a social state of law organised in a unitary and decentralized republic, and with autonomous territorial entities. Colombia's political branches are the executive, the legislative and the judiciary. It has also several autonomous bodies.[3] The executive is headed by the President of the Republic, who is the Head of State and Government (currently Ivan Duque Marquez); the legislature is divided into the Senate and the Chamber of Representatives, and the judiciary comprises the Constitutional Court, the Supreme Court of Justice, the State Council, and other minor courts.[4]

Colombia is partitioned in 32 departments (Fig. 10.1), a capital district (Bogota), and multiple districts, municipalities and indigenous territories.[5]

---

[1] Colombia, Ficha país, Oficina de Información Diplomática del Ministerio de Asuntos Exteriores, Unión Europea y Cooperación, Gobierno de España, September 2019, p. 1, http://www.exteriores. gob.es/Documents/FichasPais/COLOMBIA_FICHA%20PAIS.pdf (accessed 13 October 2019).

[2] Regiones,   Colombia.com,   https://www.colombia.com/colombia-info/informacion-general/ geografia/regiones/ (accessed 13 October 2019).

[3] Art. 113, Political Constitution of the Republic of Colombia, 1991.

[4] Art. 113–116, Political Constitution of the Republic of Colombia, 1991.

[5] Art. 286, Political Constitution of the Republic of Colombia, 1991.

Fig. 10.1 Colombia's departments (2019). Image obtained from Hablemos de Culturas https://hablemosdeculturas.com/departamentos-de-colombia/ (accessed 18 October 2019)

10.1.2   Economic Situation

With an estimated GDP of USD 347 billion (2019), Colombia is the fourth economy in Latin America.[6] In spite of economic downturns[7] in the last decade, Colombia currently maintains a solid macroeconomic framework as a result of a full-fledged inflation-targeting regime, a flexible exchange rate, a Fiscal Rule for the central government, and a Medium-Term Fiscal Framework.[8] In the first half of 2019, Colombia's economic growth accelerated to 3.3% thanks to robust private consumption and stronger investment, and it is expected to accelerate further to 3.6% in 2020.[9]

10.2   History of Space Activities

Colombia's space activities have taken many years to materialise. For many decades, the country has been incapable to develop a national space sector. The reasons behind this situation are the absence of a long-term strategic vision and an appropriate governance framework for the activation of the space ecosystem, and the existence of entry barriers to the development of the sector due to market failures.[10] According to Colombia's National Planning Department, the country has lost the opportunity to take advantage of the benefits of the space and satellite sectors to invigorate economic growth and strengthen social development.[11] For example, during the Cold War period, Colombia tried to acquire two communications satellites and one EO satellite without success. Another failed attempt was the 1976 Declaration of Bogota by which Colombia and other Equatorial countries tried in vain to claim sovereignty rights over the GEO situated over their respective territories.[12] Finally, in 2002, Colombia created the Colombia Space Commission (CSC)[13] that published a draft of a National Plan on Satellite Navigation that was never formalised in an official document.[14]

---

[6]Colombia, Ficha país, Oficina de Información Diplomática del Ministerio de Asuntos Exteriores, Unión Europea y Cooperación, Gobierno de España, September 2019, p. 2, http://www.exteriores.gob.es/Documents/FichasPais/COLOMBIA_FICHA%20PAIS.pdf (accessed 13 October 2019).

[7]Ibid.

[8]The World Bank in Colombia, The World Bank, 10 October 2019, https://www.worldbank.org/en/country/colombia/overview (accessed 18 October 2019).

[9]Ibid.

[10]Colombia prepara su primera Política Nacional Espacial, El Espectador, 9 October 2017, https://www.elespectador.com/noticias/ciencia/colombia-prepara-su-primera-politica-nacional-espacial-articulo-717173 (accessed 20 September 2019).

[11]Ibid.

[12]Ibid.

[13]Comisión Colombiana del Espacio.

[14]Colombia prepara su primera Política Nacional Espacial, El Espectador, op. cit.

Despite these setbacks, there has been some important progress in recent years on specific space projects. In 2007 the University Sergio Arboleda launched into outer space the CubeSat satellite LIBERTAD I and is currently developing the satellite LIBERTAD II.[15] In 2013, the Presidential Program for the Colombian Space Development was created and a year later the Presidency decreed the Department on Space Projects responsible for the development of the Strategic Plan on Space Development.[16] More recently, in 2017, a private initiative established the Colombian Space Agency to boost the development of the Colombian space sector, and in 2018 the Colombian Air Force launched into outer space its first satellite, FACSAT-1.[17] All of these recent initiatives show the gradual, although slow, development of a national space sector that could be reinforced in the coming years by the creation of a public space body.[18]

10.3   Space Development Policies and Emerging Programs

As noted above, the CSC has not yet elaborated the National Space Policy (NSP). Despite the lack of an NSP, there are currently different space actors with diverse goals that contribute to the development of Colombia's space activities.

10.3.1   Space Actors

The main space actors that contribute to Colombia's space development are the Colombian Space Commission, the Presidential Program for Colombian Space Development, the Colombian Air Force, and the Colombian Space Agency.

10.3.1.1   The Colombian Space Commission

Created in 2006, the Colombian Space Commission (CSC) serves as an advisory, coordination, orientation and planning inter-sectoral body that gives guidance for the implementation of the NSP. It also coordinates the creation of space plans, programs

---

[15]Ibid.

[16]See Portilla José, El limbo de la Comisión Colombiana del Espacio, Semana, 9 September 2016, https://www.semana.com/tecnologia/articulo/el-limbo-de-la-comision-colombiana-del-espacio/486295 (accessed 22 September 2019).

[17]Ibid.

[18]Despacio, así va la conquista colombiana del espacio, Semana, 20 May 2019, https://www.semana.com/contenidos-editoriales/el-tiempo-vuela/articulo/colombia-necesita-una-politica-de-desarrollo-espacial-segun-raul-joya/616617 (accessed 20 September 2019).

and projects.[19] The CSC is composed of 15 members of different national bodies, agencies and research centres, and is headed by the Vice-President of the Republic.[20] The CSC has also an Executive Secretary and a Technical Committee on Space Affairs.[21]

CSC's work is articulated in several strategic fields that are in the CSC's action plan: formulation of space policies and planning (including diagnostics, studies, plans, legislation and regulations); training for the generation and use of space science and technology; use of EO's space sciences and technology, and adoption of international space instruments.[22]

The CSC was envisaged as a preliminary body for the creation of a Colombian Space Agency. However, the agency does not yet exist and the CSC has been inactive for several years,[23] even though the 2018–2022 NDP states that the national government, in the framework of the CSC, will design an NSP (see 10.4.1).[24]

10.3.1.2   Presidential Program for Colombian Space Development

The Presidential Program for Colombian Space Development (PPDEC)[25] is a body created in 2016 under the Administrative Department of the Presidency of the Republic. It is responsible for leading, coordinating, strengthening and giving impulse to Colombian space development and its integration into the international arena through multiple space technology plans, projects and programs.[26]

PPDEC's functions include the proposal of an NSP, strategies and actions to promote Colombian space development; the guidance and promotion of the Strategic Plan on Space Development and the implementation of instruments related to this Plan; the promotion, coordination and encouragement of aerospace academic programs, scientific development and industry; the promotion and support of space

---

[19] Art. 2, Decreto 2442 de 2006, Creación de la Comisión Colombiana del Espacio, Bogota, D.C., 18 July 2006, http://www.ideam.gov.co/documents/11769/138916/Decreto+2442+del+18+de+julio+del+2006.pdf/113d8fdd-3d08-4490-9a16-c5b7538526ad (accessed 21 September 2019).

[20] Ibid., Art. 3.

[21] Ibid., Arts. 6–8.

[22] ¿Qué es la Comisión Colombiana del Espacio?, Instituto de Hidrología, Metereología y Estudios Ambientales, http://www.ideam.gov.co/web/ecosistemas/comision-colombiana-espacio-grupo-observacion-tierra (accessed 21 November 2019).

[23] During its first years of work, the CSC had planned to acquire a geostationary satellite and an EO satellite. Nevertheless, during the term of president Santos, these projects were cancelled. See Portilla José, El limbo de la Comisión Colombiana del Espacio, Semana, 9 September 2016, https://www.semana.com/tecnologia/articulo/el-limbo-de-la-comision-colombiana-del-espacio/486295 (accessed 22 September 2019).

[24] Bases del Plan Nacional de Desarrollo 2018-2022: Pacto por Colombia, pacto por la equidad, Departamento Nacional de Planeación, Bogotá, 2019, p. 582 https://colaboracion.dnp.gov.co/CDT/Prensa/PND-2018-2022.pdf (accessed 22 September 2019).

[25] Programa Presidencial para el Desarrollo Espacial Colombiano.

[26] Art. 1, Decreto 2615 de 2013 por el que se crea el Programa Presidencial para el Desarrollo Espacial Colombiano, 15 November 2013.

scientific and technological research, and innovation; and human training on space matters.[27]

10.3.1.3 Colombian Air Force

The Military Aviation School Marco Fidel Suarez (EMAVI)[28] of the Colombian Air Force is responsible for the operation of the EO satellite FACSAT-1, launched in 2018. Moreover, EMAVI organises public events on different space issues, such as the First International Symposium on the Synergy of the Colombian South West to boost the aerospace industry, held in Cali in 2018.[29]

10.3.1.4 The Colombian Space Agency

The Colombian Space Agency (CSA)[30] was established on 4 October 2017 as a private, non-profit corporation that aims to join efforts between academia, the government and the private sector and develop the aerospace services and technological industry in Colombia.[31] Its purpose is to promulgate an aerospace vision that enables Colombia to evolve in the application and development of satellite technologies with the aim of promoting productivity, efficiency, growth and integration of the different economic sectors and contributing to building a more inclusive, prosperous and better country.[32]

The CSA is chaired by an Executive Director[33] and it has a board of directors composed of different public and private entities, such as the National Planning Department, the Planetarium of Bogota, the Mercantile Exchange of Colombia, the Embassy of France and the Colombian Aeronautic Industry Corporation.[34] The Agency has also several strategic allies including national and international governments, multilateral entities, space agencies, financial entities, education institutions and the private sector.[35]

---

[27]Ibid., Art. 2.

[28]Escuela Militar de Aviación Fidel Suárez (EMAVI).

[29]La FAC apuesta por potenciar la industria aeroespacial en Colombia, Infoespacial.com, 31 October 2018, Cali, http://www.infoespacial.com/latam/2018/10/31/noticia-debate-impulso-industria-espacial-colombiana.html (accessed 20 September 2019).

[30]Agencia Espacial de Colombia.

[31]History, Colombian Space Agency, http://www.agenciaespacialdecolombia.org/en/ (accessed 20 September 2019).

[32]Ibid.

[33]Pilar Zamora Acevedo is ACS' founder and current executive director.

[34]Board of Directors, Colombian Space Agency, http://www.agenciaespacialdecolombia.org/en/ (accessed 20 September 2019).

[35]Strategic Allies, Ibid.

The Agency carries out different actions for the development and use of space technology and its applications in order to contribute to Colombia's social development and to strengthen its national defence. These actions include the promotion and development of space technology having an impact in the agricultural, education and health sectors, the promotion of satellite connectivity projects, the use of space technology to improve Colombians' financial inclusion, and the use of satellite space data for the prevention of irreversible damages to the environment, and the protection of the national territory.[36]

10.4 Current and Future Perspectives on Legal Framework Related to Space Activities

To establish the foundation of its national space capacities, Colombia has adopted several institutional regulations and ratified different international space related instruments.

10.4.1 National Space Regulations and Instruments

Colombian space legislation is based on Decree 2442/2006 that establishes the Colombian Space Commission (see 10.3.1.1),[37] Decree 2516/2013 that creates the Presidential Program for Colombian Space Development under the Administrative Department of the Presidency of the Republic (see 10.3.1.2),[38] and Law N° 1955/2019 on the establishment of the 2018–2020 NDP: "Pact for Colombia, Pact for Equity" (NDP), which sets the digital transformation of Colombia as one of the NDP's transversal strategic objectives.[39]

---

[36] Action Areas, Colombian Space Agency, http://www.agenciaespacialdecolombia.org/en/action-areas/ (accessed 20 September 2019).

[37] Decreto 2442 de 2006, Creación de la Comisión Colombiana del Espacio, Bogota, D.C., 18 July 2006, http://www.ideam.gov.co/documents/11769/138916/Decreto+2442+del+18+de+julio+del+2006.pdf/113d8fdd-3d08-4490-9a16-c5b7538526ad (accessed 21 September 2019).

[38] Decreto 2615 de 2013 por el que se crea el Programa Presidencial para el Desarrollo Espacial Colombiano, 15 November 2013, https://www.funcionpublica.gov.co/eva/gestornormativo/norma_pdf.php?i=67836 (accessed 22 September 2019).

[39] Ley N° 1955 por el cual se expide el Plan Nacional de Desarrollo 2018-2022. "Pacto por Colombia, Pacto por la Equidad", Congreso de Colombia, 25 May 2019, https://colaboracion.dnp.gov.co/CDT/Prensa/Ley1955-PlanNacionaldeDesarrollo-pacto-por-colombia-pacto-por-la-equidad.pdf (accessed 22 September 2019).

Furthermore, according to the bases of the 2018–2022 NDP,[40] an integral document of the NDP that details its strategic objectives, one of the NDP's programs is the implementation of a national policy to develop the space sector.[41] In this sense, the national government, in the framework of the CSC, will design an NSP for the development of the space sector by taking into account the following points: the revision and planning of a strengthening strategy of the current governance framework by proposing measures for its articulation with productivity and competitiveness policies, defining roles of the public entities, and articulating the interests of the public, private and academic sectors; the establishment of a roadmap for the identification of Colombia's potentialities in order to define strategic lines that the country could use to orientate its public and private efforts; and the definition of solutions to eliminate market barriers and flaws to foster the undertaking and investment in the space domain and the adoption of space technologies by the public and private sectors.[42]

10.4.2 Colombia and International Space Instruments

Colombia has ratified several international space and telecommunication treaties. Among the UN Space Treaties, Colombia has ratified the Liability Convention and the Registration Convention, and has signed the OST and ARRA. Other space related treaties ratified by Colombia include the Treaty Banning Nuclear Weapon Tests in the Atmosphere, in Outer Space and Under Water (NTB) and the ITU Convention and Constitution (Table 10.1).

---

[40]Bases del Plan Nacional de Desarrollo 2018-2022: Pacto por Colombia, pacto por la equidad, Departamento Nacional de Planeación, Bogotá, 2019, p. 582, https://colaboracion.dnp.gov.co/CDT/Prensa/PND-2018-2022.pdf (accessed 22 September 2019).

[41]The 2014–2018 NDP recognised the economic relevance of the space sector. See Colombia prepara su primera Política Nacional Espacial, El Espectador, op. cit.

[42]The NSP will be established within the framework of the National Policy of Productive Development (Política Nacional de Desarrollo Productivo – CONPES 3866 of 2016), Bases del Plan Nacional de Desarrollo 2018-2020, op. cit., p. 582.

Table 10.1 International space related treaties respected by Colombia (2019)

International space treaties ratified by Colombia[a]

Treaty[b]	Adoption	Ratification (R)/signature (S)
OST	1967	S
ARRA	1968	S
LIAB	1972	R
REG	1975	R
MOON	1979	–
NTB	1963	R
BRS	1974	R
ITSO	1971	R
IMSO	1976	R
ITU	1992	R

[a]Status of International Agreements relating to activities in outer space as at 1 January 2019, UNOOSA Doc, A/AC.105/C.2/2019/CRP.3, http://www.unoosa.org/documents/pdf/spacelaw/treatystatus/AC105_C2_2019_CRP03E.pdf (accessed 4 October 2019)
[b]Treaty on Principles Governing the Activities of States in the Exploration and Use of Outer Space, including the Moon and Other Celestial Bodies (OST); Agreement on the Rescue of Astronauts, the Return of Astronauts and the Return of Objects Launched into Outer Space (ARRA); Convention on International Liability for Damage Caused by Space Objects (LIAB); Convention on Registration of Objects Launched into Outer Space (REG); Agreement Governing the Activities of States on the Moon and Other Celestial Bodies (MOON); Treaty Banning Nuclear Weapon Tests in the Atmosphere, in Outer Space and under Water (NTB); Convention Relating to the Distribution of Programme-Carrying Signals Transmitted by Satellite (BRS); Agreement Relating to the International Telecommunications Satellite Organisation (ITSO); Convention on the International Mobile Satellite Organization (IMSO); International Telecommunication Constitution and Convention (ITU)

10.5   Cooperation Programs

In Colombia, it is the CSA that has adopted most of the cooperation agreements on space related issues with international partners.

CSA's strategic international partners include foreign agencies such as the South African National Space Agency (SANSA), the German Aerospace Center (DLR) and the Ecuadorian Civilian Space Agency (EXA). There are also foreign governments such as the French and Swedish governments. The regional organisation Comunidad Andina (CAN) and foreign companies such as Thales and Eutelsat[43] are of importance as strategic allies.

In October 2018, CSA, EXA and the Astrobotic Company[44] agreed to launch a future lunar exploration program consisting in launching the ARTEMIS-CLASS 2U

---

[43]Strategic Allies, Colombian Space Agency, http://www.agenciaespacialdecolombia.org/en/ (accessed 20 September 2019).

[44]Astrobotic Technology Inc. is a space robotic company that seeks to make space accessible to the world, see About Astrobotic, https://www.astrobotic.com/about (accessed 30 October 2019).

cubesat into lunar orbit by using the Astrobotic Lander PEREGRINE in order to carry out a satellite technology experiment. After this first phase of the Colombo-Ecuadorian Lunar Program, there will be a monitoring of this satellite to send payloads in other exploration missions of the Lander PEREGRINE in the future.[45]

10.6 Satellite Capacities

Satellite LIBERTAD I was the first ever satellite launched into outer space by Colombia. The University of Sergio Arboleda developed this satellite, which reached LEO in 500 km altitude in April 2007. Its main objective was to send telemetric data on its functioning, measure the temperature of the satellite's external borders and relevant system data (such as the electric circuit data).[46] Although the satellite still orbits the Earth, its batteries ran out some years ago. However, the same university is currently working on the development of LIBERTAD II, a satellite designed to take images from space that could be used by several national institutions.[47]

In 2018, the Colombian Air Force launched into Earth's orbit the satellite FACSAT-1 from India. FACSAT-1 was launched by the Indian Space Research Organisation (ISRO) aboard a PSLV[48] rocket. This is an EO nanosatellite whose images are used for different purposes, from urban development to natural disasters management (Table 10.2). The satellite is operated by EMAVI from the control centre in the city of Cali, which was established with the support of the Danish company GomSpace.[49]

[45]Colombia y Ecuador participarán en la exploración lunar de Astrobotic, Infoespacial.com, 8 October 2018, Bogotá, http://www.infoespacial.com/latam/2018/10/08/noticia-colombia-ecuador-participaran-exploracion-lunar-astrobotic.html (accessed 20 September 2019).

[46]Joya Raúl, Libertad 1 - Primer satélite colombiano en el espacio, Universidad Sergio Arboleda, https://www.usergioarboleda.edu.co/satelite-libertad-1/ (accessed 5 December 2019).

[47]Vida, Se cumplen 10 años del lanzamiento del satélite Libertad 1, El Tiempo, 5 May 2017, https://www.eltiempo.com/vida/ciencia/libertad-1-diez-anos-del-lanzamiento-del-satelite-colombiano-84636 (accessed 5 December 2019).

[48]The Polar Satellite Launch Vehicle (PSV) is an expendable medium-lift launch vehicle designed and operated by ISRO.

[49]Colombia lanza hoy su FACSAT-1, Infoespacial.com, 29 November 2018, http://www.infoespacial.com/latam/2018/11/29/noticia-colombia-lanza-facsat1.html (accessed 20 September 2019).

Table 10.2 Colombian Air Force's satellite FACSAT-1[a]

Colombian Air Force's satellite FACSAT-1	
State of Registry	Colombia
Date and territory or location of launch	29 November 2018 at 0427 h, 30 s UTC; Satish Dhawan Space Centre, Sriharikota, India
Basic orbital parameters	Nodal period: 94 min, 37 s Inclination: 97.4760° Apogee: 505.9466593 km Perigee: 478.5982954 km
General function	3U amateur nanosatellite equipped with an optical camera with a 30-metre ground sample distance. Intended for capacity building in space technology by the Colombian Air Force.
Space object owner or operator	Colombian Air Force
Launch vehicle	PSLV-C43

[a]Data obtained from Information furnished in conformity with the Convention on Registration of Objects Launched into Outer Space, Note verbale dated 7 March 2019 from the Permanent Mission of Colombia to the United Nations (Vienna) addressed to the Secretary-General, UNCOPUOS, ST/SG/SER.E/885, 20 March 2019, https://cms.unov.org/dcpms2/api/finaldocuments?Language= en&Symbol=ST/SG/SER.E/885

10.7   Capacity Building and Outreach

A few Colombia's universities offer space related programs. The University of Antioquia and the University of San Buenaventura currently offer aerospace engineering programs at the undergraduate and postgraduate levels, and the Universidad Libre has a specialised course on Space Law in its undergraduate general Law program (Table 10.3).

Table 10.3 Space programs in Colombia (2019)

Space programs in Colombia[a]

University	Program	Education level	Characteristics
University of Antioquia	Aerospace Engineering (Bachelor's degree)	Undergraduate	At the end of the sixth semester the student must choose one of two specialisations: applied aviation or space flight. The former includes courses on aircraft operations, aviation human factors, and aircraft maintenance and airworthiness. The latter is composed of courses on dynamic and attitude control, space missions' analysis and design, and engineering on space vehicles[b]
University of San Buenaventura	Aerospace Engineering (Master's degree)	Postgraduate	This four semester program offers three specialisations: aerospace maintenance and management, energy and propulsion, and aerospace vehicles design[c]
Universidad Libre	Course on Space Law (Bachelor's level)	Undergraduate	This one-year program focuses on the generalities of outer space and the use of space for telecommunications purposes[d]

[a]There are several Colombian universities offering programs on telecommunications engineering. A detailed list is available at https://carrerasuniversitarias.com.co/carreras/ingenieria-de-telecomunicaciones (accessed 22 September 2019)

[b]Ingeniería Aeroespacial, Universidad de Antioquia, http://www.udea.edu.co/wps/portal/udea/web/inicio/institucional/unidades-academicas/facultades/ingenieria/programas-academicos/programas-pregrado/ingenieria-aeroespacial (accessed 22 September 2019)

[c]Maestría en Ingeniería Aeroespacial, Universidad de San Buenaventura, https://www.usbbog.edu.co/facultades/facultad-de-ingenieria/posgrados/maestria-en-ingenieria-aeroespacial/ (accessed 22 September 2019)

[d]Universidad Libre, Facultad de Derecho, Contenido Programático de la Asignatura-Derecho Espacial, http://www.unilibre.edu.co/derecho/images/stories/pdfs/2013/optativas/AREA%20DERECHO%20PUBLICO/DERECHO-ESPACIAL.pdf (accessed 22 September 2019)

Appendix

See Table 10.4.

Table 10.4 Institutional contact information in Colombia (2019)

Institution	Details
Colombian Air Force	Avenue El Dorado Cra. 54 # 26-25, Bogota, Colombia Website: http://www.fac.mil.co/ Email: webmaster@simfac.mil.co
Colombian Space Agency	Cr 11 # 93°-53, office 202, Bogota, Colombia Website: http://www.agenciaespacialdecolombia.org Email: info@agenciaespacialdecolombia.org
University Sergio Arboleda	74 Street # 14-14, Bogota, Colombia Website: https://www.usergioarboleda.edu.IEEco/

Chapter 11
Ecuador

Contents

Abstract Ecuador has made important progress in the development of its space
sector in recent years, mainly since the creation of the national space agency that
has made an extraordinary contribution to putting the country in the international
space arena. This chapter presents a description of Ecuador's main space activities.
Section one is a general overview of the South American country; section two presents
Ecuador's space history; section three and four focus on the country's space policies,
legislation and actors; section five describes the recent space cooperation programs
adopted by Ecuador and other international partners; section six concerns Ecuador's
space activities, and sections seven and eight briefly mention the country's space
technology development and its capacity building and outreach activities.

© Springer Nature Switzerland AG 2020   319
A. Froehlich et al., *Space Supporting Latin America*, Studies in Space Policy 25,
https://doi.org/10.1007/978-3-030-38520-0_11

11.1   General Country Overview

The Republic of Ecuador (Ecuador) is located in northwest South America. It borders Colombia to the north, Peru to the south and east, and the Pacific Ocean to the west. The capital is San Francisco de Quito (Quito). Ecuador's land surface is 283,561 km^2 and its population is 17,180,000 inhabitants (2019). The official language is Spanish, together with Quechua and Shuar. The Ecuadorian currency is the USD.[1]

Geographically, Ecuador has four regions: the Costa Region, in the west, has fertile plains, hills, sedimentary basins and low elevations. The Sierra or Inter Andean Region is covered by high mountains, volcanoes and peaks, and there are several natural parks with a rich variety of flora and fauna. The East or Amazonian Region, a zone with wet-tropical forests, has many hills in the west Andes that go down to the Amazonia plain. Finally, the Galapagos or Insular Region is constituted by an archipelago composed of 13 volcanic islands, six small islands and 107 listed rocks and islets.[2]

11.1.1   Political System

Ecuador is a constitutional, democratic, unitary, plurinational and secular state and decentralized republic.[3] The country is territorially organised in regions, provinces, cantons and rural parishes[4] (Fig. 11.1).

The President of the Republic (currently Lenin Moreno) exercises the executive function and is the head of state and government; the National Assembly (unicameral) is in charge of the legislative function, and the National Court of Justice, the provincial courts of justice and other tribunals and bodies exercise the judicial function. There is also the National Electoral Council and the Electoral Court, which exercise the electoral function; and the Citizen's Participation Council, the Ombudsman and the General Comptroller of the state, which are responsible for civic participation, human rights protection, transparency and accountability, respectively.[5]

---

[1] Ecuador, EcuRed, https://www.ecured.cu/Ecuador (accessed 5 November 2019).

[2] Regiones naturales de Ecuador, Wikipedia, https://es.wikipedia.org/wiki/Regiones_naturales_de_Ecuador (accessed 5 November 2019).

[3] Art. 1, Constitución de la República del Ecuador, Official Registry N° 449, 20 October 2008.

[4] The metropolitan autonomous districts, the Galapagos province and the indigenous and pluricultural territorial circumscriptions have special regimes, Art. 242, Ibid.

[5] Título IV, Participación y Organización del Poder, Ibid.

Fig. 11.1 Ecuador's provinces (2019)

11.1.2   Economic Situation

Between 2007 and 2014, Ecuador experienced economic growth thanks to the boom in oil prices and poverty reduction. However, the recent decrease in oil prices has highlighted problems, such as an inefficient public sector, large macroeconomic imbalances, a lack of stabilisation mechanisms and limited private investment.[6] To tackle some of these problems, in 2019, the International Monetary Fund (IMF) approved an agreement with Ecuador to provide support to government economic reforms proposed in the 2018–2021 Prosperity Plan. The same year, the World Bank authorized a Development Policy Loan. Currently, the government is implementing significant structural reforms to adapt its economy to a challenging international context.[7]

11.2   History of Space Activities

Ecuador's space history started during the space race when the USA, through NASA, agreed with Ecuador on the installation on Ecuadorian territory of a monitoring station for NASA's satellites. Nevertheless, for almost fifty years, Ecuador did not really possess a national space sector. It was not until the first years of the current century that a private initiative decided to develop space activities for the benefit of Ecuadorian society. The Government, through the Ecuadorian Air Force, agreed to collaborate with this private initiative. The subsequent creation of an Ecuadorian space agency and a space plan to be implemented by this civilian institution marked the beginning of the Ecuadorian space sector that has achieved significant progress to date (Table 11.1).

---

[6]The World Bank In Ecuador, The World Bank, 15 October 2019, https://www.worldbank.org/en/country/ecuador/overview (accessed 6 November 2019).
[7]Ibid.

Table 11.1 Ecuador's space history

Year	Ecuador's space history[a]
1957	NASA installs on the side of the Cotopaxi volcano the Mini Track satellite monitoring station to monitor and control USA satellite orbits
1977	Creation of the Centre for Natural Resources Integrated Survey by Remote Sensors (CLIRSEN)[b]
1982	CLIRSEN is responsible for the maintenance of NASA's facilities and equipment left by the Agency in 1981
1989	The Cotopaxi facilities become a receiving, recording and processing station for satellite data
2003	Formation of the project "EX SOMINUS AD ASTRA" or project ESAA, later known as the project "Ecuador to Space", with the first Ecuadorian mission: the Scientific Suborbital Mission ESAA—01
2006	First Ecuadorian to undergo professional training as an astronaut in the Yuri Gagarin Cosmonaut Training Center (GCTC), Russia. Announcement of the first Ecuadorian space mission: ESAA—01
2006	First contact between the Ecuadorian Air Force (FAE)[c] and the ESAA—01 project
2007	The FAE general commander recognises the ESAA project as the first Ecuadorian astronautic initiative and Ronnie Nader as the first Ecuadorian astronaut
2007	Announcement of the creation of the Ecuadorian Space Program for the next ten years
2007	ESAA and FAE carry out the first Ecuadorian aerospace exercise (a suborbital flight take-off simulation) aboard a MIRAGE FIJE plane
2007	Creation of the first Ecuadorian space agency: the Ecuadorian Civilian Space Agency (EXA)[d]
2008	EXA and FAE carry out the DÉDALO project
2008	Beginning of the POSEIDÓN project by EXA and FAE
2008	The International Astronautical Federation (IAF) accepts EXA as a national space agency
2009	Development of the HERMES and the "A Satellite in the Classroom" projects
2012	Creation of the Ecuadorian Space Institute (IEE)[e] by Executive Decree
2013	Launch of satellite PEGASO, the first Ecuadorian satellite, from China
2013	Launch of satellite KRYSAOR from China
2014	Satellites PEGASO and KRYSAOR make contact
2018	The American consortium Irvine CubeSat STEM Program (ICSP) commissions EXA to design, build and test a laser transmitter for Earth orbit communications
2018	EXA and the Colombian Space Agency signs an agreement to launch the first Latin American manned mission into outer space
2018	Start of the Ecuadorian lunar program
2018	Launch of IRVINE01 and IRVINE02 satellites into outer space
2019	Dissolution of the IEE by the President of the Republic
2019	Sacebit UK announces that EXA and DEREUM were selected to build a walking robot to land on the moon in 2021

[a]La Historia Espacial del Ecuador, Agencia Espacial Civil Ecuatoriana, http://exa.ec/ (accessed 14 November 2019)
[b]Centro de Levantamientos Integrados de Recursos Naturales por Sensores Remotos (CLIRSEN)
[c]Fuera Aérea Ecuatoriana (FAE)
[d]Agencia Espacial Civil Ecuatoriana (EXA)
[e]Instituto Espacial Ecuatoriano (IEE)

11.3   Space Development Policies and Emerging Programs

The lack of a governmental space plan has not prevented the emergence of a civilian space program called the Ecuadorian Civilian Space Plan (ECSP). Moreover, there are several space actors that contribute to some extent to the implementation of this plan.

11.3.1   The Ecuadorian Civilian Space Plan

The Ecuadorian Civilian Space Plan was presented for the first time on 29 August 2007.[8] ECSP's objective is to provide access to space to the Ecuadorian people by Ecuadorian hands by training at least one national astronaut for the implementation of scientific, technical and educational missions, and carrying out experiments from different institutions in each of these missions.[9]

The ECSP is created, organized and operated by the Ecuadorian Project "To Space", and is composed of three phases[10]:

1. The ESSA—01 suborbital phase (ESAA—01A suborbital and ESAA—01B suborbital missions). Its objective is to reach an altitude of 100 km and carry out two Ecuadorian scientific experiments in each mission.
2. The ESAA—02 orbital phase, whose objective is to reach an altitude of 400 km to carry out four experiments in a ten-day stay aboard the International Space Station (ISS) or other similar facility.
3. The ESAA—03 moon landing phase. Its objective is to reach an altitude of 384,000 km to land on the moon and plant the Ecuadorian flag on the Earth's natural satellite. The installation of a radio transmitter on the moon is also projected.

11.3.2   Space Actors

There are three main actors that contribute to the development of Ecuador's space activities: the Ecuadorian Civilian Space Agency, the Ecuadorian Air Force, and the Ministry of Telecommunications and the Information Society (MTSI).[11]

---

[8]ECSP's early objectives were the creation of a civil space agency, the development of a microgravity laboratory plane, the launch into outer space of five manned missions, the launch of the first Ecuadorian satellite to Earth orbit, and reach the Moon in 2020; see Nader Ronnie, El Programa Espacial Civil Ecuatoriano, Reporte al 2010, Agencia Espacial Civil Ecuatoriana, 2010, https://www.defensa.gob.ec/wp-content/uploads/downloads/2013/04/POWER-POINT-EL-PROGRAMA-ESPACIAL-DE-EXA.pdf (accessed 14 November 2019).

[9]El Objetivo, Agencia Espacial Civil Ecuatoriana, http://exa.ec/ (accessed 15 November 2019).

[10]Ibid.

[11]Ministerio de Telecomunicaciones y de la Sociedad de la Información (MTSI).

11.3.2.1   The Ecuadorian Civilian Space Agency

The Ecuadorian Civilian Space Agency (EXA), the first ever Ecuadorian space agency, was created in 2007 as an independent civil body. The Agency was recognised internationally as a national space agency on 28 September 2008, and is member of the IAF.[12] EXA is responsible for the implementation of the ECSP (see 11.3.1) and the development of scientific research on planetary and space sciences, and to foster science education in Ecuador. EXA has its own astronaut, Ronnie Nader, who is currently the president of EXA's Board of Directors.[13] Since its inception, EXA has carried out several significant space projects alone or in collaboration with the FAE.

11.3.2.2   The Ecuadorian Air Force

Ecuador is aware of is strategic geographical situation due to the great array of space activities carried out by different countries over its subjacent territory. The geostationary orbit (GEO) and the electromagnetic spectrum are of vital importance to the country's national defence and development, and for this reason the development of public policies on the matter is crucial for the country.[14] To this end, the national defence sector participates and possesses EO and aerospace capacities.[15] In this regard, the FAE has been an important Ecuadorian space actor, namely because it has worked together with EXA in the development of some space projects, such as the PEGASO and POSEIDON satellite programs (see Sect. 11.6).

11.3.2.3   The Ministry of Telecommunications and the Information Society

The Ministry of Telecommunications and the Information Society (MTSI) is the national body responsible for the coordination of the telecommunications sector policy. In particular, the MTSI carries out the administrative procedures before the ITU for the allocation of geostationary or other satellite orbits for the benefit of Ecuador.[16]

---

[12]¿Quiénes somos?, Agencia Espacial Civil Ecuatoriana, http://exa.ec/ (accessed 14 November 2019).

[13]Ibid.

[14]Política de la Defensa Nacional: Libro Blanco, Ministerio de Defensa Nacional, Quito, 2018, p. 20, available at https://www.defensa.gob.ec/wp-content/uploads/2019/01/Pol%C3%ADtica-de-Defensa-Nacional-Libro-Blanco-2018-web.pdf.

[15]Ibid., p. 60.

[16]Other functions of the MTSI are available at https://www.telecomunicaciones.gob.ec/objetivos/.

11.4   Current and Future Perspectives on the Legal Framework Related to Space Activities

Despite its significant space achievements and ambitions, Ecuador does not possess specific national regulations on the matter. On the contrary, the telecommunications sector is governed by two national laws.

11.4.1   National Space Regulations and Instruments

Currently, there are no space regulations in Ecuador. Some years ago, the IEE was created by Executive Decree 1246 of 19 July 2012[17] with the main objective of exercising Ecuador's rights over its corresponding GEO slots.[18] However, six years later, the IEE was dissolved by the President of the Republic as part of an economic package, and its functions were moved to the Geographic Military Institute by Executive Decree N° 174 of 11 April 2019.[19]

Concerning the telecommunications sector, the Organic Law on Telecommunications (LOT) of 18 February 2015[20] regulates the use of orbital resources and satellite services. Moreover, this law determines that the use of the radio electric spectrum associated with satellite networks, as well as the provision of services related to these networks, are administered, regulated and controlled by the state.[21] There is also the General Regulations of the Organic Law on Telecommunications, adopted on 26 January 2016, which develops and applies the LOT.[22]

11.4.2   Ecuador and International Space Instruments

Ecuador is legally bound by several international legal instruments related to the space sector. Specifically, Ecuador has ratified three of the five UN Space Treaties: the OST, ARRA and the Liability Convention. It has also ratified several instruments related

---

[17] Other objectives included scientific space research; the coordination of space projects and programs with the National Development Goals; the development of space technology, and the promotion of the peaceful use of outer space, Ibid.

[18] Ibid.

[19] Ibid.

[20] Ley Orgánica de Telecomunicaciones, Registro Oficial, Año II, N° 439, Quito, 18 February 2015, https://www.telecomunicaciones.gob.ec/wp-content/uploads/downloads/2016/05/Ley-Org%C3%A1nica-de-Telecomunicaciones.pdf (accessed 15 November 2019).

[21] Chapter IV, Ibid.

[22] Reglamento General a la Ley Orgánica de Telecomunicaciones, Decreto Ejecutivo 864, Registro Oficial Suplemento 676, 25 January 2016, https://www.telecomunicaciones.gob.ec/wp-content/uploads/2016/02/Reglamento-Ley-Organica-de-Telecomunicaciones.pdf (accessed 15 November 2019).

Table 11.2 International space treaties respected by Ecuador

International space treaties respected by Ecuador

Treaty[a]	Adoption	Ratification (R)/signature (S)
OST	1967	R
ARRA	1968	R
LIAB	1972	R
REG	1975	–
MOON	1979	–
NTB	1963	R
BRS	1974	–
ITSO	1971	R
IMSO	1976	R
ITU	1992	R

[a]Treaty on Principles Governing the Activities of States in the Exploration and Use of Outer Space, including the Moon and Other Celestial Bodies (OST); Agreement on the Rescue of Astronauts, the Return of Astronauts and the Return of Objects Launched into Outer Space (ARRA); Convention on International Liability for Damage Caused by Space Objects (LIAB); Convention on Registration of Objects Launched into Outer Space (REG); Agreement Governing the Activities of States on the Moon and Other Celestial Bodies (MOON); Treaty Banning Nuclear Weapon Tests in the Atmosphere, in Outer Space and under Water (NTB); Convention Relating to the Distribution of Programme-Carrying Signals Transmitted by Satellite (BRS); Agreement Relating to the International Telecommunications Satellite Organisation (ITSO); Convention on the International Mobile Satellite Organization (IMSO); International Telecommunication Constitution and Convention (ITU)

to the telecommunications sector, such as the ITU Convention and Constitution (Table 11.2).

11.5 Cooperation Programs

Space cooperation between Ecuador (mostly through EXA) and other partners includes space technology exchange agreements, launch services conventions, and joint space projects. Concerning Ecuador's partners, they are usually governments, foreign space agencies and industries.

The launch into Earth orbit of Ecuador's first satellites (see Sect. 11.6) was possible thanks to the cooperation agreements with China, which launched the PEGASO satellite from the Jiuquan launch centre, and Russia, which launched the KRYSAOR satellite from the Dombvarovski cosmodrome. Moreover, a cooperation agreement between the Technological Equinoctial University (UTE)[23] and the Russian Southwest State University (SWSU) developed the UTE-UESOR and ECUADOR-UTE satellites in 2017 and 2019, respectively.[24]

---

[23]Universidad Tecnológica Equinoccial (UTE).

[24]Ecuador en el Espacio, Latam Satelital, 15 July 2017, http://latamsatelital.com/ecuador-en-el-espacio/ (accessed 15 November 2019).

These space cooperation programs include contracts signed by EXA and the American consortium ICSP for the development of the satellite programs IRVINE01 and IRVINE02. These missions were the first American satellites with Ecuadorian components provided by EXA (50% and 60%, respectively). In particular, the second contract consisted of the design, construction and test by EXA of a laser emitter for Earth orbit communications to be integrated in satellite IRVINE02.[25] This satellite was launched into outer space in 2018 by the American company SPACE-X.[26]

Another important cooperation agreement is the joint venture adopted in 2018 between EXA and the Colombian Space Agency, both civilian non-governmental space agencies, for the implementation of the Colombo-Ecuadorian Lunar Program (CELP).[27] Among its goals, the CELP aims to send into Earth orbit the first Colombian astronaut. In addition, both space agencies have signed a contract with the ASTROBOTIC company to use its PEREGRINE Lander to explore the moon, marking the beginning of the CELP and the Ecuadorian Lunar Program.[28]

Moreover, in the same year EXA signed an MoU with the company RBC SIGNALS[29] for collaboration on an Optical Communication System for LEO and Lunar/Deep Space programs, including the CELP.[30] As part of the MoU, both partners will also collaborate on the Irvine CubeSat STEM Program. In this context, EXA holds a twelve-year contract for the provision of space and laser technology for the IRVINE CubeSat STEM Program.[31]

11.6 Satellite Capacities

Ecuador reached a significant development in its space sector with the design, construction and launch of EXA's first satellites: NEE-01 PEGASO and NEE-02 KRYSAOR (see Table 11.3). This achievement was followed by the launch of two satellites developed by an Ecuadorian University in collaboration with a Russian university.

---

[25] La historia de Ecuador al espacio, Agencia Espacial Civil Ecuatoriana, http://exa.ec/index.html (accessed 15 November 2019).

[26] Ibid.

[27] Ibid.

[28] Román Víctor, Colombia y Ecuador firman convenio para enviar misión a la Luna, El Espectador, 6 October 2018, https://www.elespectador.com/noticias/ciencia/colombia-y-ecuador-firman-convenio-para-enviar-mision-la-luna-articulo-816510 (accessed 15 November 2019).

[29] RBC Signals is a global space communications provider serving satellite operators with an improved model for the delivery and processing of data spacecraft in orbit, see RBC Signals, http://rbcsignals.com/ (accessed 15 November 2019).

[30] RBC Signals and Ecuadorian Civilian Space Agency (EXA) Announce Collaboration for Optical Communication System, RBC Signals, Seattle, Washington, 4 October 2018, http://rbcsignals.com/rbc-signals-and-ecuadorian-civilian-space-agency-exa-announce-collaboration-for-optical-communication-system/ (accessed 15 November 2019).

[31] Ibid.

Table 11.3 Ecuador's satellites launched into outer space[a]

Ecuador's satellites launched into outer space[b]				
Satellite	Date	Orbit	Features	Focus
NEE-1 PEGASO	25 April 2013	• Inclination: 98.05° • Apogee: 660 km • Perigee: 636 km	• Launch vehicle: CHANG ZHENG 2D • Launch site: Jiuquan Launch Centre, China • Launching States: China and Ecuador • Operator of the object launched: EXA	• Status: active • General functions: monitor potential threats from near space objects and space debris
NEE-02 KRYSAOR	21 November 2013	• Inclination: 98.7° • Apogee: 890 km • Perigee: 720 km	• Launch vehicle: DNEPR RS-20B • Launch site: Dombvarovski cosmodrome • Launching States: Ecuador and Russia • Operator of the object launched: EXA	• Status: active • General functions: repeat NEE – 01 PEGASO's signal, and monitor potential threats by near space objects and space debris

[a]Ecuador's satellites are not registered with the United Nations
[b]Data obtained from Ecuadorian Civilian Space Agency, http://exa.ec/index-en.html, https://es.wikipedia.org/wiki/NEE-01_Pegaso, and https://www.ecured.cu/Sat. %C3%A9lite_NEE-02_KRYSAOR (accessed 15 November 2019)

11.6.1 Satellite NEE-01 PEGASO

Ecuador launched into outer space its first satellite NEE-01 PEGASO on 25 April 2013. Its first transmission was captured by the HERMES-A ground station ten days later. Unfortunately, two weeks after its launch, PEGASO had an orbital problem that caused the loss of its signal. Almost eight months later, EXA's personnel recovered the PEGASO's signal thanks to the NEE-02 KRYSAOR satellite.[32]

PEGASO was developed to function as an orbital sentinel to monitor potential threats from near space objects, and to control space debris.[33]

11.6.2 Satellite NEE-02 KRYSAOR

On 21 November 2013, Ecuador launched its NEE-02 KRYSAOR satellite, a PEGASO class satellite. KRYSAOR's main function is to be a backup to PEGASO's signal, which was repeated for the first time on 25 January 2015. Since then, KRYSAOR has received PEGASO's signals every time both satellites orbit in parallel to a maximal range of 2,000 km.[34]

---

[32]Satélites, Agencia Espacial Civil Ecuatoriana, http://exa.ec/ (accessed 6 November 2019).
[33]Ibid.
[34]Ibid.

KRYSAOR's mission is practically the same as that of PEGASO's. However, the former has a camera with a better resolution than the latter's.[35]

11.6.3   University Satellites

In recent years, Ecuadorian scientists and experts from the Faculty of Engineering Sciences of the UTE have designed, developed and launched into Earth orbit two nanosatellites. The first satellite, UTE-UESOR was launched in 2017 from the Baikonur cosmodrome, Kazakhstan. UTE-UESOR's main functions were the collection of scientific data to study the Earth and the monitoring of the density of cosmic particles.[36]

In July 2019, the UTE launched into orbit its second satellite, ECUADOR-UTE, from the Vostochny cosmodrome, Russia. These two satellites were developed in collaboration with the SWSU with the support of the Russian space system (the satellites were launched by two SOYUS 2.1B rockets). ECUADOR-UTE measures the Earth's magnetic field and monitors the main parameters of the own satellite.[37]

11.7   Space Technology Development

Most Ecuadorian space projects have been developed by EXA. A few of these projects have been developed by EXA together with FAE and several foreign institutions. There are also satellite projects that have been developed by the UTE (see 11.6.3). Some of the most important national space projects are the following: DÉDALO, POSEIDÓN, HIPERIÓN, HERMES and HERMES-DELTA.

The DÉDALO Project—the first Latin American microgravity plane. The DÉDALO project was an inter-institutional project carried out by EXA and FAE to develop a national microgravity plane.[38] The first DÉDALO's mission, EXA/FAE-01, was accomplished on 10 April 2008 when three mission members (two aboard a MIRAGE F1JE plane and one from ground) obtained 301 seconds of zero gravity thanks to a gravimeter created by EXA. On 6 May 2008, EXA and FAE carried out

---

[35] Ibid.

[36] Ecuador en el Espacio, Latam Satelital, 15 July 2017, http://latamsatelital.com/ecuador-en-el-espacio/ (accessed 15 November 2019).

[37] La UTE pone en órbita su segundo nanosatélite, EFE, Quito, 5 July 2019, https://www.expreso.ec/ciencia-y-tecnologia/ute-lanzamiento-orbita-nanosatelite-ciencia-KI2956158 (accessed 15 November 2019).

[38] Proyecto Dédalo: el primer avión latinoamericano de microgravedad, Fuerza Aérea Ecuatoriana, Agencia Espacial Civil Ecuatoriana, Power Point Presentation, available at Proyectos, Agencia Espacial Civil Ecuatoriana, http://exa.ec/index.html (accessed 14 November 2019).

the EXA/FAE-02 mission aboard a T-39 SABRELINER plane, which was named "FUERZA-G UNO "CÓNDOR" (FG-1 CÓNDOR).[39]

POSEIDÓN Project—the youngest human being in the world to experience microgravity. This project consisted of a biometric study of the child and youth physiology in microgravity, as well as a study of fluid dynamics in a weightless environment.[40] The study was carried out in 2008 by two Ecuadorian boys of ten and seven years, respectively, aboard the FG-1 CÓNDOR, a record that the Guinness World Records awarded in June 2008 as the youngest person in the world experiencing microgravity.[41]

HIPERIÓN Project—a field study to determine the type and quantity of UV radiation reaching Ecuadorian territory. The EXA's planetary sciences division carried out a study of the weakening of the ozone layer thanks to the work of Ecuadorian engineers and scientists, and the use of national weather stations and foreign satellite data.[42]

HERMES Project—the first "bridge" between the Internet and the Earth orbit. In September 2009, EXA announced the finalisation of the HERMES project aimed at connecting Internet with the Earth orbit in a stable and permanent way through a space flight control station called HERMES-A/MINOTAURO. The project has enabled students and scientists from all over the world to access satellites and spaceships by using a computer with an Internet connexion.[43]

HERMES-DELTA Project—"A Satellite in the Classroom" program created in 2009 to give schools real time access to meteorological satellites. The HERMES-DELTA System allows students to connect to several satellites (four US satellites: NOAA-15, NOAA-17, NOAA-18 and NOAA-19) to observe the weather in a Latin American region with a range of 6,000 km.[44]

---

[39]This is considered the first laboratory plane for microgravity developed in Latin America. With this project, Ecuador became the third country in the world to develop a plane of this kind by its own and without foreign support. Ibid.

[40]Proyecto Poseidón: el ser humano más joven del mundo en experimentar microgravedad, Fuerza Aérea Ecuatoriana, Agencia Espacial Civil Ecuatoriana, Power Point Presentation, available at Proyectos, Agencia Espacial Civil Ecuatoriana, http://exa.ec/index.html (accessed 14 November 2019).

[41]Ibid.

[42]El Informe Hiperión, 22 October 2008, Agencia Espacial Civil Ecuatoriana, Power Point Presentation, available at Proyectos, Agencia Espacial Civil Ecuatoriana, http://exa.ec/index.html (accessed 14 November 2019).

[43]Agencia Espacial Civil Ecuatoriana anuncia la creación del primer "puente" entre internet y la órbita terrestre: las Naciones Unidas invitan a EXA a entrenar científicos de todo el mundo para aprender a usarlo, Agencia Espacial Civil Ecuatoriana, http://exa.ec/ (accessed 15 November 2019).

[44]Agencia Espacial Civil Ecuatoriana anuncia el programa "Un satélite en el aula" para dar accesso en tiempo real a satélites meteorológicos para escuelas y colegios ecuatorianos, Agencia Espacial Civil Ecuatoriana, http://exa.ec/ (accessed 15 November 2019).

11.8 Capacity Building and Outreach

The development of the ECSP has necessitated the training of specialists and professionals in the space field, namely in the satellite sector. From the training of the Ecuadorian astronaut, Ronnie Nadar, in Russian territory to the dissemination of the use of satellite technology through the program "A satellite in the classroom" of EXA, Ecuador has continuously searched for opportunities to increase its capital human in the space sector.

EXA has also cooperated with several Ecuadorian educational institutions to provide them space knowledge to train future space professionals. In 2006, the Agency selected the Educational Unit Nuestra Señora de Alborada (Guayaquil), the Cotopaxi Academia (Quito) and the Las Catalinas College (Cuenca) to offer its advice and support on space matters.[45]

Another example of space capacity building is the program led by the UTE for the development of its satellite ECUADOR-UTE (see 11.6.3) for which several students of Mechatronics Engineering received a scholarship to go to the SWSU in Kursk, Russia.[46]

Appendix

See (Table 11.4).

Table 11.4 Institutional contact information in Ecuador (2019)

Institution	Details
Ecuadorian Civilian Space Agency (EXA)	Website: http://exa.ec/index.html
Ministry of National Defence	La Exposición Street S4-71 and Benigno Vela, 170403, Quito, Ecuador Website: https://www.defensa.gob.ec/
Ministry on Telecommunications and the Society of Information (MTSI)	Avenue 6 December N25-75 and Avenue Colón, 17052, Quito, Ecuador Website: https://www.telecomunicaciones.gob.ec/
Technological Equinoctial University (UTE)	Rumipamba Street s/n, Quito, Ecuador Website: https://www.ute.edu.ec/

---

[45]La agencia EXA busca sus futuros talentos en los colegios del país, El Telégrafo, 17 March 2016, https://www.eltelegrafo.com.ec/noticias/septimo/1/la-agencia-exa-busca-sus-futuros-talentos-en-los-colegios-del-pais (accessed 16 November 2019).

[46]La UTE pone en órbita su segundo nanosatélite, EFE, Quito, 5 July 2019.

Chapter 12
Mexico

Contents

Abstract The entry of Mexico in the space arena started at the beginning of the space race in 1957. Since then, Mexico's interest in space has considerably expanded to nowadays encompass multiple issues, from the strengthening of space science and technology to the formation of a national space industry. Moreover, Mexico's active participation in international space fora and the adoption of international space cooperation agreements shows its increased enthusiasm to take advantage of the

© Springer Nature Switzerland AG 2020   333
A. Froehlich et al., *Space Supporting Latin America*, Studies in Space Policy 25,
https://doi.org/10.1007/978-3-030-38520-0_12

benefits of this domain. This chapter presents a descriptive analysis of Mexico's space sector. The first sections focus on a general overview of the country and the history of Mexico's space activities. Then, the economic perspective of the sector is analysed particularly in relation to the institutional space budgets and the Mexican space industry. Next, the main characteristics of Mexico's space and satellite policies and legal frameworks, as well as the space actors responsible to implement them, are presented. Mexico's participation in the consolidation of international space law, and the current space cooperation agreements celebrated by Mexico are also noted. The last sections focus on issues concerning Mexico's satellite capabilities, space technology development, and capacity building and outreach, with special emphasis on space education.

12.1 General Country Overview

With its 1,964,375 km^2 of territory,[1] Mexico (officially the United Mexican States) is a country geographically located in the North America region. It has borders with the United States of America in the north, and with Guatemala and Belize in the south. The country is composed of 32 federal entities, one of which is the capital Mexico City. Mexico has an estimated population of 126,577,691[2] and Spanish is Mexico's main language, although there are more than 60 indigenous languages.[3]

12.1.1 Political System

Mexico is a democratic, representative and secular republic, and a federation composed of 32 states or federal entities (Fig. 12.1).[4] The supreme power of the Federation is divided in the executive, legislative and judicial powers. At the federal level, the executive power is headed by the President, who is also the Head of State, and Commander in Chief of the Armed Forces. The president is directly elected every six years with no possibility of re-election. The legislative power constitutes the Congress, which is composed of two Chambers: the Deputy Chambers (500 members) and the Senate (128 members). The judicial power is made up of the Supreme

---

[1] Mexico's Exclusive Economic Zone covers 3,149,920 km^2, see Mexico's General Consulate of El Paso, Ministry of Foreign Affairs, https://consulmex.sre.gob.mx/elpaso/index.php/2016-03-16-21-05-21/2016-03-16-21-07-37 (accessed 23 April 2019).

[2] Demographic indicators of the Mexican Republic in 2019, National Population Council, http://www.conapo.gob.mx/work/models/CONAPO/Mapa_Ind_Dem18/index_2.html (accessed 23 April 2019).

[3] México en breve, United Nations Development Program, http://www.mx.undp.org/content/mexico/es/home/countryinfo.html#Poblaci%C3%B3n (accessed 23 April 2019).

[4] Art. 40, Political Constitution of the United Mexican States, 5 February 1917, last reform 27 January 2016 http://www.diputados.gob.mx/LeyesBiblio/htm/1.htm (accessed 23 April 2019).

Mexico's political division

Source: diymapo.net (c)

Fig. 12.1 Mexico's political division

Court of Justice of the Nation, the Electoral Tribunal, collegiate and unitary courts, and district courts.[5]

Federal entities also have the aforementioned branches. The executive is headed by a governor; the legislative is composed of one single chamber (local congress), and the judicial branch is headed by a Supreme Court of Justice.

On 1 July 2018 the left-winger Andrés Manuel López Obrador (known by his initials "AMLO"), candidate of the coalition "Juntos Haremos Historia" formed by the political parties MORENA (National Regeneration Movement), PT (Workers Party) and PES (Social Encounter Party),[6] won the presidential elections. MORENA obtained an absolute majority in both the Senate and the Chamber of Deputies. The President has ruled Mexico since December 2018.[7] His stated priorities are to tackle corruption and reverse decades of free-market economic policy.[8]

---

[5] Ibid., articles 50, 80 and 94, respectively.

[6] Movimiento Regeneración Nacional (MORENA), Partido del Trabajo (PT), and Partido Encuentro Social (PES).

[7] Mexican political system, Ministry of Foreign Affairs, Mexican Government, https://globalmx.sre.gob.mx/index.php/en/democracy-and-rule-of-law/mexican-political-system (accessed 23 April 2019).

[8] Mexico country profile, BBC News, 3 December 2018, https://www.bbc.com/news/world-latin-america-18095241 (accessed 23 April 2019).

12.1.2   Economic Situation

In terms of GDP measured at purchasing power parity, Mexico is the world's 11th largest economy.[9] Several structural reforms have increased Mexico's growth in the last decades. Tax policy, telecom, competition policy, labour market, energy and other recent structural reforms have shown some progress in many Mexico's regions and sectors, namely in terms of productivity growth. The telecommunications reform, for example, has notably increased growth in the sector and reduced services prices (up to 25%).[10]

However, the benefits of these reforms have not been inclusive enough and disparities across regions have increased, especially between the highly productive north region and the less productive south.[11] In addition, several firms have had no benefit from these reforms and they constantly struggle to perform better with limited success.[12]

Mexico's foreign trade and investment have been constantly increasing as demonstrated by the signature of twelve free trade agreements with 46 countries. The country has also benefited from its integration into global value chains, mostly as an assembler of manufactured inputs, and it has achieved the reduction of trade barriers in key sectors such as media and telecoms.[13] Nevertheless, other sectors suffer from stringent local regulations, weak legal institutions, rooted informality, corruption and insufficient financial development.

Mexico's economic growth is also hindered by other factors, such as the high concentration of incomes, poverty, insecurity, financial exclusion, insufficient educational achievement, gender gap, weak rule of law, and persistent levels of corruption and crime.[14]

12.1.3   Social Development

According to the OECD, living standards in Mexico have increased in the last years in many areas, such as health, jobs and education. However, all these standards are lower than the OECD average.[15]

---

[9]OECD Economic Surveys: Mexico, January 2017, p. 10, at http://www.oecd.org/eco/surveys/Mexico-2017-OECD-economic-survey-overview.pdf.

[10]Ibid., p. 38.

[11]Ibid., pp. 14–15.

[12]Ibid., p. 39.

[13]Ibid., p. 42.

[14]Ibid., p. 14.

[15]Cf. Mexico, Better Life Index, OCDE, http://www.oecdbetterlifeindex.org/countries/mexico/ (accessed 24 April 2019).

In the field of employment, about 61% of Mexicans aged 15–64 have a paid job. Roughly 79% of men are employed, compared to 45% of women.[16] Mexicans work 2,246 h per year (more than 480 h above the OECD average) but their productivity is 20%.[17]

Concerning health, life expectancy at birth is 75 years, 78 years for women and 72 for men. In terms of water quality, 67% of people are satisfied with the quality of their water.[18]

In the field of education, 37% of Mexicans adults aged 25–64 have finished upper secondary education. Concerning the quality of the education system, an average student scored 416 points in reading literacy, maths and science according to the OECD's Programme for International Student Assessment (PISA), a lower result in comparison to the 486 points OECD average.[19]

Subjective well-being is marginally above the OECD average. In general, Mexicans are satisfied with their life. On a scale from 0 to 10, general satisfaction with life was rated 6.6 on average, while the OECD average is 6.5.[20]

12.2 History of Space Activities

Mexican space history (Table 12.1) started in the same year as the beginning of the space race between the United States and the Soviet Union, in 1957. Since then, the development of the space sector in Mexico has increased considerably in different areas, from space science, technology and industry to space policy and law. The creation of the Mexican Space Agency and the Federal Institute of Telecommunications, and the expansion of space industries in the country in the last decade are not only the result of this evolution but the building blocks for the future development of the Mexican space sector.

---

[16]Ibid.

[17]"En México trabajamos más horas, pero somos menos productivos", Forbes México, 9 January 2018 https://www.forbes.com.mx/mexico-trabajamos-mas-horas-pero-menos-productivos/ (accessed 24 April 2019).

[18]Mexico, Better Life Index, op. cit.

[19]Ibid.

[20]Ibid.

Table 12.1 Mexico's space history (1957–2019)

Year	Mexico's space history
1957	Mexico's space activities' history[a] begins on 28 December 1957[b] when a group of the Physics School of the Autonomous University of San Luis Potosí launches "FÍSICA 1", an 8 kg scientific rocket with a length of 1.7 m, which reached an altitude of 2,500 m
1959	The Ministry of Communications and Transportation (MCT)[c] promotes the experimentation of the construction of small rockets to perform high atmosphere measurements. Two liquid propellant rockets are built: SCT1 and SCT2. The former is launched in 1959 reaching 4,000 while the latter is launched in 1960 reaching 25,000 m
1960	A tracking station is built in Guaymas, Sonora (Northwest region) in collaboration with NASA. This station serves mainly to track American satellites and other space objects.
1962	Creation of the National Commission of Outer Space (CONEE)[d] responsible for carrying out investigations in all disciplines of the space sector.
1962	The National Autonomous University of Mexico (UNAM)[e] establishes the Outer Space Department to study the Solar System's physics.
1968	Transmission of the television signal of the 1968 Olympic Games and the 1970 World Cup thanks to the Mexican telecommunications networks.
1975	Launching of the MITL I rocket probe from a mobile platform, which exceeds 100 km before falling to the ground[f]
1976	CONEE disappears after launching small rockets to attract rains to deserted zones.
1977	Creation of the Mexican Communications Institute.
1985	ITU grants Mexico's first geostationary orbital positions and associated frequencies. Mexican Government acquires its first satellite system: the MORELOS System which is composed of satellites MORELOS 1 and 2, and a satellite control centre located in Mexico City[g]
1985	Rodolfo Neri Vela becomes Mexico's first astronaut as a mission specialist for the joint NASA/ESA mission STS-61B aboard the Space Shuttle ATLANTIS. He helps to place in orbit the MORELOS 2 satellite and conducts some experiments for the Mexican Government[h]
1993	Launching of satellite SOLIDARIDAD I
1994	Launching of satellite SOLIDARIDAD II
1996	Dissolution of the Mexican Communications Institute[i]
1997	Privatisation of fixed satellite services
2009	Launching of the Mexican satellite system MEXSAT
2010	Promulgation of the Mexican Space Agency Act that gives birth to the Mexican Space Agency (AEM)[j]
2011	AEM starts operations
2012	Launching of satellite BICENTENARIO

(continued)

Table 12.1 (continued)

Year	Mexico's space history
2013	Creation of the Federal Telecommunications Institute (IFT)[k]
2015	Launching of satellite MORELOS 3
2016	The MCT receives overall control of MEXSAT which is made up of the BICENTENARIO and MORELOS 3 satellites, two control centres located in Mexico City and Hermosillo (state of Sonora), their associated networks and some test stations[l]
2019	Launching of AZTECHSAT-1, the first Mexican CubeSat satellite, which was developed by students of the Popular University of the State of Puebla (UPAEP)[m]

[a]Most of the information was taken from Mexican Space Agency et al., Orbit Plan: 2.0: Roadmap for Mexico's Space Industry, ProMéxico, Mexico City, 2017, https://www.gob.mx/cms/uploads/attachment/file/414932/Plan_Orbita_2.0.pdf, and *Acuerdo mediante el cual se dan a conocer las Líneas Generales de la Política Espacial de México*, Executive Power, Ministry of Communications and Transportation, Official Journal of the Federation, 13 July 2011, https://www.gob.mx/cms/uploads/attachment/file/73124/Lineas_Generalas_Politica_Espacial_de_Mexico.pdf
[b]In 1949 the Geophysics Institute of UNAM and the Ministry of Communications and Transportation created a team to build rockets to be launched a decade later, in Ávila, Norma, "La Agencia Espacial Mexicana", *La Jornada*, 8 October 2006
[c]Secretaría de Comunicaciones y Transportes
[d]"Comisión Nacional del Espacio Exterior" (CONEE), Red Universitaria del Espacio, National Autonomous University of Mexico, http://www.astroscu.unam.mx/congresos/rue/Antecedentes_Coheteria.html (accessed 18 April 2019)
[e]Universidad Nacional Autónoma de México (UNAM)
[f]Nájar, Alberto, "Después de 36 años, México busca volver a la carrera espacial", *BBC Mundo*, 21 July 2011, https://www.bbc.com/mundo/noticias/2011/07/110720_ciencia_mexico_agencia_espacial_an. (accessed 18 April 2019)
[g]Ruiz Esparza, Gerardo, El Sistema Satelital Mexicano "MEXSAT": Pilar Fundamental de la Reforma de Telecomunicaciones, El Financiero, 12 May 2015, https://www.elfinanciero.com.mx/opinion/gerardo-ruiz-esparza/el-sistema-satelital-mexicano-mexsat-pilar-fundamental-de-la-reforma-de-telecomunicaciones (accessed 18 April 2019)
[h]Rodolfo Neri Vela, International Space Hall of Fame, New Mexico Museum of Space History, http://www.nmspacemuseum.org/halloffame/detail.php?id=111 (accessed 18 April 2019)
[i]Rivera Parga, José Ramón, Space exploration: an opportunity to increase national power of Mexican State, Revista del Centro de Estudios Superiores Navales, Mexico City, 2017, Vol. 38, No. 4, October-December
[j]Agencia Espacial Mexicana (AEM)
[k]Instituto Federal de Telecomunicaciones (IFT)
[l]Mendieta, Susana, "SCT recibe control total del Sistema Satelital Mexicano", Milenio, 25 August 2016, https://www.milenio.com/negocios/sct-recibe-control-sistema-satelital-mexicano
[m]AZTECHSAT-1, UPAEP, https://upaep.mx/aztechsat (accessed 5 January 2020)

12.3 Economic Perspective of Space Programs

On the one hand, the development of Mexico's space sector implies not only political will but also significant financial resources. To carry out its functions, the Mexican Space Agency and the Federal Telecommunications Institute have specific assigned budgets. Nevertheless, these budgets are usually limited and nowadays both institutions suffer from important financial cuts. On the other hand, space industry and other related industries have considerably increased in the last decades to the benefit of Mexico's development.

12.3.1   Institutional Space Budgets

The AEM and the IFT are the main Mexican bodies with competences in space related issues. Neither of them has more than a decade of existence.[21] While AEM's budget has been modest, IFT's budget has been more substantial. However, nowadays both institutional budgets have been the object of important reductions that could impair or delay the development of the Mexican space sector.

12.3.1.1   The Mexican Space Agency's Budget

In its first year of operations, there was no specific budget allocated to AEM by the Mexican congress. It was the MCT that used its own resources to finance the first activities of the Agency.[22] From 2012 to date, congress has assigned specific budgets to AEM every year but they have been very inconsistent. For example, in 2012 the Agency received MXN 60 million; in 2015, AEM's budget was almost doubled with MXN 112 million[23] but decreased to MXN 62 million by 2019 (Table 12.2).

Variability in the allocation of financial resources can be a problem for the Agency to achieve its objectives and to create long-term space projects. José Hernández, a NASA astronaut of Mexican origin, has qualified AEM as a "paper institution" and stressed not only that AEM must have tangible projects[24] but an estimated budget of MXN 1 billion to operate.[25] In a similar way, Mexican astronaut Neri Vela has

---

[21] The AEM was not the first Mexican space body; the National Commission of Outer Space was created in 1962 but disappeared in 1976. In the same way, the IFT is the successor of the Federal Telecommunications Commission; see Sect. 12.2.

[22] Rodríguez Ivet, "Agencia espacial no alcanzó presupuesto", Expansión, 2 December 2010 https://expansion.mx/manufactura/2010/12/02/agencia-espacial-mexicana (accessed 16 April 2019).

[23] AEM has also external resources to complement its budget. For example, in 2015 AEM's budget allocated by the Congress was MXN 111,983,200 but additional financial resources increased it to MXN 123,483,200. See http://www.apartados.hacienda.gob.mx/presupuesto/temas/pef/2015/docs/09/r09_jzn_feie.pdf.

[24] Rodríguez Yazmín, "Agencia Espacial Mexicana es una institución de papel: astronauta José Hernández", El Universal, 25 October 2018, https://www.eluniversal.com.mx/nacion/sociedad/agencia-espacial-mexicana-es-una-institucion-de-papel-astronauta-jose-hernandez (accessed 16 April 2019).

[25] He stated that "AEM does not have the budget to design, build and launch a useful communications satellite for the country capable of monitoring the Earth with different cameras and sensors. Until now contributions ranges 100 million pesos but it is required an investment 10 times higher", Valadéz Blanca, "La AEM require 10 veces más presupuesto: José Hernández", Milenio, 16 December 2014, https://www.milenio.com/cultura/aem-requiere-10-presupuesto-jose-hernandez (accessed 16 April 2019) (no official translation).

Table 12.2 Space and telecommunication institutional budgets (2011–2019)

AEM's Budget by year[a]		IFT's Budget by year
2011	Not specified	Not applicable
2012	MXN 60,000,000[b]	Not applicable
2013	MXN 97,772,029	Not applicable
2014	MXN 112,811,973	MXN 2,000,000,000[c]
2015	MXN 123,483,200	MXN 2,000,000,000[d]
2016	MXN 98,807,913	MXN 2,000,000,000[e]
2017	MXN 92,482,883[f]	MXN 1,980,000,000[g]
2018	MXN 77,821,489[h]	MXN 1,998,000,000[i]
2019	MXN 62,246,743[j]	MXN 1,500,000,000[k]

[a]The value of the Mexican peso against the USD dollar from 2011 to 2019 decreased by 58.20% (MXN 12.11 per dollar in 2011 to MXN 19.55 per dollar in 2019). See figures at Historical data USD/MXN, investing.com, https://es.investing.com/currencies/usd-mxn-historical-data
[b]AEM's budgets from 2012 to 2016 can be found here: http://www.diputados.gob.mx/LeyesBiblio/abro/pef_2012/PEF_2012_abro.pdf
[c]See Presupuesto de Egresos de la Federación para el Ejercicio Fiscal 2014, Official Journal of the Federation, 3 December 2013, http://www.diputados.gob.mx/LeyesBiblio/abro/pef_2014/PEF_2014_orig_03dic13.pdf
[d]See Presupuesto de Egresos de la Federación para el Ejercicio Fiscal 2015, Chamber of Deputies of the Congress of the Union, 3 December 2014, http://www.diputados.gob.mx/LeyesBiblio/abro/pef_2015/PEF_2015_abro.pdf
[e]See Presupuesto de Egresos de la Federación para el Ejercicio Fiscal 2016, Chamber of Deputies of the Congress of the Union, 27 November 2015, http://www.diputados.gob.mx/LeyesBiblio/abro/pef_2016/PEF_2016_abro.pdf
[f]See Presupuesto de Egresos de la Federación para el Ejercicio Fiscal 2017, Official Journal of the Federation, 30 November 2016, http://www.diputados.gob.mx/LeyesBiblio/abro/pef_2017/PEF_2017_orig_30nov16.pdf
[g]Ibid.
[h]See Presupuesto de Egresos de la Federación para el Ejercicio Fiscal 2018, Official Journal of the Federation, 29 November 2017, https://www.dof.gob.mx/nota_detalle.php?codigo=5506080&fecha=29/11/2017
[i]Ibid.
[j]See Presupuesto de Egresos de la Federación para el Ejercicio Fiscal 2019, Chamber of Deputies of the Congress of the Union, 28 December 2018, http://www.diputados.gob.mx/LeyesBiblio/pdf/PEF_2019_281218.pdf
[k]Ibid.

commented that the Agency requires an annual budget of USD 100 million to create projects and prepare astronauts. He highlighted that AEM's yearly budget is mostly directed to the payment of salaries preventing the financing of projects in universities and technological institutions, as well as investing capital in industries.[26]

[26]Ramírez Rebeca, "Agencia Espacial necesita presupuesto de 100 mdd", Vanguardia, 21 October 2017, https://vanguardia.com.mx/articulo/agencia-espacial-necesita-presupuesto-de-100-mdd (accessed 16 April 2019).

It is difficult to believe that in the short term AEM will achieve significant progress. A reduction of AEM's budget to almost the same amount as in its first year of operations demonstrates that the space sector, at least at the governmental level, is not considered a national priority. Current political developments do not indicate that this downward trend will change in the near future.[27]

12.3.1.2   Federal Telecommunications Institute's Budget

With the adoption of the constitutional telecommunications reform and the creation of the IFT in 2013, the Mexican telecommunications sector received a strong financial impulse. Due to its vast competences and functions, as well as its high number of civil servants, IFT's budget is much higher than AEM's budget. IFT's yearly budget was relatively constant from 2014 to 2018 reaching roughly MXN 2 billion per year. However, in 2019 the Institute lost 25% of its global budget, an important financial cut that could prevent IFT from adequately fulfilling all its functions.[28] Therefore, as in the case of AEM, it is difficult to see significant progress in the telecommunications and broadcasting sectors of the country in the coming years.

12.3.2   The Mexican Space Industry

The space industry in Mexico is diverse and is constantly growing. However, there is currently no consolidated space industry. Possible reasons for this are the lack of public and private funding and genuine interest by all relevant actors to give impulse to the space industry and to place it on the international competitive arena.[29] Although the Mexican space industry is small in comparison to other Mexican industries or foreign space industries,[30] there is a consensus that the sector has an enormous

---

[27]In 2019, science and technology sectors have suffered significant financial reductions. For example, the National Council of Science and Technology (Consejo Nacional de Ciencia y Tecnología) budget decreased by 8% in comparison with the last year. Cf. Contreras Raúl, "Ciencia y tecnología clave para el desarrollo", Excélsior, 16 February 2019, https://www.excelsior.com.mx/opinion/raul-contreras-bustamante/ciencia-y-tecnologia-clave-del-desarrollo/1296782 (accessed 18 April 2019).

[28]In January 2019, IFT brought a constitutional appeal against the Budget of Expenditures of the Federations' Decree declaring that it infringed its autonomy and prevented it from having minimum resources to operate, Guadarrama José de Jesús, "IFT va por controversia constitucional por recorte presupuestal", Excélsior, 10 January 2019, https://www.excelsior.com.mx/nacional/ift-va-por-controversia-constitucional-por-recorte-presupuestal/1289467 (accessed 18 April 2019).

[29]Cf. Blanco Martha, E. Carrera (Clúster): "Hay que ver el espacio como la vía de desarrollo de México", Infoespacial.com, 25 March 2019 http://www.infoespacial.com/latam/2019/03/25/noticia-carrera-cluster-espacio-desarrollo-mexico.html (accessed 17 April 2019).

[30]Orbit Plan: 2.0: Roadmap for Mexico's Space Industry, ProMéxico, Mexico City, 2017, https://www.gob.mx/cms/uploads/attachment/file/414932/Plan_Orbita_2.0.pdf.

potential to be exploited.[31] Space industry is necessary for the Mexican economy particularly through the creation of jobs, technology transfer, and the promotion and training of human capital. Moreover, space industry responds to social needs in areas related to health, telecommunications, agriculture, natural disasters, and security.[32]

12.3.2.1   Industrial Capacities

Mexico already has some complementary industries with the potential to be an important development factor for the country and for the space sector.[33] Orbit Plan 2.0, a document prepared by several Mexican institutions (see Sect. 12.4.2.2), divides the industrial capacities of the country in five industries or technologies: (1) space industry; (2) electronic industry; (3) aerospace industry; (4) automobile industry, and (5) information technologies.

Space Industry

The creation of MXSpace by several Mexican enterprises interested in the space sector foresees the expansion of the space industry in many areas. MXSpace is a civil association that contemplates actions to develop, test and fly satellite systems for remote sensing purposes, namely to increase understanding and improve the use of resources for Mexican industries and populations.

The initiative follows a strict calendar divided in several steps that go from micro-satellite systems integrated tests to rocket firing in preparation for the first launch in the following years.[34] MXSpace supports enterprises in the development of space technological areas such as the development and construction of satellites, launch vehicles, engines, optical and telemetric systems, space operations and information services. One of the principal goals of MXSpace is to support the exploitation of space resources allocated to Mexico to address issues of national interest, namely the administration of water resources, disaster prevention and response, exploitation of natural resources, monitoring and national security, and development of geographical and commercial information services.[35]

MXSpace and AEM have worked together to promote the development of the Mexican space industry, collaboration that was strengthened with a cooperation

---

[31] See "Inauguran en Querétaro encuentro 'Industry Day' de oportunidades comerciales en el espacio", Mexican Space Agency, 19 August 2018, https://www.gob.mx/aem/es/articulos/inauguran-en-queretaro-encuentro-industry-day-de-oportunidades-comerciales-en-el-espacio-170963?idiom=es (accessed 22 April 2019).

[32] Orbit Plan 2.0, op. cit., p. 10.

[33] Ibid.

[34] MXSPACE, Mexican Space Initiative, http://mxspace.mx/nosotros/ (accessed April 17, 2019).

[35] Ibid.

agreement in 2015. Moreover, in 2018 MXSpace launched the "MXSpace Magazine" with the intention of becoming a national communication model of the space sector, an achievement that was celebrated by AEM.[36]

MXSpace's projects[37] include the construction of a private laboratory for the development of useful loads operated by Aisystems; the manufacturing of printed circuit boards for space uses by Simple Complexity; the development of femtosatellites by Thumbsat de México; the development of a launcher for small loads to be launched to LEO by Ketertech and Datiotec, and the development of ground segment infrastructure and capabilities by Latitude 19:36.[38]

Electronic Industry

There are several relevant manufacturing companies in electronic components and semiconductors in Mexico such as Kyocera, Skyworks and Intel. Moreover, some of the most important international companies that provide manufacturing electronic services are located in Mexico, including Foxconn, Pegaton, Flex, Jabil, New Kinpo Group, Sanmina, Celestca, Benchmark Electronics and Universal Scientific Industrial.[39]

Aerospace Industry

For a long time, Mexico has been recognized worldwide for its high level aerospace industry. Foreign companies such as Bombardier, Safran Group, General Electric, HoneyWell and Eurocopter work in the country in areas related to maintenance services, manufacturing, repairs and operations, engineering and design, and auxiliary services for commercial and military aircraft.[40]

Automotive Industry

Mexican territory is ideal for the establishment of automotive companies. Global companies such as General Motors, Ford, Chrysler, Volkswagen, Nissan, Honda,

---

[36] "Presentan en AEM revista MXSPACE Magazine", MXSpace, 26 February 2018, http://mxspace.mx/2018/02/presentan-en-aem-revista-mxspace-magazine/ (accessed April 17, 2019).

[37] Orbit Plan 2.0, op. cit., p. 34.

[38] Latitud 19:36 is an industrial group that aims to become a strategic provider of satellite control services for the national and global space industry. See Latitud 19:36, MXSPACE Iniciativa Espacial Mexicana, http://mxspace.mx/portfolio/latitud-1936/ (accessed 28 September 2019).

[39] Orbit Plan 2.0, op. cit., p. 34.

[40] Ibid.

BMW, Toyota, Volvo and Mercedes-Benz have together approximately 25 productive complexes in 14 Mexican states.[41]

Information Technologies

There is in Mexico a great array of small, medium and large companies providing information technologies (IT) services, such as IBM, Cisco, Microsoft, Ericsson and Softek. Concerning the "Internet of Things", the Mexican cities of Guadalajara and Puebla are becoming highly competitive poles in the area with their projects "Ciudad Creativa Digital" and "Capital Mundial de Innovación y Diseño", respectively.[42]

12.3.2.2   Regional Development

AEM has identified four Mexican states with strong development opportunities in the space sector: Querétaro, Jalisco, Coahuila, and Hidalgo (Fig. 12.2).

Querétaro's industry specializes in advanced materials, aeronautics and automobiles. Jalisco's industries are strong in the electronic, aeronautic, manufacture, engineering, software development, design, and administration sectors. Coahuila has

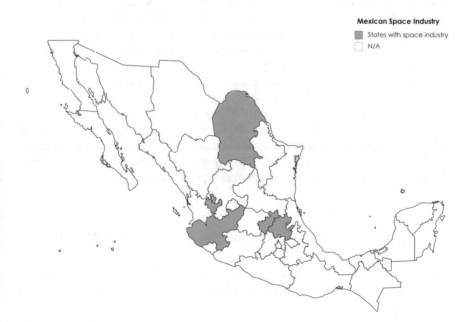

Fig. 12.2   Mexican states with strong space related industry

---

[41] Ibid.
[42] Ibid, p. 35.

consolidated industries in the automobile, manufacture and metallurgical sectors, and it also has important human capital training in engineering, IT and technical development, while Hidalgo possess talent in the metallurgical, chemistry, electronic, propulsion, and textile industries, as well as experience in space related sectors such as aerospace and manufacture.[43]

12.3.2.3   Strengths, Weaknesses, Opportunities and Threats

Orbit Plan 2.0 uses an analysis called SWOT (Strengths, Weaknesses, Opportunities and Threats) to understand the current state of Mexico's space industry and to provide a perspective on its future development.[44]

On the one hand, Mexican strengths in this sector are qualified human capital (mainly engineers), specialized space research (Mexico has an important network of national laboratories and observatories), and the aeronautic sector (there are more than 300 national and foreign aeronautic related companies).[45] To capitalize on these strengths, Mexico has to take advantage of developmental and internationalization opportunities. Orbit Plan 2.0 identified several opportunities, such as the integration of productive chains and the development of suppliers; the connection between Mexican space networks and international space programs; the creation of links with the world space industry, and the exploration of new markets of advanced technological level and satellite solutions for the Internet of Things. Opportunities at the international level are the reform of the international regulatory framework related to space services outsourcing to the private sector, as well as the inclusion of the Mexican diaspora working in the space field.[46]

On the other hand, Mexico must reduce its weaknesses by, for instance, updating the strategy of the sector to improve coordination between the relevant space actors, and strengthening the execution of planned strategies through an efficient use of resources. Regarding identified threats, the unforeseen consequences of global political, financial and economic crises could impair the development of the Mexican space industry, especially when they lead to regressive policies preventing technology transfers or reducing national budgets for science and technology.[47]

12.4   Space Development Policies and Emerging Programs

In the last decade Mexico has laid the ground for strengthening its space sector not only to improve the country's development but also to become a significant space

---

[43] Ibid, p. 38.
[44] Ibid, p. 63.
[45] Ibid, p. 44.
[46] Ibid.
[47] Ibid.

actor at the international level. Mexico has adopted its space and satellite policies and put in practice its National Space Activities Programme. The telecommunication sector has also been improved. Moreover, the creation of new institutions responsible to promote and develop these sectors is a milestone in Mexico's space history. In addition, the joint efforts of the Mexican government, industry and academia that have materialized in the Orbit Plan document to foster the national space industry is an important step to improve national space capabilities in the following years.

12.4.1   Space Policies and Main Priorities

Mexico has two main policies on space related issues: a space policy and a satellite policy.[48] The two policies are of recent creation—neither of them has more than a decade of existence—and they aim to strengthen Mexico's space sector to contribute to the well being of all Mexicans and the development of the country. Each policy has its own priorities to achieve this development.

12.4.1.1   Space Policy

One year after the adoption of the Mexican Space Agency Act (see Sect. 12.5.1.1), the MCT approved the General Guidelines of the Mexican Space Policy based on the proposals submitted by AEM's Government Board and prepared after four advisory and roundtable fora in which many national and international experts in the space field, higher education institutions and public research centres participated.[49]

The Guidelines[50] define the Mexican Space Policy as a "State policy that transcends political or economic conjunctures" whose purpose is to turn Mexican aerospace[51] scientific, technological and industrial development towards new opportunities and niches, to place Mexico into the international space competition arena,

---

[48]There is also a telecommunications policy implemented by the MCT. Due to its comprehensive character and for reasons of space, this policy will not be analyzed in this study. For a detailed explanation of the telecommunications policy see OECD Telecommunication and Broadcasting Review of Mexico 2017, 31 August 2017, https://www.oecd.org/publications/oecd-telecommunication-and-broadcasting-review-of-mexico-2017-9789264278011-en.htm; see also Soria Gerardo, "Política y regulación de telecomunicaciones", El Economista, 11 July 2018, https://www.eleconomista.com.mx/opinion/Politica-y-regulacion-de-telecomunicaciones-20180711-0009.html (accessed 24 April 2019).

[49]*Acuerdo mediante el cual se dan a conocer las Líneas Generales de la Política Espacial de México*, Executive Power, Ministry of Communications and Transportation, Official Journal of the Federation, 13 July 2011, https://www.gob.mx/cms/uploads/attachment/file/73124/Lineas_Generalas_Politica_Espacial_de_Mexico.pdf.

[50]The term "Guidelines" is used to refer to the overall instrument to distinguish them from the "guidelines", which are the guidelines (parameters) per se.

[51]It seems that the Guidelines use the terms "space" and "aerospace" as synonyms. For example, the terms "aerospace industries" and "space industries" or "aerospace applications" and "space

and to create new and better jobs. At the same time, the Guidelines are an opportunity for AEM to advance the national and regional leadership of Mexico in the space sector.

Taking into account the benefits of the development of space activities for Mexican society, the Guidelines enumerate nine strategic objectives that are part of the Mexican space policy (Table 12.3).

In addition to these objectives, the development of space activities is related to the public policy axes of the National Development Plan (NDP), an important instrument adopted every six years (every presidential turn) which determines the public policies that the Government carries out to improve the development of the country.

For example, from 2007 to 2012, the period when AEM and the Guidelines were created, the five NDP's axes were: rule of law and security; competitive and job-creating economy; equal opportunities; environmental sustainability, and democracy and responsible foreign policy.[52]

To attain the aforementioned objectives, and according to the relevant NDP axes, AEM implements the guidelines, which are the following (Table 12.4).

Table 12.3 Mexican space policy's strategic objectives

Mexican space policy's strategic objectives	
1	To create an institutional framework for space development in Mexico
2	To ensure that space activities have a significant role in the protection and security of the population
3	To articulate the public and private sectors in the space areas related to the protection of national sovereignty, and to generate a leadership with national capacities for the creation of satellite systems with their own infrastructure and technology
4	To establish the criteria for the promotion of space activities in Mexico according to their scientific, social and economic impact
5	To promote productive chains linking the industrial to the academic and services sectors so as to increase national competitiveness
6	To strength the international cooperation policy by the adoption of agreements that benefit space activities and guarantee good technology transfers, as well as the full integration of Mexico into the world space community
7	To stimulate the growth of a self-sustainable space industry and with technological capacity to compete globally
8	To promote the expansion and consolidation of a space knowledge culture into society, namely children and young people
9	To guarantee and preserve in the Mexican space policy the values related to human development, security and peace

applications" are used in similar contexts. The NSAP 2015 refers to the term "aerospace sector" as including both space and aeronautic sectors.

[52] Federal Executive Power, National Development Plan 2007–2012, Mexico, 2007, http://pnd.calderon.presidencia.gob.mx/pdf/PND_2007-2012.pdf.

Table 12.4 Mexican space policy's guidelines

Mexican space policy's guidelines	
1	State conduct of the space sector
2	State autonomy in the space sector
3	Protection of national sovereignty and security
4	Protection of population
5	Environmentally sustainability
6	Research, scientific and technological development and innovation
7	Development of the productive sector
8	Training of human resources
9	Coordination, regulation and certification
10	International cooperation
11	Disclosure of aerospace activities
12	Funding
13	Organisation and management

Although the guidelines state that they must be revised at least once every four years, there is no evidence[53] that they have been modified to date, even when the NDP has changed two times since then.[54]

National Space Activities Program

The National Space Activities Program (NSAP) is an institutional program elaborated by AEM and adopted by the MCT. It is through the execution of the NSAP that the Agency implements the Mexican Space Policy. The NSAP aligns to the NDP and establishes the axes, objectives, strategies and action lines that AEM must follow when carrying out its activities. Moreover, AEM coordinates its efforts with other public, private and social entities of the space sector.

---

[53] AEM's official website only contains the 2011 Guidelines.

[54] The current National Development Plan does not contain specific issues on the development of the space or telecommunications sector, see Plan Nacional de Desarrollo 2019–2024, México, Gobierno de la República, 30 April 2019, https://lopezobrador.org.mx/wp-content/uploads/2019/05/PLAN-NACIONAL-DE-DESARROLLO-2019-2024.pdf (accessed 30 May 2019).

National Space Activities Program 2011–2015

The first NSAP, adopted in February 2012 for the period 2011–2015,[55] was originally a confidential and private instrument and it was intended only for internal use by AEM.

It contemplated three space activities whose development would be prioritized by the country: communications, EO, ocean and atmosphere observation, and generation of science and technology. It also established a medium-term forward-looking approach to the Mexican space sector to transform Mexico into a country with first class space scientific and technological activities.

The NSAP was structured in five strategic activities axes: human capital training in the space field; scientific research and space technological development; industrial and business development and competitiveness in the space sector; international affairs, standards and security space issues, and financing, organization and information technologies on space issues.

The NSAP detailed the level of progress in the form of objectives that each axis had to reach per year by considering the NDP's axes and the Mexican Space Policy General Guidelines.

National Space Activities Program 2015

On 14 April 2015, the NSAP[56] was updated[57] with some important differences from its predecessor. First, the new NSAP was published in the Official Journal of the Federation, which gave it legal validity as a public and official institutional program of AEM. Second, the 2015 NSAP was not explicitly adopted for a specific period as the 2011–2015 NSAP was, although some AEM's documents refer to this program as the 2013–2018 NSAP.[58] That seems to indicate that the NSAP could be implemented by the Agency even after the conclusion of the presidential term or the NDP for which it was created, at least if the new administration does not derogate it. Third, the NSAP is aligned not only to the NDP (2013–2018)[59] but also to the Communications and

---

[55] Ministry of Communications and Transportation, Programa Nacional de Actividades Espaciales 2011–2015, Mexican Space Agency, February 2012, https://www.gob.mx/cms/uploads/attachment/file/73432/PNAE_2011-2015.pdf. It is possible that the NSAP was initially created for the period 2011–2012 and extended later to 2015 to give it continuity after the change of Administration, as suggested by the Orbit Plan Roadmap 2013 (see Sect. 12.4.2.1), p. 9.

[56] Acuerdo por el que se expide el Programa Nacional de Actividades Espaciales, Official Journal of the Federation, 14 April 2015, https://www.gob.mx/cms/uploads/attachment/file/73432/PNAE_2011-2015.pdf.

[57] It was proposed by AEM's Director, approved by the AEM's Government Board and adopted by the Ministry of Communications and Transportation.

[58] Cf. Ministry of Communications and Transportation, 2013–2018 NSAP Results report, 2018, https://www.gob.mx/cms/uploads/attachment/file/442785/Programa_Nacional_de_Actividades_Espaciales_2013-2018.pdf.

[59] The 2013–2018 NDP's national goals were: (1) Mexico in peace; (2) Mexico with an inclusive approach; (3) Mexico with quality education; (4) Prosperous Mexico, and (5) Mexico with global responsibility. For example, regarding the "Mexico with global responsibility" goal, and according to

Table 12.5 National Space Activities Program's strategic objectives 2015

2015 National Space Activities Program's strategic objectives
To advance the development of a space infrastructure aimed to deal with the social necessities related to security, protection of the population, disaster response, broadband connectivity, and environmental care
To support the development of the national space sector by promoting innovation, public and private investment, business creation, job creation, and increasing competitiveness
To foster the creation of national capacities and strategic competences in the space area by promoting education, strengthening research, and articulating different actors in the application of space sciences and technologies
To put Mexico in an important place in the international community regarding the peaceful, efficient and sustainable use of space, and to encourage international cooperation to face global problems related to this field

Transportation Sectorial Programme (2013–2018), which includes space infrastructure as part of the communications and transportation sector. Finally, the NSAP 2015 is much more detailed than the previous Plan and it contains important modifications such as a documented diagnostic of the Mexican space sector and several financial indicators.

A significant characteristic of the NSAP is the use of the OECD's recommendations for the development of the space sector as a structural guide to strengthen Mexican space capabilities. This means that the Agency implements its activities according to the three blocks defined by OECD's recommendations: the construction of user-oriented sustainable space infrastructure, the promotion of public use, and stimulation of the participation of the private sector in space activities. In addition, by using the "space economy"[60] concept, which includes the value chains directly or indirectly related to space activities, the NSAP takes the OECD's findings on Mexican economic activities related to the national space sector.

The NSAP 2015 has four strategic objectives (Table 12.5). Each of these objectives has its own strategies and action lines, all of them detailed in the NSAP 2015, which consists of practical steps that AEM must take to comply not only with NSAP's

the NSAP's objectives, AEM must become an actor with global responsibility that interacts with the space community so that it becomes one of the most important actors in the global space community. The space sector is absent in the current 2019–2024 NDP, see Plan Nacional de Desarrollo 2019–2024, Official Journal of the Federation, 12 July 2019, https://www.dof.gob.mx/nota_detalle.php?codigo=5565599&fecha=12/07/2019 (accessed 9 September 2019).

[60]Space economy "is the segment of a country's economy resulting from activities such as exploration, exploitation and use of the outer space. These activities, which employ objects launched and placed in space, include: scientific research; technological development; design, fabrication, manufacturing and operation of telecommunication systems; global positioning systems; and Earth and cosmos observation", Mexican Space Agency et al., Orbit Plan: Roadmap for Mexico's Space Industry, ProMéxico, Mexico City, October 2012, p. 13 https://www.gob.mx/cms/uploads/attachment/file/73145/PLAN_DE_ORBITA_2013_INGLES.pdf.

objectives but also with the sectorial strategies and objectives, as well as the NDP's objectives.[61]

12.4.1.2 Satellite Policy

Although Mexico ventured in the satellite sector several decades ago (see Sect. 12.2) becoming one of the first countries to have its own satellite system, it was not until 2018 that it adopted a comprehensive policy on the matter. After organizing a public consultation in 2017,[62] the MCT enacted the Satellite Policy of the Federal Government (SPFG) on 5 May 2018[63] to guarantee the intrinsic social and economic benefits derived from satellite technology (see Sect. 12.4.3.1).

The SPFG strengthens the satellite capacity reserved to the state that was established in the Mexican legislation as part of the measures adopted in the privatization process of the first national satellites with the aim of guaranteeing the availability of satellite capacity for government services associated with national security and social services. National bodies such as the Ministry of National Defence, the Ministry of the Navy, and the Federal Police use the satellite capacity reserved to the state.[64]

The SPFG lays the ground for the development and continuity of progress made in relation to the availability of national satellite capabilities and services, at least for the next 15 years, through the following axes (Table 12.6).

Table 12.6 Satellite policy's axes (2018)

Axes of the satellite policy of the Federal Government	
1	Satellite technologies and services for social inclusion
2	Satellite technologies and services for economic development
3	Satellite technologies and services for national security
4	Satellite technologies and services for technological and knowledge development
5	Satellite technologies and services for international cooperation

---

[61] There are in total twelve strategies and 36 action lines. In addition, the NSAP establishes that AEM will take the necessary actions to comply with three cross strategies of the NDP: the Program for a Close and Modern Government; the Program to Democratize Productivity; and the National Program for Equal Opportunities and Non Discrimination against Women (PROIGUALDAD – Programa Nacional para la Igualdad de Oportunidades y No Discriminación contra las Mujeres).

[62] Consulta de la Política Satelital Mexicana, Mexican Space Agency, https://www.gob.mx/participa/consultas/politica-satelital-mexicana (accessed 18 April 2019).

[63] It also received feedback from several public institutions and space actors, as well as advice from Euroconsult, a world leader satellite consulting company. Acuerdo que establece la política en material satelital del Gobierno Federal, Ministry of Communications and Transportation, Official Journal of the Federation, 5 May 2018, http://dof.gob.mx/nota_detalle.php?codigo=5522574&fecha=15/05/2018.

[64] Ibid.

Moreover, through this instrument Mexico seeks to directly and indirectly promote investment in the satellite sector.[65]

12.4.2 Space Long Term Strategies and Objectives

Some of Mexico's space long-term strategies have been developed in the last decade by the Mexican Government together with academia and the space industry. These strategies are defined in instruments consisting of roadmaps to develop the Mexican Space Industry.

12.4.2.1 Orbit Plan: Roadmap for Mexico's Space Industry[66]

In October 2012 AEM, ProMéxico[67] and a working group of representatives and deputies from academia, industry and the government (triple helix) created the *Orbit Plan: Roadmap for Mexico's Space Industry* document, a first roadmap to develop a national space industry that takes into account the AEM's Act, the General Guidelines of Mexico's Space Policy and the NSAP.

The Orbit Plan analysed various aspects of the space economy (characteristics of the global market, market segments, entrepreneurial structure) to appreciate the current situation of space activities in the world. It also presented an overview of the Mexican space sector and analysed its strengths, weaknesses, opportunities and threats in order to identify areas to be improved to consolidate a world-class space sector. Therefore, the Orbit Plan listed several strategic milestones to be achieved within a period of five years (no later than 2017), and included some projects and a work plan for their accomplishment.

The strategic milestones[68] determined by the Orbit Plan were (Table 12.7).

---

[65]One of the most relevant measures of the 2013 Telecommunications Reform was the opening of direct foreign investment up to 100% in satellite communications, Acuerdo que establece la política en material satelital del Gobierno Federal, Ministry of Communications and Transportation, Official Journal of the Federation, 5 May 2018, http://dof.gob.mx/nota_detalle.php?codigo=5522574&fecha=15/05/2018.

[66]Mexican Space Agency et al., Orbit Plan: Roadmap for Mexico's Space Industry, ProMéxico, Mexico City, October 2012, https://www.gob.mx/cms/uploads/attachment/file/73145/PLAN_DE_ORBITA_2013_INGLES.pdf.

[67]ProMéxico is the federal government agency responsible for coordinating strategies aimed at strengthening Mexico's participation in the international economy by supporting the export process and the internationalization of companies established in the country and coordinating actions to attract foreign investment.

[68]A strategic milestone is a goal or objective based on the prospective analysis of trends. For a milestone to be considered strategic, it must be SMART, this is Specific, Measurable, Aggressive but attainable, Relevant and Time framed. Each milestone had several projects and specific completion dates. There were 27 projects in total.

Table 12.7 Strategic milestones (Orbit Plan, 2012)

Orbit Plan—strategic milestones	
Milestone I	Validation, Standardization and Certification Centre for world-class space product testing laboratories
Milestone II	To establish a business with the technological capacity to design and develop space projects under a Public-Private Partnership model, that is both the core and interface with participants in space projects
Milestone III	Integrating a low-orbit multifunction satellite platform with 50% of critical technologies developed in Mexico
Milestone IV	Creation of a Public-Private Partnership institute to coordinate the triple helix for innovation in advanced materials with aerospace applications
Milestone V	Mexico will have a 1% share in the space industry (one billion dollars)

Despite its good intentions and the important efforts of those who elaborated it, the Orbit Plan did not meet its targets due to budget limitations[69] and because its goals were too ambitious to be achieved in the short term. Then, it soon became necessary to update it.[70]

12.4.2.2 Orbit Plan 2.0: Roadmap for Mexico's Space Industry[71]

Elaborated in 2017 by the same actors, Orbit Plan 2.0 is the updated version of the 2012 Orbit Plan. It includes recent developments in the space market; follows up the agreements, projects and activities of the first version; defines specific strategic lines that take into account the new trends, and proposes concrete actions to develop and strengthen the space industry in the following years. It incorporates elements not only to boost the development of the space industry but also to contribute to the development of sectors such as education, security, telecommunications, environmental sustainability, and health.

Contrary to the first version, the Orbit Plan 2.0 determines long-term strategic milestones and projects that respond to an integral vision of what its creators foresee for the Mexican space sector for the next decades.

As in the first version, Orbit Plan 2.0 uses a SWOT analysis[72] of the current situation of Mexico's space sector and a prospective analysis to identify and study international trends in this field (see Sect. 12.3.2). As a result, four thematic axes

---

[69]Rivera Parga, José, op. cit., p. 8.

[70]The Orbit Plan itself envisaged this possibility when stating that *"Due to its strategic nature, the orbit plan is a living document that requires frequent feedback and updating. It also demands flexibility to adapt to the major changes in the space market"*, p. 61.

[71]Orbit Plan: 2.0: Roadmap for Mexico's Space Industry, ProMéxico, Mexico City, 2017, https://www.gob.mx/cms/uploads/attachment/file/414932/Plan_Orbita_2.0.pdf.

[72]The SWOT (strengths, weaknesses, opportunities and threats) analysis shows the Mexican space industry's current situation compared to the rest of the world and identifies areas that must be leveraged or improved to consolidate a world-class space sector, Ibid., p. 63.

were identified from which four strategic milestones and 22 projects were defined (Table 12.8). Each milestone has a specific deadline and every project has a proposed term to be accomplished.

The four thematic axes are the following:

1. Innovation and opportunity niches for the industry and convergent services (EO, civil protection).
2. Self-determination in the development of space activities and cooperation for the strengthening of the Mexican space sector.
3. Giving impulsion to both the value chain of the space field and industrial development.
4. Promotion of digital access and development of applications and solutions.

Table 12.8 Strategic milestones (Orbit Plan 2.0, 2017)

Summary of strategic milestones[a]		
Thematic axe	Milestone	Strategic projects and action lines
Markets, industries and convergent services (EO, civil protection)	In 2035, Mexico will meet the needs of public and convergent private markets linked to the space sector, and will be positioned among the three world leaders in global share market with a 40% participation, including the use of space technologies to take care of the population and deal with climate change	Inventory update of industrial capacities, innovation and services; technological surveillance of capacities and state of the art
		Expansion of the innovation network capabilities and validation of solutions for the space sector
		Consolidation of the national archive of images management and satellite data
		Strengthening of the National Technical Committee for the Standardization of Space as the standardization body of the space sector, and coordination with the Mexican Certification Entity for the accreditation of laboratories related to the space sector
		Establishment of the Mexican cluster of organizations, industries and services related to space
		Development of a strategy of design, manufacture and put small satellites in service

(continued)

Table 12.8 (continued)

Summary of strategic milestones[a]		
Thematic axe	Milestone	Strategic projects and action lines
		Conduct of a study of strategic and opportunity markets for the country in space matters
		Development of specialized strategies for the strategic markets defined in the study
Self-determination through sovereignty and cooperation for the development of the Mexican's space sector	By 2036, Mexico will guarantee access to space by strengthening the decision-making capacity for the preservation and expansion of orbits and their convergent radio spectrum resources, and the establishment of two additional orbital positions	Clear definition of national bodies competences in the identification and monitoring of satellite replacement dates, and establishment of mechanisms and policies that guarantee a program for the conservation and expansion of satellite orbital positions and the necessary frequencies
		Policy and bases of planning and coordination to guarantee the management of orbits and the associated radio spectrum
		Update of the comprehensive catalogue of the capacities of associated laboratories
		Creation of the first network of venture capital investors for space infrastructure projects
Competitive participation in the space market in components, goods and services	By 2026, Mexico will have an important role in the development of components, goods and services, with a participation of roughly 1% of the global market	Technological prospective analysis to link the national IT sector to the space sector
		Planning and coordination policy and basis to give certainty and promote activities for the development of launchers in Mexico
		Creation of a multidisciplinary management group for the identification and analysis of strategic and opportunities markets for the country in space matters, development of specialized strategies based on the identified high value markets, and development of a pilot program for launchers in the national territory

(continued)

Table 12.8 (continued)

Summary of strategic milestones[a]

Thematic axe	Milestone	Strategic projects and action lines
		Establishment of an offsets systems scheme for the space sector
		Development of a strategy to increase the national content in satellites and ground segments
		Development of a strategy of international alliances
Digital access and content development and solutions	By 2026, Mexico will have developed the necessary space infrastructure to expand the Latin American connectivity range by 25%	Creation of a Latin American civil protection network by satellite technology
		Development of a strategy for the national integration of 45% for applications of the communications systems for national security
		Analysis of potential sites for the location of spaceports within the national territory and availability of these areas
		Development of spaceports by using national technological and innovation capacities

[a]This table was obtained for the Orbit Plan 2.0, op. cit., pp. 81–82

12.4.3 Space Actors

The main governmental actors participating in the Mexican space sector are the MCT, AEM, the IFT, and Telecomunicaciones de México. Other actors not directly related to space issues are also relevant when using space technologies or services to fulfil their obligations.

12.4.3.1 Ministry of Communications and Transportation

The MCT is a federal entity in charge of the promotion of transport systems through public policies and strategies with the aim of supporting increasing coverage and services availability. It elaborates and conducts the Federal telecommunications and broadcasting policies.[73]

---

[73]Ley Orgánica de la Administración Pública Federal, Official Journal of the Federation, 29 December 1976.

   The MCT is also responsible for the coordination of satellite orbit procedures with the competent international organizations, organizations from other countries, and national concessionaires or foreign operators; it establishes policy to promote the availability of capacity and sufficient satellite network services for national security, and the Federal Government's social services, needs, objectives, and goals; it manages and oversees the efficient use of wholly-owned or licenced, acquired or reserved allocated state satellite capacity; it ensures continuity of the satellite services provided by the state under long-term policies, and promotes investment in the country's infrastructure, satellite, and telecommunications and broadcasting services.[74]

12.4.3.2   Mexican Space Agency

AEM is the national body responsible for the implementation of the Mexican space policy and the NSAP. It is as a decentralized and sectorised public organism with legal personality and technical and administrative autonomy. It was created in 2010 by the AEM Act and started its operations on 1 November 2011. AEM belongs to the Communications and Transportation Sector, which is coordinated by the MCT. AEM's headquarters are based in Mexico City.

   AEM's mission is to use space science and technology to address Mexico's population needs and to create high quality jobs by fostering innovation and development of the space sector, to contribute to Mexico's competitiveness and position in the international community, and to use space in a pacific, efficient and responsible way.

   AEM has convened important cooperation agreements with several governmental entities, universities, research centres, space industries, international organizations and national space agencies of other states.[75] In recent years, the Agency has organised a great array of national and international space events, such as the 67th International Astronautical Congress in 2017[76] and the World Space Week[77] which positioned Mexico in third place worldwide in science dissemination and technology in 2018.[78]

---

[74]Ley Federal de Telecomunicaciones y Radiodifusión, Official Journal of the Federation, 14 July 2014. http://www.ift.org.mx/sites/default/files/contenidogeneral/asuntos-internacionales/federaltelecommunicationsandbroadcastinglawmexico.pdf.

[75]See Sect. 12.6.

[76]67th IAC, International Astronautical Federation, 2016 http://www.iafastro.org/events/iac/iac2016/ (accessed 7 April 2019).

[77]Semana Mundial del Espacio'18 México, https://haciaelespacio.aem.gob.mx/sme-mexico/2018/ (accessed 25 May 2019).

[78]"Mexico, third place in science dissemination and technology worldwide", Notimex, 1 March 2019, http://www.notimex.gob.mx/ntxnotaLibre/671333/mexico-third-place-in-science-dissemination-and-technology-worldwide.

12.4.3.3 Federal Telecommunications Institute

The IFT is an autonomous agency with its own legal personality and equity focused on the efficient development of broadcasting and telecommunications. It is responsible for the regulation, promotion and oversight of the use, development and operation of the radio spectrum, satellite orbits, satellite services, public telecommunications networks and the provision of broadcasting and telecommunications services, as well as access to active, passive and other essential inputs guaranteeing the right to access to information and universal access to those services. The Institute is also the authority on economic competition in the broadcasting and telecommunications sectors.

IFT grants, revokes and authorizes assignments or changes in the control, ownership or operation of companies with broadcasting and telecommunication concessions, including geostationary orbital positions and associated frequencies.[79]

12.4.3.4 Telecomunicaciones de México

Telecomunicaciones de México (Telecomm) is a decentralized organism of the Federal Public Administration with legal personality, created in 1986. Similar to AEM, Telecomm is part of the Communications and Transportation Sector of the MCT. It provides telegraphy, radiotelegraphy and telecommunications public services. Telecomm functions related to space consist of the installation, operation and use of transmitting and receiving ground stations and satellite radio communication systems; the occupation and exploitation of geostationary orbital positions and satellite orbits assigned to Mexico, with their respective frequency bands and rights of emission and reception of signals; the use and exploitation of radio frequency bands; the establishment of the interconnection of telecommunication systems under its charge with other national or foreign operators of public telecommunications networks, including the Mexsat Satellite System; and participation in technological and industrial research on satellite and telecommunication matters.[80]

12.4.3.5 Other Actors

Apart from the aforementioned bodies, there are several governmental actors whose activities are necessary for the development of the space sector, namely the ministries of National Defence (SEDENA); Navy (SEMAR); Environment and Natural Resources (SEMARNAT); Interior (SEGOB); Agriculture and Rural Development

---

[79]Ley Federal de Telecomunicaciones y Radiodifusión, Official Journal of the Federation, 14 July 2014.

[80]Organic Statute of Mexico's Telecommunications, Official Journal of the Federation, 14 February 14 2018, http://www.telecomm.gob.mx/gobmx/wp-content/uploads/2016/04/Estatuto%20Org%C3%A1nico%20de%20Telecomm%20(14%20DE%20FEBRERO%20DE%202018).pdf.

(SADER); the National Commission for the Knowledge and Use of Biodiversity (CONABIO); the National Forestry Commission (CONAFOR); Mexican Petroleum (PEMEX), and the National Institute of Statistics and Geography (INEGI).[81]

12.5 Perspectives on the Legal Framework Related to Space Activities

At the national level, Mexico has adopted several regulations to carry out its space and telecommunications activities, all of them regulating to some extent the functions of the relevant national bodies. At the international level, Mexico has ratified the most important treaties on the field and has fulfilled its international obligations on the creation of a national registry of its space objects launched into outer space.

12.5.1 National Space Regulations and Instruments

Mexico's specialized space norms include the Mexican Space Agency Act, AEM's internal regulations and the Mexican Norm on CubeSats. The telecommunications sector is mainly regulated by the Federal Telecommunications and Broadcasting Act.

12.5.1.1 Mexican Space Agency's Act

On 30 July 2010, the then president Felipe Calderon promulgated the Mexican Space Agency Act that gave birth to AEM.[82] The Act establishes the AEM's objectives, general structure, competences and functions, and matters concerning the Agency's budget and assets. One of the most important characteristics of the Act is the enumeration of the Mexican Space Policy instruments that are implemented by the Agency.

---

[81] Secretaria de la Defensa Nacional (SEDENA); Secretaría de Marina (SEMAR); Secretaría de Medio Ambiente y Recursos Naturales (SEMARNAT); Secretaría de Gobernación (SEGOB); Secretaría de Agricultura y Desarrollo Rural (SADER); Comisión Nacional para el Conocimiento y Uso de la Biodiversidad (CONABIO); Comisión Nacional Forestal (CONAFOR); Petróleos Mexicanos (PEMEX); Instituto Nacional de Estadística y Geografía (INEGI). See Orbit Plan 2.0, p. 32.

[82] Ley que crea la Agencia Espacial Mexicana, Official Journal of the Federation, 30 July 2010, https://www.gob.mx/cms/uploads/attachment/file/73063/Ley_que_crea_la_AgenciaEspacialMexicana.pdf (accessed 21 May 2019).

12.5.1.2   Mexican Space Agency Internal Regulations

The Organic Statute of AEM was adopted by AEM's Government Board on 27 November 2012.[83] It details the structure and organization of the Agency as well as the competences and functions of all of its decisional, administrative and technical bodies. Together with the Organic Statute there is the Organization Manual of the Agency, adopted by the Government Board on 5 December 2013, which coordinates and deeply details the functions and competences of the administrative and technical bodies of the Agency.

12.5.1.3   Mexican Norm on CubeSats

In August 2018, the Ministry of Economy published the Declaration of validity of the Mexican Norm NMX-AE-001-SCFI-2018 which determines the conditions for the creation of CubeSat satellites in the country. This norm, which became effective on 22 October 2018, also classifies the different types of CubeSats based on its main technical characteristics. Moreover, the norm was adopted in conformity with the International Norm ISO/DIS 17770, Space Systems-Cube Satellites (CubeSat), ISO 2015, with some appropriate modifications.[84]

12.5.1.4   Federal Telecommunications and Broadcasting Act

The Federal Telecommunications and Broadcasting Act entered into force on 14 July 2014. It mainly regulates the use and exploitation of the radio spectrum; public telecommunication networks; access to active and passive infrastructure; orbital resources; satellite communication, and the provision of telecommunications and broadcasting public services of general interest. According to the Act, the Mexican state maintains the original, inalienable and imprescriptible domain over the radio spectrum, but the use and exploitation of the radio spectrum and orbital resources can be allowed according to the relevant legal provisions.[85]

---

[83]*Acuerdo No 4/II/ORD./11.04.12/S adopting the Organic Statute of the Mexican Space Agency* (last reform on 10 April 2015), https://www.gob.mx/cms/uploads/attachment/file/73035/Estatuto_Organico_AEM.pdf.

[84]Publican la primera Norma mexicana para impulsar sector industrial de satélites miniaturizados, Mexican Space Agency, 26 August 2018, https://www.gob.mx/aem/prensa/publican-la-primera-norma-mexicana-para-impulsar-sector-industrial-de-satelites-miniaturizados-172274; see also Declaratoria de vigencia de la Norma Mexicana NMX-AE-001-SCFI-2018, Official Journal of the Federation, Ministry of the Interior, 22 August 2018, http://dof.gob.mx/nota_detalle.php?codigo=5535554&fecha=22/08/2018 (accessed 10 December 2019).

[85]Ley Federal de Telecomunicaciones y Radiodifusión, Cámara de Diputados del Congreso de la Unión, 14 July 2014, http://www.diputados.gob.mx/LeyesBiblio/pdf/LFTR_140219.pdf (accessed 7 April 2019).

12.5.2   Mexico and International Space Instruments

Mexico's active engagement in respect of international space law has been concretized in the ratification of the most important space and telecommunications treaties, the creation of a national registry of its space objects launched into outer space, as well as its formal engagement to comply with almost all international and European guidelines concerning space debris mitigation.

12.5.2.1   Space Treaties and Other Instruments

Mexico is one of the few states in the world that have ratified the five Space Treaties.[86] Mexico has also ratified other international agreements related to space or telecommunications such as the Treaty Banning Nuclear Weapon Tests in the Atmosphere, in Outer Space and Under Water (NTB), and the ITU Convention and Constitution[87] (Table 12.9).

Table 12.9 International space related treaties ratified by Mexico

International space treaties ratified by Mexico		
Treaty	Adoption	Mexico's ratification
OST	1967	R
ARRA	1968	R
LIAB	1972	R
REG	1975	R
MOON	1979	R
NTB	1963	R
BRS[a]	1974	R
ITSO[b]	1971	R
IMSO[c]	1976	R
ITU	1992	R

[a]Convention Relating to the Distribution of Programme-Carrying Signals Transmitted by Satellite, 21 May 1974
[b]Agreement Relating to the International Telecommunications Satellite Organization, 20 August 1971
[c]Convention on the International Mobile Satellite Organization, 3 September 1976

---

[86]At 1 April 2018, only 14 states have ratified the five Space Treaties: Australia, Austria, Belgium, Chile, Kazakhstan, Kuwait, Lebanon, Mexico, Morocco, the Netherlands, Pakistan, Peru, Turkey and Uruguay. COPUOS, Status of International Agreements relating to activities in outer space at a 1 January 2018, A/AC.105/C.2/2018/CRP.3, 9 April 2018 http://www.unoosa.org/documents/pdf/spacelaw/treatystatus/AC105_C2_2018_CRP03E.pdf.
[87]Ibid.

12.5.2.2 National Registry

According to Article II of the Registration Convention, Mexico created a national registry to entry its space objects launched into outer space. The registry was created on 1 March 2013,[88] although the first Mexican launched space object dates from 1985. To date, the country has launched 13 space objects (Table 12.11). AEM maintains the registry[89] and informs the UN Secretary-General through the United Nations Office of Outer Space Affairs (UNOOSA) on all space objects launched into outer space by Mexico as the State of Registry in accordance with Article IV of the Registration Convention.[90]

12.5.2.3 Space Debris Mitigation Instruments

Mexico has not yet adopted a national regulatory framework on space debris mitigation. However, it is aligned with the UNCOPUOS Space Debris Mitigation Guidelines and it supports the IADC Guidelines on Space Debris Mitigation and agrees with ITU Recommendations ITU-R.S 1003 on the environmental protection of the geostationary satellite orbit, the standards of the European Code of Conduct for Space Debris Mitigation and ISO 24113 for Space Systems: Space Debris Mitigation Requirements.[91]

12.6 Cooperation Programs

Nowadays, Mexico has adopted more than thirty space cooperation agreements with national space agencies, international governmental and non-governmental organizations, as well as space, telecommunications and technology companies. An important number of these agreements are valid for a few years, although it is likely that many of them will be extended for similar periods. There are also several agreements whose validity is undefined. Apart from the cooperation agreements with the UN, Mexico's space cooperation programmes are concentrated in four regions: Latin America, North America, Europe, and Asia-Pacific (Table 12.10).

---

[88] Nota verbal de fecha 31 de octubre de 2013 dirigida al Secretario General por la Misión Permanente de México ante las Naciones Unidas (Viena), COPUOS, ST/SG/SER.E/INF/28, 11 December 2013, http://www.unoosa.org/pdf/reports/regdocs/SERE_INF_028S.pdf.

[89] AEM Act, article 4, para. XIII.

[90] Mexican launched space objects Registry, 2018, https://www.gob.mx/cms/uploads/attachment/file/354275/RegistrodeObjetos__2018.pdf (accessed 6 February 2019).

[91] Compendium of space debris mitigation standards adopted by States and international organizations, Mexico, UNCOPUOS A/AC.105/C.2/2014/CRP.15, 18 March 2014, p. 31, http://www.unoosa.org/pdf/limited/c2/AC105_C2_2014_CRP15E.pdf (accessed 26 April 2019).

Table 12.10 Current agreements with space agencies, international organisations and other partners (2019)

Current agreements with space agencies, international organizations and other partners[a]

Country/international organisation/company	Agreement	Institution or company	Date of signature	Validity
Argentina	Framework Agreement between the Mexican Space Agency and the National Commission on Space Activities concerning space cooperation for peaceful purposes	CONAE—Argentina National Commission on Space Activities	29/07/2016	5 years
Argentina	Memorandum of Understanding between the Mexican Space Agency and Sur Technological Entrepreneurs, S.R.L.	Sur Technological Entrepreneurs, S.R.L.	02/11/2018	5 years
Austria	Memorandum of Understand between Mexican Space Agency and Space Generation Advisory Council	SGAC—Space Generation Advisory Council	01/04/2019	Undefined
Belgium	Memorandum of Understanding between the Mexican Space Agency and the Belgian Federal Office for Science Policy concerning space cooperation for peaceful purposes	BELSPO—Belgian Federal Office for Science Policy	18/02/2019	5 years

(continued)

Table 12.10 (continued)

Current agreements with space agencies, international organizations and other partners[a]

Country/international organisation/company	Agreement	Institution or company	Date of signature	Validity
Belgium	Letter of Intent between the Mexican Space Agency and LUCIAD, N.V.	LUCIAD N.V.	19/02/2019	Undefined
China	Memorandum of Understanding between the Mexican Space Agency and the China National Space Administration concerning space cooperation for peaceful purposes	CNSA—China National Space Administration	21/05/2015	5 years
Ecuador	Letter of Intent between the Mexican Space Agency and the Ecuadorian Civilian Space Agency concerning cooperation in exploration and utilization of outer space for peaceful purposes	EXA—Ecuadorian Civilian Space Agency	05/10/2016	Undefined
Europe	Letter of Intent between the European Space Agency and the Mexican Space Agency	ESA—European Space Agency	08/06/2018	Undefined
Europe	Letter of Intent Concerning Collaboration on the International Lunar Prize Competition	Airbus—Airbus Defence and Space	07/09/2018	Undefined

(continued)

Table 12.10 (continued)

Current agreements with space agencies, international organizations and other partners[a]

Country/international organisation/company	Agreement	Institution or company	Date of signature	Validity
Europe	GNSS Sensor Station Network Extension and Data Exchange	ESA—European Space Agency (ESOC—European Space Operations Centre)	31/08/2018	3 years
France	Extension of the Framework Agreement between the Mexican Space Agency and the French National Center for Space Studies concerning space cooperation for peaceful purposes	CNES—French National Center for Space Studies	04/11/2019	5 years
France	Specific Cooperation Agreement between the Mexican Space Agency and the French National Center for Space Studies regarding French-Mexican space cooperation for the study of the environment, climate and oceans	CNES—French National Center for Space Studies	16/07/2015	5 years

(continued)

Table 12.10 (continued)

Current agreements with space agencies, international organizations and other partners[a]

Country/international organisation/company	Agreement	Institution or company	Date of signature	Validity
France	Specific cooperation agreement between the National Center for Space Studies (French Republic) and the Mexican Space Agency, concerning cooperation for the organization of a workshop on the use of satellite technology for forestry	CNES—French National Center for Space Studies	26/06/2019	1 year
Germany	Mutual Agreement between the German Aerospace Center (DLR) and the Mexican Space Agency concerning the transfer of ownership in respect of the "Antenna System for Reception of Payload Data from Earth Observation Satellites"	DLR—German Aerospace Center	31/12/2013	Undefined
Germany	Extension of the Memorandum of Understanding between the Mexican Space Agency and the Deutsches Zentrum für Luft- und Raumfahrt e. V. concerning Space Cooperation for Peaceful Purposes	DLR—German Aerospace Center	02/10/2018	1 year

(continued)

Table 12.10 (continued)

Current agreements with space agencies, international organizations and other partners[a]

Country/international organisation/company	Agreement	Institution or company	Date of signature	Validity
Greece	Memorandum of Understanding between the Mexican Space Agency and the Hellenic Space Agency concerning space cooperation for peaceful purposes	HSA—Hellenic Space Agency	01/02/2019	5 years
India	Memorandum of Understanding between the Mexican Space Agency and the Indian Space Research Organisation of the Republic of India concerning space cooperation for peaceful purposes	ISRO—Indian Space Research Organisation	22/10/2014	5 years
International Space Education Board	Confirmation by Members of the International Space Education Board	ISEB—International Space Education Board	18/12/2017	3 years
Israel	Framework Agreement between the Mexican Space Agency and the Israel Space Agency concerning cooperation in the exploration and use of outer space for peaceful purposes	ISA—Israel Space Agency	14/09/2017	5 years

(continued)

Table 12.10 (continued)

Current agreements with space agencies, international organisations and other partners[a]

Country/international organisation/company	Agreement	Institution or company	Date of signature	Validity
Israel	Non-Disclosure Agreement between the Mexican Space Agency and Israel Aerospace Industries Ltd. (IAI/MBT)	IAI/MBT—Israel Aerospace Industries	13/06/2013	Undefined
Italy	Memorandum of Understanding between the Mexican Space Agency and the Italian Space Agency (ASI) on peaceful space cooperation	ASI—Italian Space Agency	27/09/2017	5 years
Italy	Joint Statement between the Italian Space Agency and the Mexican Space Agency related to a space partnership	ASI—Italian Space Agency	16/06/2015	Undefined
Japan	Memorandum of Understanding for cooperation on applications in space	NEC Corporation	26/07/2015	Undefined
Paraguay	Letter of Intent between the Mexican Space Agency and the Space Agency of Paraguay concerning cooperation in the peaceful uses of outer space	AEP—Space Agency of Paraguay	18/01/2018	Undefined

(continued)

Table 12.10 (continued)

Current agreements with space agencies, international organizations and other partners[a]

Country/international organisation/company	Agreement	Institution or company	Date of signature	Validity
Poland	Memorandum of Understanding between the Mexican Space Agency and the Polish Space Agency concerning technical and scientific cooperation in the exploration and use of outer space for peaceful purposes	POLSA—Polish Space Agency	24/04/2017	5 years
Russia	Letter of Intent between the Mexican Space Agency and the State Space Corporation Roscosmos on cooperation in the Field of exploration and use of outer space for peaceful purposes	ROSCOSMOS—State Corporation for Space Activities	29/09/2016	Undefined
Ukraine	Memorandum of Understanding between the State Space Agency of Ukraine and the Mexican Space Agency concerning space cooperation for peaceful purposes	SSAU—State Space Agency of Ukraine	26/09/2017	5 years

(continued)

Table 12.10 (continued)

Current agreements with space agencies, international organizations and other partners[a]

Country/international organisation/company	Agreement	Institution or company	Date of signature	Validity
United Kingdom	Memorandum of Understanding between the United Kingdom Space Agency and the Mexican Space Agency concerning cooperation in the exploration and use of outer space for peaceful purposes	UKSA—United Kingdom Space Agency	30/05/2013	Undefined
United Kingdom	Preliminary Memorandum of Understanding between Ecometrica and the Mexican Space Agency concerning development of a constellation of small satellites and downstream services to support climate change resilience and mitigation	ECOMETRICA	20/03/2019	January 2020

(continued)

Table 12.10 (continued)

Current agreements with space agencies, international organizations and other partners[a]

Country/international organisation/company	Agreement	Institution or company	Date of signature	Validity
United Nations	Cooperation agreement between the United Nations and the Mexican Space Agency to facilitate cooperation and host a regional support office for implementing the activities of the United Nations Platform for Space-Based Information for Disaster Management and Emergency Response programme	UNOOSA—United Nations Office for Outer Space Affairs	28/09/2016	3 years
United Nations	Memorandum of Understanding between the United Nations and the Mexican Space Agency	UNOOSA—United Nations Office for Outer Space Affairs	28/09/2016	5 years
United States of America	Non-reimbursable Space Act Agreement between the Mexican Space Agency and the National Aeronautics and Space Administration for collaboration on the AZTECHSAT-1 Cubesat Communications Technology Demonstration	NASA—National Aeronautics and Space Administration	15/03/2019	2 years

(continued)

Table 12.10 (continued)

Current agreements with space agencies, international organizations and other partners[a]

Country/international organisation/company	Agreement	Institution or company	Date of signature	Validity
United States of America	Reimbursable Space Act Agreement between Mexican Space Agency and the National Aeronautics and Space Administration of the United States for participation in the NASA international Internship Program	NASA—National Aeronautics and Space Administration	17/07/2018	5 years
United States of America	Letter of Intent between the Mexican Space Agency and Astrobotic Technology, Inc. to join efforts in shaping the development and delivery of the first charge of Latin America to the lunar surface	Astrobotic Technology, Inc.	01/06/2015	Undefined
United States of America	Nondisclosure Agreement between the Mexican Space Agency and ViaSat, Inc.	Viasat Inc.	18/06/2018	2 years
United States of America	Space Systems Company Bilateral Nondisclosure Agreement	Lockheed Martin Corporation—Lockheed Martin Space Systems Company	19/09/2018	3 years

(continued)

Table 12.10 (continued)

Current agreements with space agencies, international organizations and other partners[a]

Country/international organisation/company	Agreement	Institution or company	Date of signature	Validity
United States of America	Letter of Support to Mission Innovation and Launch Opportunity (MILO) and the Mexican Space Agency	MILO—Mission Innovation and Launch Opportunity	10/01/2018	Undefined
United States of America	Memorandum of Understanding between the Mexican Space Agency and the Secure World Foundation (SWF)	SWF—Secure World Foundation	02/14/2019	10 months
Venezuela	Cooperation agreement in space science, technology and innovation for the use of outer space for peaceful purposes between the Mexican Space Agency and the Bolivarian Agency for Space Activities of the Government of the Republic of Venezuela	ABAE—Bolivarian Agency for Space Activities of the Bolivarian Republic of Venezuela	17/09/2015	5 years

[a]General Coordination of International Affairs and Space Security, Mexican Space Agency, https://www.gob.mx/cms/uploads/attachment/file/293626/Current_Agreements_with_Space_Agencies_and_International_Organizations.pdf; Memorandum of Cooperation between the Hellenic and the Mexican Space Agency, 14 Current Agreements with Space Agencies and International Organizations, General Coordination of International Affairs and Space Security, Mexican Space Agency, 17 May 2019, https://www.gob.mx/cms/uploads/attachment/file/462340/Current_Agreements_with_Space_Agencies_and_International_Organizations_Descargar_documento.pdf (accessed 26 May 2019). See an interactive map at https://www.google.com/maps/d/viewer?mid=1PmDrtLbTKSeUixI6Z-YpMSohk8mJtV_p&ll=12.915772333477822%2C-14.38333781056076&z=2

12.6.1 Mexico and Latin America

Mexico's space cooperation with other Latin American countries has materialized through the adoption of several bilateral agreements with space agencies and
companies and its participation in the Americas Space Conference.

12.6.1.1 Bilateral Agreements

The AEM currently has space cooperation agreements with four Latin American
countries: a framework agreement with the Argentina National Space Activities
Commission (CONAE); a letter of intent with the Ecuadorian Civilian Space Agency
(EXA); a letter of intent with the Space Agency of Paraguay (AEP), and a cooperation
agreement with the Bolivarian Agency for Space Activities of the Government of
the Bolivarian Republic of Venezuela (ABAE). All these instruments aim to promote
the peaceful use of outer space between the relevant parties.

Moreover, AEM has also signed a memorandum of understanding (MoU) with the
Sur Technological Entrepreneurs, S.R.L., also known as SpaceSUR, an Argentinian
space company specialized in engineering and software solutions for the complete
aerospace cycle. The MoU defines joint work areas such as EO applications, capacity
building and diffusion, observations satellites, and development of alliances with the
Mexican industry.[92]

12.6.1.2 Multilateral Cooperation

Mexico has participated in Latin American space and telecommunications fora,
namely in the Americas Space Conference. The proposal for the creation of a regional
space agency has also been largely supported by Mexico.

Americas Space Conference

The Americas Space Conference (CEA) is the main regional and multilateral forum
related to the space sector. It was created in 1990 with the support of the UN and
its aim is to promote the development of space activities in Latin America and the
Caribbean. Mexico participates actively in CEA Conferences. For example, Mexico
organized the VI CEA entitled "Space and Development: Space Activities at the Service of Humanity and the Americas' Development" in Pachuca, state of Hidalgo, in
November 2009. In this Conference, Mexico assumed the Secretariat Pro-Tempore
and promoted a high level political dialogue in favour of a Latin American space
agenda, a dialogue with space actors of other regions, the negotiation of common

---

[92] AEM y SpaceSUR firman Memorando de Entendimiento, SpaceSUR, https://www.spacesur.com/
2018/11/28/aem-y-spacesur-firman-memorando-de-entendimiento/ (accessed 27 May 2019).

positions in the space sector in international fora, and the positioning of regional space priorities. The "Declaration of Pachuca" was adopted by the VI CEA. It is an important step towards consolidating the CEA as an intergovernmental and interagency coordination process among its members.[93]

Latin American Space Agencies Alliance's Initiative

In November 2013, the Brazilian Space Agency proposed the creation of a Latin-American Space Agencies Alliance (ALAS) with the aim of establishing a regional forum and promoting the creation of shared space programs and projects to address the necessities and demands of the region's countries. This proposal was welcomed by some Latin American countries, including Mexico,[94] and was seriously considered in an international symposium celebrated in Mexico in December 2014.[95] However, the initiative has received less attention than other regional cooperation frameworks such as CEA and the proposal for the creation of a Latin American Space Agency.

Latin American Space Agency's Proposal

Within the framework of the 67th International Astronautical Congress held in Guadalajara, Mexico, in 2016, AEM's director mentioned the possibility of creating a regional space agency.[96] This Latin American agency would not be linked to sending spaceships to space but to face common problems such as natural disasters surveillance, the fight against climate change, meteorology, reduction of the digital gap, and national security issues.[97] In 2018, AEM's director highlighted that

---

[93] Since its creation, CEA has dealt different subjects such as space technology, environment, long distance education, disasters management, and space legislation. To date, CEA has met seven times and adopted the equal number of declarations: I San José, Costa Rica, 1990; II Santiago, Chile, 1993; III Punta del Este, Uruguay, 1996; IV Cartagena de Indias, Colombia, 2002; V, Quito, Ecuador, 2006; VI Pachuca, México, 2010; VII Managua, Nicaragua, 2015, and VIII, Caracas, Venezuela, 2017, in Mexican Space Agency, Conferencia Espacial de las Américas, 22 March 2016, https://www.gob.mx/aem/acciones-y-programas/conferencia-espacial-de-las-americas-cea.

[94] Leonardi, Ivan, "Brasil propone una alianza Latinoamericana de agencias espaciales", Mundo-Geo, 11 November 2013, https://mundogeo.com/es/blog/2013/11/11/brasil-propone-una-alianza-latinoamericana-de-agencias-espaciales/.

[95] United Nations, "Informe del Simposio de las Naciones Unidas y México sobre Tecnología Espacial Básica: Logro de una tecnología espacial accesible y asequible", UNCOPUOS, A/AC.105/1086, 23 December 2014, http://www.unoosa.org/pdf/reports/ac105/AC105_1086S.pdf.

[96] Argentina y México van por un NASA "latinoamericano", El Economista, 30 September 2016 (accessed 3 April 2019), https://www.eleconomista.com.mx/arteseideas/Argentina-y-Mexico-van-por-un-NASA-latinoamericano-20160930-0044.html.

[97] Zazueta Oliver, "Idean agencia espacial de Latinoamérica", Reforma, 29 September 2016 (accessed 3 April 2019). https://www.reforma.com/aplicacioneslibre/articulo/default.aspx?id=950894&md5=15a50a7ce2dcddd0547f60b0c16c5806&ta=0dfdbac11765226904c16cb9ad1b2efe.

this idea needs dialogue, reflexion and consideration.[98] The idea of a regional space Agency was also proposed by the crew of the first Latin American mission of The Mars Society in which students from six Latin American countries, including Mexico participated.[99]

ReLaCa-Espacio Network

Although representatives of Mexican institutions and universities have not yet participated in ReLaCa-Espacio's activities, a regional network of universities and institutions that aims to foster technology, policy and law space research, the invitation is open.[100]

Other Regional Fora

Mexico has participated in other regional space related fora such as the 3rd International Space Forum (ISF) at Ministerial Level—The Latin American and Caribbean Chapter held in Buenos Aires, Argentina on 1 November 2018, where it called for international cooperation between universities of the Latin America and Caribbean region to form human capital specialized in the space sector to face common problems[101]; and the First Latin American Congress on Aerospace Science and Technology held in San Luis Potosi (Mexico's central region) in September 2012, where members of the government, academia and industry talked about the future of the Mexican space sector.[102]

---

[98]"Expertos coinciden en necesidad de Latinoamérica de impulsar alianza espacial", EFE, Montevideo, 21 May 2018, (accessed 3 April 2019), https://www.efe.com/efe/america/tecnologia/expertos-coinciden-en-necesidad-de-latinoamerica-impulsar-alianza-espacial/20000036-3623068.

[99]The Colombian researcher Camilo Guzmán considers that the regional space agency would reduce the costs of space technologies to the benefit of sustainable development by preventing disasters and deforestation, avoiding epidemics, and strengthening telemedicine, in "América Latina necesita cooperación especial para desarrollo regional", Mexican Space Agency, 67th IAC, Guadalajara, 23 September 2016 https://www.aem.gob.mx/iac2016-NOTIMEX/notas/243962.html (accessed 3 April 2019).

[100]In the preliminary programme of the IV International ReLaCa-Espacio Meeting held on 24 May 2019 in Asuncion, Paraguay, a lecture by a representative of the Mexican Space Agency on the "Pillars for the Global Governance on Outer Space Activities in the XXI century" was scheduled. However, this participation was removed in the final programme, see Preliminary Programme, IV Encuentro International de la Red Latinoamericana y del Caribe del Espacio, Asuncion, Paraguay, 24 May 2019, http://www.aepdiri.org/images/2019/archivos/Programa_24_mayo_2019_Asuncion.pdf (accessed 26 May 2019).

[101]Mexican Space Agency, "Exhorta AEM a cooperación especial en región Latinoamericana y del Caribe", 14 November 2018, https://www.gob.mx/aem/prensa/exhorta-aem-a-cooperacion-espacial-en-region-latinoamericana-y-del-caribe-182081.

[102]Grajeda, Genaro, "1er Congreso Latinoamericano de Ciencia y Tecnología Aeroespacial", Reconoce Mx, http://www.reconoce.mx/1er-congreso-latinoamericano-de-ciencia-y-tecnologia-aeroespacial/ (accessed 7 April 2019).

12.6.2 Mexico and Europe

The European experience in the space sector has led Mexico to create ties with ESA to give impulse to its space activities. Furthermore, Mexico has signed important bilateral agreements with several European space agencies and companies.

12.6.2.1 Bilateral Agreements

On one hand, Mexico has many space cooperation agreements with eight space agencies in Europe: a mutual agreement and an MoU with the German Aerospace Center (DLR); an MoU with the Belgian Federal Office for Science Policy of Belgium; a framework agreement and two specific cooperation agreements with the French National Center for Space Studies (CNES); an MoU with the Hellenic Space Agency of Greece (HSA); an MoU and a joint statement with the Italian Space Agency (ASI); an MoU with the Polish Space Agency (POLSA); an MoU with the United Kingdom Space Agency (UKSA); an MoU with the State Space Agency of Ukraine (SSAU), and a letter of intent with the State Corporation for Space Activities of the Federation of Russia (ROSCOSMOS).

In general, these cooperation agreements aim to promote the peaceful use and exploration of outer space. Specifically, the mutual agreement between DLR and AEM concerns the transfer of ownership in respect of the "Antenna System for Reception of Payload Data from Earth Observation Satellites", and the specific cooperation agreements between AEM and CNES relate to Franco-Mexican space cooperation for the study of the environment, climate and oceans,[103] and for the organization of a workshop on the use of satellite technology for forestry.

AEM also has cooperation agreements with other European governmental bodies, NGO's and private companies, including a letter of intent with Luciad, NV, a Belgian international supplier of geographic information system tools; a preliminary MoU with ECOMETRICA, a British company specialized in downstream space and sustainability reporting, regarding the development of a constellation of small satellites and downstream services to support climate change resilience and mitigation; and an MoU with the Space Generation Advisory Council (SGAC), a space non-governmental organisation and network based in Austria (see Sect. 12.10.2.1).

12.6.2.2 Multilateral Cooperation

Space cooperation between Mexico and Europe comprises a letter of intent between AEM and ESA, and an agreement for a Global and Navigation Satellite System (GNSS) sensor station, a network extension and the exchange of data between AEM

---

[103]"Avances en cooperación espacial entre México y Francia", Latam Satelital, 22 March 2016, http://latamsatelital.com/avances-en-cooperacion-espacial-entre-mexico-y-francia/ (accessed 7 April 2019).

and ESA-European Space Operations Centre. A letter of intent has also been signed between AEM and AIRBUS Defence and Space concerning the International Lunar Prize Competition.

Also, as mentioned above (see Sect. 12.6.1.2), the ECSL of ESA has given impulse to the newly created ReLaCa-Espacio Network to promote cooperation between Latin America's universities and research institutions in space related issues, such as technology, policy and law. Mexico has not yet participated in its activities but the invitation is still open.

12.6.3 Mexico and North America

Mexico's relations with the United States of America and Canada are extremely important. The North American Free Trade Agreement (NAFTA) and the new trade agreement negotiated in 2018[104] by the three countries are a good example of this trilateral cooperation. Thus, the geographic situation of Mexico could be seen as an excellent opportunity for the development of its space sector. For years, Mexico has celebrated many telecommunications and broadcasting bilateral agreements with its northern neighbours, and nowadays, AEM and the United States have adopted many space bilateral agreements. In contrast, Mexico and Canada have not yet adopted any space agreement (apart from satellite services and telecommunications agreements) nor has there been a regional agreement on the matter.

12.6.3.1 Bilateral Agreements

Without any doubt, relations between Mexico and the United States of America (USA) are extremely important for the development of both countries. Space is one of the multiple sectors of this bilateral cooperation. Mexico has a strong interest in improving its space capabilities, technology and human capital by taking advantage of the long experience of USA's space sector. NASA and AEM have two cooperation agreements in force: one is for collaboration on the AZTECHSAT-1 Cubesat Communications Technology Demonstration, and the other for the participation in the NASA international Internship Program.

The objective of the first agreement is to launch the AZTECHSAT-1 nanosatellite from the International Space Station (ISS) in 2019. This project, on which more than 70 people from both countries are working, also aims to build satellite development

---

[104]See "Resultados de la modernización del acuerdo comercial entre México, Estados Unidos y Canadá", Mexican Government, 1 October 1018, https://www.gob.mx/tlcan/acciones-y-programas/resultados-de-la-modernizacion-del-acuerdo-comercial-entre-mexico-estados-unidos-y-canada?state=published (accessed 24 April 2019).

capabilities for future Mexican generations.[105] The second agreement allows Mexican students to do research internships at NASA to acquire knowledge in space science and technology in order to integrate national telecommunications, TIC's or aerospace sectors.[106]

Apart from the aforementioned space agreements, Mexico and the USA have adopted approximately 20 bilateral instruments on satellite services and the use of bands for broadcasting, among other services. Mexico and Canada have also adopted bilateral cooperation agreements on satellite services and telecommunications.[107]

Space cooperation agreements between AEM and U.S. companies have been adopted in recent years. These include a letter of intent with Astrobotic Technology, Inc. for the development and delivery of the first charge of Latin America to the lunar surface; a nondisclosure agreement with ViaSat, Inc., an American communications company; a nondisclosure agreement with Lockheed Martin Space Systems Company; a letter of support with Mission and Innovation and Launch Opportunity (MILO); and an MoU with Secure World Foundation (SWF), a private operating foundation dedicated to the secure and sustainable use of space.

12.6.3.2 Multilateral Cooperation

There is currently no regional space cooperation agreement between Mexico, the USA and Canada.

12.6.4 Mexico and the Asia-Pacific

The active participation of AEM in recent years has led Mexico to adopt bilateral cooperation agreements with some of the most important space agencies of the Asia-Pacific region. Mexico's growing interest in this region has been strongly manifested with the recent invitation of the Organization for Asia-Pacific Space Cooperation to Mexico to participate in the sessions of this space regional forum.

---

[105] "Lanzarán nanosatélite mexicano "Aztechsat-1" desde Estación Espacial Internacional, en 2019", Mexican Space Agency, 25 February 2019, https://www.gob.mx/aem/prensa/lanzaran-nanosatelite-mexicano-aztechsat-1-desde-estacion-espacial-internacional-en-2019-148785?idiom=es (accessed 7 April 2019); see also Kulu Erik, "AzTechSat-1", Nanosats Database, 2019, https://www.nanosats.eu/sat/aztechsat-1 (accessed 19 April 2019).

[106] "Publica AEM Convocatoria para Estancias de Investigación en NASA", Mexican Space Agency, 21 February 2019, https://www.gob.mx/aem/prensa/publica-aem-convocatoria-para-estancias-de-investigacion-en-nasa-191611?fbclid=IwAR3BLxydRgYayBqm-AaGKmoJNtwq86mi7WSv3rHbFpw0I79EcwXKzcxlGz4 (accessed 7 April 2019).

[107] The Ministry of Communications and Transportation keeps the record of these instruments. See http://www.sct.gob.mx/informacion-general/normatividad/telecomunicaciones/tratados-y-acuerdos-Internacionales.

12.6.4.1 Bilateral Agreements

AEM has celebrated bilateral space agreements with some of the major spacefaring nations of the Asia-Pacific region, all of them concerning space cooperation for peaceful purposes: an MoU with China National Space Administration (CNSA), an MoU with the Indian Space Research Organisation (ISRO) and a framework agreement with the Israel Space Agency (ISA). In addition, AEM and the Japan Aerospace Exploration Agency (JAXA) have created a cooperation agenda to improve and diversify knowledge, investment and experience exchanges in the aerospace sector to promote Mexico's development in this area.[108] Also, Mexico and New Zealand have started talks to cooperate in the aerospace sector of both countries and to strengthen the nano satellites launch sector of New Zealand.[109]

Space cooperation with Asian companies includes a non-disclosure agreement between AEM and Israel Aerospace Industries Ltd., an Israeli company that produces aerial and astronautic systems for both civil and military usage, and an MoU for cooperation on applications in space with NEC Corporation, a Japanese provider of information technology services and products.

12.6.4.2 Multilateral Cooperation

Although space cooperation between Mexico and the Asia-Pacific region is modest, it has a strong potential for the coming years as demonstrated, for example, by the recent invitation to Mexico to participate in the Asia-Pacific Space Cooperation Organisation forum.

Asia-Pacific Space Cooperation Organisation

In July 2015, Mexico was invited to join the Asia-Pacific Space Cooperation Organisation (APSCO), an international intergovernmental organization based in Beijing, China, as a member observer in a first stage. Later, in December 2018 and in the framework of APSCO's tenth anniversary and its ninth International Symposium held in Beijing, AEM was invited to join APSCO's forums to take advantage of space and joint scientific development, creating new possibilities to collaborate in different aerospace areas. In May 2019, APSCO invited Mexico to become a permanent member of the Organization. This cooperation contemplates support with

---

[108]It has also been considered the possibility to make joint launches of nanosatellites from the Kibo laboratory of the ISS, "Esquema de cooperación AEM-JAXA, diversifica y enriquece transferencia de conocimientos", Mexican Space Agency, 20 February 2016, https://www.gob.mx/aem/prensa/esquema-de-cooperacion-aem-jaxa-diversifica-y-enriquece-transferencia-de-conocimientos-20548 (accessed 7 April 2019).

[109]Anuncian colaboración espacial México y Nueva Zelanda, Mexican Space Agency, 3 August 2019, https://www.gob.mx/aem/prensa/anuncian-colaboracion-espacial-mexico-y-nueva-zelanda-211751 (accessed 9 September 2019).

real-time satellite images in case of disasters, monitoring of seismic activity, agricultural productivity projects, and use of telecommunications infrastructure for health purposes.[110] The cooperation will also promote knowledge transfer, technology, and the generation of new knowledge with the aim of participating in Asia-Pacific space projects that will benefit Mexican youth.[111]

Centre for Space Science and Technology Education in Asia and the Pacific

In July 2015, Mexico integrated in the activities of the Centre for Space Science and Technology Education in Asia and the Pacific (CSSTEAP) to train Mexicans in the major disciplines of space science and technology.[112] CSSTEAP is a regional centre affiliated to the UN that promotes space education, training, application programmes and regional cooperation.

12.6.5 Mexico and the United Nations

The universal character of the UN and its significant role in space issues have led Mexico to participate in a productive and continuous manner in the work of the organization, namely in the activities of UNCOPUOS and its scientific and technical, and legal subcommittees. Mexico's participation in the work of the ITU is also important.

12.6.5.1 Bilateral Agreements

Mexico has concluded two bilateral agreements in space matters with the UN: an MoU to create a regional centre for space science and technology in Mexico, and a cooperation agreement to facilitate cooperation and host a regional support office for implementing the activities of the UN-SPIDER programme.

---

[110]Invitan a AEM a ser miembro permanente de la APSCO, Mexican Space Agency, 10 May 2019, https://www.gob.mx/aem/prensa/invitan-a-aem-a-ser-miembro-permanente-de-la-apsco-199663 (accessed 27 May 2019).

[111]"Mexico participates in international space forum in Beijing", El Universal, 12 December 2018 https://www.eluniversal.com.mx/english/mexico-participates-international-space-forum-beijing (accessed 3 April 2019).

[112]"Se integra México a actividades de cooperación internacional Asia-Pacífico en materia especial", Mexican Space Agency, 27 July 2015 https://www.gob.mx/aem/prensa/se-integra-mexico-a-actividades-de-cooperacion-internacional-asia-pacifico-en-materia-espacial-19686 (accessed 3 April 2019).

Regional Centre for Space Science and Technology Education

In 1990, the UNOOSA's Programme on Space Applications considered the necessity to establish Regional Centres for Space Science and Technology Education (CRECTEALC) to allow developing countries to build regional capacity in space science and technology and its applications. These centres should be created on the basis of affiliation to the UN through a cooperation agreement. Two regional centres were established in Brazil and Mexico for Latin America and the Caribbean. The agreement between Mexico and the secretariat of CRECTEALC for the operation of the Centre in Mexico was adopted on 23 October 2002.[113]

The objective of the centre is to enhance the capabilities of member states, at regional and international levels, in various disciplines of space science and technology to advance their scientific, economic and social development. The centre provides postgraduate education, research and application programmes with emphasis on remote sensing, satellite communications, satellite meteorology and space science for university educators and research and application scientists.[114]

UN-SPIDER

UNOOSA and AEM signed a Cooperation Agreement in September 2016 to incorporate AEM as a Regional Support Office of the UN Platform for Space-Based Information for Disaster Management and Emergency Response programme (UN-SPIDER programme) under the UNOOSA framework. AEM has supported UN-SPIDER through the provision of experts to conduct technical advisory missions to Central American countries. In addition, AEM is a partner in the SEWS-D project, a Latin American and the Caribbean project which aims to strengthen national drought policies by using space-based information on a permanent basis.[115] AEM coordinates activities between the parties involved and provides support in managing the project.[116]

Even though the Cooperation Agreement was only signed in 2016, UN-SPIDER had previously worked with Mexico. For example, in 2011 UN-SPIDER and CRECTEALC conducted a theoretical and practical training course in Mexico to strengthen the knowledge and skills of professionals of the Mexican government on the use of radar imagery in case of floods, namely the peculiarities concerning their

---

[113]CRECTEALC, INAOE, http://crectealc.inaoep.mx:8080/CREC/ See Sect. 12.6.4.2.

[114]UNCOPUOS, Regional centres for space science and technology education (affiliated to the United Nations), UNGA A/AC.105/782, 14 March 2002.

[115]"SEWS-Drought in Latin American and the Caribbean", UN-SPIDER, Knowledge Portal, http://www.un-spider.org/projects/SEWS-D-project-caribbean. See also "AEM conducts teleconference with SEWS-D partners", Knowledge Portal, UN-SPIDER, 13 October 2016, http://www.un-spider.org/news-and-events/news/aem-conducts-teleconference-sews-d-partners (accessed 26 April 2019).

[116]Mexico Regional Support Office, Knowledge Portal, UN-SPIDER, http://www.un-spider.org/network/regional-support-offices/mexico-regional-support-office (accessed 26 April 2019).

acquisition by specific satellites and the features displayed in the images emerging from specific characteristics of microwaves.[117] Cooperation between UN-SPIDER and CRECTEALC has been continuously strengthened.[118]

12.6.5.2 Institutional Participation

Mexico's role in the UN organs and specialized agencies has been renowned and active. Mexico has been a member of UNCOPUOS and the ITU for decades and constantly participates in their work by proposing and adopting, together with other states, standards, legal rules, recommendations and other instruments related to space and telecommunications.

UNCOPUOS

Mexico is one of the founding members of UNCOPUOS.[119] Since the creation of UNCOPUOS on 13 December 1958, Mexico has actively participated in the work of the UN Committee and of its Scientific and Technical and Legal Subcommittees.

UNOOSA

Mexico has given to UNOOSA information concerning its space activities to fulfil its international obligations or on a voluntary basis. For example, as part of its international obligations as the State of Registry according to the 1975 Registration Convention, Mexico has duly notified to the UN Secretary-General the information of its space objects launched into outer space[120] as well as the creation of its national registry.[121] Concerning notification on a voluntary basis, Mexico has responded to

---

[117]"Mexico-Regional Training course", Knowledge Portal, UN-SPIDER, October 2011, http://www.un-spider.org/advisory-support/training-activities/mexico-regional-training-course (accessed 26 April 2019).

[118]"UN-SPIDER visits Mexico Campus of the Regional Centre for Space Science Technology Education for Latin America and the Caribbean", Knowledge Portal, UN-SPIDER, 18 July 2017, http://www.un-spider.org/news-and-events/news/un-spider-visits-mexico-campus-regional-centre-space-science-technology (accessed 26 April 2019).

[119]Question of the peaceful use of outer space, UNGA RES 1348 (XIII), 13 December 1958, http://www.unoosa.org/pdf/gares/ARES_13_1348E.pdf.

[120]Mexico, Online Index of Objects Launched into Outer Space http://www.unoosa.org/oosa/osoindex/search-ng.jspx?lf_id=#?c=%7B%22filters%22:%5B%7B%22fieldName%22:%22en%23object.launch.stateOrganization_s%22,%22value%22:%22Mexico%22%7D%5D,%22sortings%22:%5B%7B%22fieldName%22:%22object.launch.dateOfLaunch_s1%22,%22dir%22:%22desc%22%7D%5D%7D (accessed 13 April 2019).

[121]"Información proporcionada de conformidad con el Convenio sobre el registro de objetos lanzados al espacio ultraterrestre", UNCOPUOS, ST/SG/SER.E/INF/28, 11 December 2013, http://www.unoosa.org/pdf/reports/regdocs/SERE_INF_028S.pdf (accessed 19 April 2019). See Sect. 12.5.2.2.

many UNOOSA questionnaires on its national space activities and other related issues.[122]

International Telecommunication Union

Mexico has been member of the International Telecommunication Union (ITU) since 1993. Mexico's engagement in ITU's instruments includes the ratification of ITU's Constitution and Convention, International Telecommunications Regulations, Radio Regulations, and many World Administrative Radio Conferences and World Radiocommunication Conferences.[123]

12.7 Satellite Capacities

As presented in Sect. 12.2, the first Mexican satellites were launched into outer space in the 1980s. Since the launching of the MORELOS System, the Mexican satellite sector has seen a positive upgrade trend in the last years (see Sect. 12.4.1.2 and Table 12.11). Nowadays, satellites have become necessary to satisfy social needs, create new economic opportunities, maintain national security and face natural disasters. Public and private satellite systems participate in the development of the country thanks to an appropriate use of Mexican orbital resources.

12.7.1 Satellite Systems

In Mexico the satellite system is composed of two subsystems, one public and one of concessions. The first system, called MEXSAT, is integrated by the BICENTE-NARIO and MORELOS 3 satellites while the second comprises the concessioned orbital positions and authorizations granted to individuals to exploit the emission and reception rights of foreign satellites.[124]

MEXSAT System provides safe communications in all Mexican territory, and control and autonomy of the Federal Government over strategic satellite security communications. Satellite BICENTENARIO provides national security and social

---

[122] See for example Mexico's answers to the UNOOSA's survey "Questions on suborbital flights for scientific missions and/or for human transportation", A/AC.105/1039/Add.11, 28 March 2018, available at http://www.unoosa.org/oosa/oosadoc/data/documents/2018/aac.105/aac.1051039add.11_0.html (accessed 24 April 2019).

[123] For a complete list of agreements see https://www.itu.int/online/mm/scripts/gensel26?ctryid=1000100411.

[124] Acuerdo que establece la política en material satelital del Gobierno Federal, Ministry of Communications and Transportation, Official Journal of the Federation, 5 May 2018, http://dof.gob.mx/nota_detalle.php?codigo=5522574&fecha=15/05/2018.

Table 12.11 Mexico's satellites launched into outer space

Mexico's satellites launched into outer space

Satellite	Launched	Orbit	Features	Focus
MORELOS 1	1985	Geostationary	• Orbital position: undefined • Launch vehicle: Space Shuttle Discovery • Launch site: Kennedy Space Centre, Florida, USA	• Owner: Telecomunicaciones de México • Status: inactive • Expected life: 9 years • General functions: coverage of the national territory with television, radio and telephony signals and data transmission
MORELOS 2	1985	Geostationary	• Orbital position: undefined • Launch vehicle: Space Shuttle Atlantis • Launch site: Kennedy Space Centre, Florida, USA	• Owner: Telecomunicaciones de México • Status: inactive • Expected life: 19 years • General functions: coverage of the national territory with television, radio and telephony signals and data transmission
SOLIDARIDAD 1	1993	Geostationary	• Orbital position: 109.2° W • Launch vehicle: ARIANE 4 • Launch site: Kouru, French Guiana	• Owner: Satélites Mexicanos, S. A. de C. V. • Status: inactive • Expected life: 7 years • General functions: commercial and social services

(continued)

Table 12.11 (continued)

Mexico's satellites launched into outer space

Satellite	Launched	Orbit	Features	Focus
SOLIDARIDAD 2	1994	Geostationary	• Orbital position: 113° W • Launch vehicle: ARIANE 4 • Launch site: Kouru, French Guiana	• Owner: Satélites Mexicanos, S. A. de C. V. • Status: inactive • Expected life: 15 years • General functions: social access and connectivity satellite services
UNAMSAT B	1996	orbit below 1000 km with an 83° inclination	• Orbital position: low orbit (1000 km) with an inclination of 83° • Launch vehicle: KOSMOS 3 M • Launch site: Plesetsk Cosmodrome, Russia	• Owner: National Autonomous University of Mexico • Status: inactive • Expected life: undefined • General functions: scientific
SATMEX 5	1998	Geostationary	• Orbital position: 116,8° W • Launch vehicle: ARIANE 4 • Launch site: Kouru, French Guiana	• Owner: Satélites Mexicanos, S. A. de C. V. • Status: inactive • Expected life: 15 years • General functions: commercial and social satellite services

(continued)

Table 12.11 (continued)

Mexico's satellites launched into outer space

Satellite	Launched	Orbit	Features	Focus
SATMEX 6	2006	Geostationary	• Orbital position: 113° W • Launch vehicle: ARIANE 5 • Launch site: Kouru, French Guiana	• Owner: Satélites Mexicanos, S. A. de C. V. • Status: active • Expected life: 15 years • General functions: commercial and social satellite services
QUETZAT 1	2011	Geostationary	• Orbital position: 77° W • Launch vehicle: PROTON BREEZE-M • Launch site: Baikonur Cosmodrome, Kazakhstan	• Owner: Quetzsat, S. A. de C. V. • Status: active • Expected life: 15 years • General functions: restricted satellite television services
BICENTENARIO	2012	Geostationary	• Orbital position: 114.8° W • Launch vehicle: ARIANE 5 • Launch site: Kouru, French Guiana	• Owner: Mexican Government/Telecomunicaciones de México • Status: active • Expected life: 15 years • General functions: commercial and social satellite services

(continued)

Table 12.11 (continued)

Mexico's satellites launched into outer space

Satellite	Launched	Orbit	Features	Focus
SATMEX 8	2013	Geostationary	• Orbital position: 116.8° W • Launch vehicle: PROTON BREEZE-M • Launch site: Baikonur Cosmodrome, Kazakhstan	• Owner: Satélites Mexicanos, S. A. de C. V. • Status: active • Expected life: 15 years • General functions: telecommunications services
E115WB (SATMEX 7)	2015	Geostationary	• Orbital position: 114.9° W • Launch vehicle: FALCON-9 v1.1 • Launch site: Cape Canaveral, Florida, USA	• Owner: Satélites Mexicanos, S. A. de C. V. • Status: active • Expected life: 15 years • General functions: communications satellite
MORELOS 3	2015	Geostationary	• Orbital position: 113.1° W • Launch vehicle: ATLAS V (421) • Launch site: Cape Canaveral, Florida, USA	• Owner: Mexican Government/operated by Telecomunicaciones de México • Status: active • Expected life: 15 years • General functions: communication satellite services; vehicle tracking; early warning to prevent emergencies and provision of support during natural disasters
E117WB (SATMEX 9/EUTELSAT 117 West B)	2016	Geostationary	• Orbital position: 117° W • Launch vehicle: FALCON-9 v.1. • Launch site: Cape Canaveral, Florida, USA • Other launching States: USA	• Owner: Satélites Mexicanos, S. A. de C. V. • Status: active • Expected life: 15 years • General functions: communications satellite

coverage services and offers fixed communication services (through extended bands C and Ku). Satellite MORELOS 3 provides mobile communication services (through band L).[125]

12.7.2 Orbital Resources

Mexican orbital resources are the geostationary and non-geostationary resources. Geostationary resources match with the Geostationary Orbital Positions (GOP's) available to Mexico, which grants exploitation licenses to satellite, services concessionaries or assigns them to the MEXSAT System. Non-geostationary resources are the positions and frequencies outside GEO, including Low, Medium and High Earth Orbits.[126]

12.7.2.1 Geostationary Orbital Positions

Mexico has two types of orbital resources: planned and not planned (coordinated).[127] Planned orbital resources assigned by the ITU are GOP's 69.2° W, 77° W, 127° W, 136° W for satellite broadcasting services, while GOP's 113° W and 116.8 ° W are for fixed satellite services. GOP's 77° W and 113° W are used by QuetSat and Telecomm, respectively. GOP's 69.2° W, 116.8° W, 127° W and 136° W are currently available.[128]

Non-planned orbital resources are GOP's whose frequency bands and coverage are different from those defined by the ITU, although they are used according to the ITU's Radio communications Regulations to avoid interferences with other positions. GOP's 113° W, 114.9° W and 116.8° W have been concessioned to Eutelsat Americas (GOP's 113° W, 114.9° W and 116.8° W) and Telecomm (113° W and 114.9° W) in different frequency bands.[129]

12.7.2.2 Foreign Orbital Resources

Foreign satellite operators interested in providing coverage with their signals or in distributing their telecommunication services in Mexico need an authorization

---

[125] Ibid.

[126] Ibid.

[127] "Recursos orbitales geoestacionarios y no geoestacionarios", Espectro Radioeléctrico, Federal Telecommunications Institute, http://www.ift.org.mx/espectro-radioelectrico/recursos-orbitales/en-mexico (accessed 19 April 2019).

[128] Acuerdo que establece la política en material satelital del Gobierno Federal, Ministry of Communications and Transportation, Official Journal of the Federation, 5 May 2018, http://dof.gob.mx/nota_detalle.php?codigo=5522574&fecha=15/05/2018.

[129] Ibid.

Table 12.12 Mexico's geostationary orbital resources

Mexico's geostationary orbital resources				
Planned	69.2° W	Broadcasting	National	Available
	77° W	Broadcasting	National	Quetsat
	116.8° W	Fixed services	National	Available
	127° W	Broadcasting	National	Available
	136° W	Broadcasting	National	Available
	113° W	Fixed services	National	Telecomm
Not-planned	113° W	Fixed services	Regional	Eutelsat Americas
		Mobile services	Regional	Telecomm
	114.9° W	Fixed services	Regional	Telecomm
				Eutelsat Americas
	116.8° W	Fixed services	Regional	Eutelsat Americas

granted by the IFT.[130] Once this authorization is obtained, they can provide services in the national territory[131] (Table 12.12).

12.8 Launch Capability and Sites

As mentioned in Sect. 12.7, all Mexican satellites have been launched from the jurisdiction of third states. Although some rockets have been launched from Mexican territory in the past, they did not reach high altitudes or Earth's orbits.[132] Nowadays, there are no spaceports, installations or launching sites on Mexican territory. However, Mexico has shown interest in developing a launch industry to access outer space by its own means. Therefore, Mexico foresees the creation of launching capabilities in the following decades through a series of projects defined in Orbit Plan 2.0.[133]

A first project is to identify and elaborate national legal rules to lay the foundations for the development, integration and operation of launchers in Mexico, namely on

---

[130]The Institute has its own Public Concessions Registry that is available at http://ucsweb.ift.org.mx/vrpc/.

[131]To know the concessions procedure before the creation of the Federal Telecommunications Institute see "Regulación satelital en México: studio y acciones", Federal Telecommunications Commission, June 2013, http://www.ift.org.mx/sites/default/files/contenidogeneral/espectro-radioelectrico/regulacionsatelitalenmexicoestudioyacciones19-06-2013-final.pdf (accessed 19 April 2019).

[132]See Sect. 12.2.

[133]That does not mean that Mexico will not continue to request launches to third States to place its satellites in orbit, at least in the short term. Project 2.2 of the Orbit Plan 2.0 related to the policy and bases of planning and coordination to guarantee the management of orbits and the associated radio spectrum takes into account some aspects related to the future management process of a launcher and the possible legal implications of launch contracts.

standardization, tests and certification issues. A second project is to create a Multi-disciplinary Management Group responsible for the coordination of the development project of a launcher pilot program in the country. This group will be made up of representatives of the triple helix (government, academia and industry) and must deliver a preliminary design of mission and funding of a satellite launcher. A third project is to identify potential and available launching sites in Mexican territory. This project will also be a common effort of the triple helix and it could be centred on the launching of small satellites in a first phase. A fourth project is to create a preliminary design of mission and funding of a spaceport to develop a future spaceport infrastructure that allows the exploration of new market niches, articulate the space sector value chain and strengthen the Mexican space ecosystem.[134] The results of these projects are planned to be attained by 2026.

12.9   Space Technology Development

Several Mexican universities, research centres and industries have developed technology that can be used directly or indirectly in space activities (see Sect. 12.3.2). From astronomy to rocketry purposes,[135] Mexico has made significant progress in developing space technology. One of the latest efforts in the sector is the University Space Program of UNAM established in 2017, which aims to promote the participation of the university community in the creation and dissemination of scientific knowledge to foster progress of space technology and its applications in Mexico.[136] Moreover, the recent creation of the Cluster Espacial Red Global Mx organization aims to join efforts in all the country to develop space technologies and applications, as well as to strengthen the national space industry.

12.9.1   Cluster Espacial Red Global Mx

Cluster Espacial Red Global Mx is a worldwide non-profit organization of Mexican professionals in Mexico and abroad that, through international cooperation, aims to promote the space sector in the country and to encourage new generations to become part of it. The creation of the Cluster in August 2018 was sponsored by the Institute for Mexicans Abroad of the Ministry of Foreign Affairs, and the Ministry of Innovation, Science and Technology of the Government of Jalisco. The Cluster will

---

[134]Projects 3.2, 3.3, 4.3 and 4.4, respectively.

[135]See Tecnología Espacial en México, ¿solo cohetes? 1ª Parte, Hacia el Espacio, Mexican Space Agency, 1 January 2015, https://haciaelespacio.aem.gob.mx/revistadigital/articul.php?interior=176 (accessed 3 June 2019).

[136]Programa Espacial Universitario de la Universidad Nacional Autónoma de México, http://peu.unam.mx/ (accessed 10 September 2019).

take advantage of the experience acquired by Mexicans abroad to develop technology, satellites, and expand the space industry to attract investment and create jobs in the country.[137]

The Cluster joins private industry, academy, government and civil society in an integral organism that designs and provides services, goods, business and models for the space industry. Main Cluster's projects are *Space City,* which foresees the creation of a space city in the Mexican state (Mexico's central region) to develop space science, technology and innovation projects; *Centros del Futuro* which consists of innovation, development and creativity space industry hubs what will be located in different Mexican states; *Talento Mx* which aims to foster Mexican youth to participate in science and technology projects; *Maya Space*, a program designed to form and promote Mexican space start-ups, and *Carrera Space*, a first generation productive chain model integrated by Mexican and Latin American entrepreneurs.[138]

12.10 Capacity Building and Outreach

The development of Mexico's space sector is a combination of the efforts of multiple actors, public, private and social. Space education through the activities of the Mexican government and universities is vital to form future professionals in the sector. Moreover, the participation of non-governmental institutions is also important to spread knowledge through the organization of space related activities.

12.10.1 Space Education

Space education in Mexico has increased progressively in the past years in different disciplines such as engineering and space sciences and technology. Public and private universities and institutions in several Mexican states contribute to the formation of new professionals and specialists to strengthen the space sector nationally and internationally. In addition, the AEM has an important role in the promotion of space education.

---

[137]"Mexicanos en el Exterior forman el clúster Espacial de la Red Global mx", Mexican Space Agency, 13 September 2018, https://www.gob.mx/aem/prensa/mexicanos-en-el-exterior-forman-el-cluster-espacial-de-la-red-global-mx-174433?idiom=es (accessed 17 April 2019).

[138]Blanco Martha, "E. Carrera (Clúster): "Hay que ver el espacio como la vía de desarrollo de México", Infoespacial.com, 25 March 2019 http://www.infoespacial.com/latam/2019/03/25/noticia-carrera-cluster-espacio-desarrollo-mexico.html (accessed 17 April 2019).

12.10.1.1 Higher Education Institutions with Aerospace Programs[139]

There are several federal and local public universities, technological institutions and private universities in Mexico with bachelor's or postgraduate degrees on space science or technology. Specifically, there are five programs on space technology and two space science programs at the bachelor's level, and ten space related programs at the postgraduate level. In addition, there is one law program that offers a course on aeronautic and space law at the bachelor's level, and some CRECTEALC's (see Sect. 12.6.5.1) space science and technology courses addressed to people with bachelor's studies (Table 12.13).

Table 12.13 Higher Education Institutions' Aerospace Programs (2019)

Higher Education Institutions' Aerospace Programs[a]			
University	Program	State	Education level
Autonomous University of Baja California (Valle de Palmas Unity)	Aerospace engineering	Baja California	Bachelor's level
Autonomous University of Baja California Mexicali Faculty of Engineering	Aerospace engineering	Baja California	Bachelor's level
Autonomous University of Chihuahua	Aerospace engineering	Chihuahua	Bachelor's level
Marist University of Guadalajara	Aerospace engineering	Jalisco	Bachelor's level
Popular Autonomous University of the State of Puebla	Aerospace engineering	Puebla	Bachelor's level
National Autonomous University of Mexico Faculty of Sciences	Earth Sciences with orientation in Space Sciences	Mexico City	Bachelor's level
National Autonomous University of Mexico Faculty of Sciences	Earth Sciences with orientation in Space Sciences	Queretaro	Bachelor's level

(continued)

[139]Space Education portal, Mexican Space Agency, April 2019, and Duarte Muñoz, Carlos et al., AEM 2S-322-01 Human Capital Offer Analysis Project, Mexican Space Agency, 2015, pp. 27–47, http://www.educacionespacial.aem.gob.mx/images/normateca/pdf/20160606_Demanda_en_la_triple_helice_DIS%20REV%20BRT.pdf.

Table 12.13 (continued)

Higher Education Institutions' Aerospace Programs[a]

University	Program	State	Education level
National Autonomous University of Mexico Faculty of Law	Law (Aeronautic and Space Law Course)	Mexico City	Bachelor's level
National Autonomous University of Mexico Faculty of Sciences	Postgraduate in Astronomical Sciences	Mexico City	Postgraduate
National Autonomous University of Mexico Faculty of Sciences	Postgraduate in Earth Sciences	Mexico City	Postgraduate
University of Guadalajara	Master's Degree in Physics (with concentration in Astrophysics)	Jalisco	Postgraduate
Aeronautical University of Queretaro	Master's Degree in Aerospace Engineering	Queretaro	Postgraduate
Metropolitan Polytechnic University of Hidalgo	Master's Degree in Aerospace Engineering	Hidalgo	Postgraduate
Centre for Research in Geography and Geomatics	Master's Degree in Space Planning	Mexico City	Postgraduate
National Polytechnic Institute School of Mechanical and Electrical Engineering	Master's Degree in Sciences in Aeronautical and Space Engineering	Mexico City	Postgraduate
National Institute of Astrophysics, Optics and Electronics	Master's Degree in Space Science and Technology	Puebla	Postgraduate
National Institute of Astrophysics, Optics and Electronics	Master's Degree in Astrophysics	Puebla	Postgraduate
National Institute of Astrophysics, Optics and Electronics	PhD in Astrophysics	Puebla	Postgraduate
Regional Centre for Space Science and Technology Education (CRECTEALC)	Courses on remote sensing and geographical information systems, and satellite communications and global positioning systems	Puebla	Bachelor's and postgraduate levels

[a]Duarte Muñoz Carlos et al., AEM 2S-322-01 Human Capital Offer Analysis Project, Mexican Space Agency, 2015, pp. 27, 33 and 34

12.10.1.2 The Mexican Space Agency and Space Education

One of AEM's functions is to promote the development of formal space education, as well as the dissemination of space research studies.[140] Therefore, AEM has launched a digital platform called "Educación Espacial"[141] (Space Education) to facilitate access to space education and to share space related information, such as courses, national and international contests, news, and interactive content. Furthermore, AEM has a digital magazine called "Hacia el Espacio" (Towards Space) that contains articles and other space related information that are freely available on its website.[142]

12.10.1.3 Space Telecommunications Programs

In the field of space telecommunications, the Research, Innovation and Development Telecommunication Centre (CIDTE)[143] of the Autonomous University of Zacatecas (Mexico's central region) develops wireless communication systems for the development of satellite broadband connectivity solutions. CIDTE has two education programmes, one related to radiofrequencies and antennas for the creation of satellite telecommunication systems, in particular for ground stations and for pico and nano satellites, and one related to the control and automation for the development of mechanical and control systems applied to space telecommunications, and for the design of new unmanned vehicles.[144]

The state of Zacatecas also will be home to the first Space Telecommunication Centre of the country, a project designed to promote TIC's such as communication between satellites, from space to land, or land to land via space in order to reduce digital gaps and consolidate Mexico as a global aerospace power. This centre will be operational in the second semester of 2019[145] (Fig. 12.3).

---

[140] Art. 4, AEM's Act.

[141] Mexican Space Agency, Educación Espacial portal, https://www.educacionespacial.aem.gob.mx/index.html#.

[142] Available at https://haciaelespacio.aem.gob.mx/revistadigital/.

[143] Centro de Investigación, Innovación y Desarrollo en Telecomunicaciones.

[144] Research and Development Areas, CIDTE's website, http://cidte.uaz.edu.mx/web/index.php/investigacion-y-desarrollo/areas-de-desarrollo/ (accessed 5 April, 2019).

[145] "Visita Subsecretaría de SCT obra del primer Centro de Telecomunicaciones Espaciales del país, en Zacatecas", Mexican Space Agency, 27 January 2019 https://www.gob.mx/aem/articulos/visita-subsecretaria-de-sct-obra-del-primer-centro-de-telecomunicaciones-espaciales-del-pais-en-zacatecas-189197?idiom=es (accessed 7 April 2019).

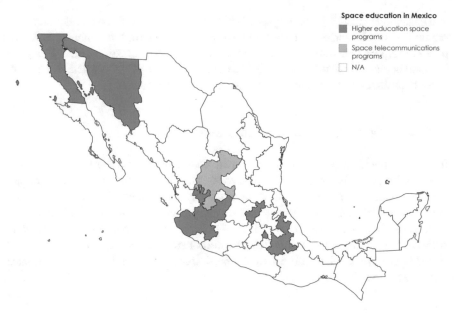

Fig. 12.3 States with space education in Mexico

12.10.2 Mexico and Non-governmental Space Institutions

Space education in Mexico would not be the same without the large space science and technology dissemination activities organised by a great array of civil society organisations. The work of the Space Generation Advisory Council in the country also contributes to this goal.

12.10.2.1 Space Generation Advisory Council

SGAC, a non-governmental, non-profit organisation and network that aims to represent young space students and professionals to the UN, space agencies, industry, and academia all over the world, has helped Mexican youth to become space sector leaders.

The international approach of SGAC Mexico was consolidated in 2016 with the establishment of two national points of contact. They have two main duties: one is to approach Mexican youth within the country and the other to approach youth abroad and liaise with scientists and technologists of Mexican and Latin American origin that are living abroad.[146] National points of contact facilitate the creation of ties between SGAC and the Mexican research institutions, universities, companies

---

[146]Space Generation Advisory Council, National Points of Contact, Mexico, accessed 29 March 2019, https://spacegeneration.org/regions/north-central-america/mexico (accessed 5 April 2019).

and space agencies.[147] For example, SGAC Mexico and AEM have collaborated to organise several courses, trainings and exchanges. In 2016, they planned and prepared the Space Generation Congress and the International Astronautical Congress in Guadalajara, Mexico.[148] On 4 April 2019, AEM and SGAC signed an MoU in order to strengthen their relations and to promote space activities.

12.10.2.2 Other Space Related Institutions

In Mexico there is a great array of non-governmental institutions, associations, societies, student organisations, amateur networks, and other groups whose activities are directly related to space issues, such as astronomy and physics,[149] space science and technology, aerospace engineering, and rocketry. Many of these institutions participate actively in the development of the Mexican space sector and help Mexican society to learn more about this field, as demonstrated, for example, in the 2018 World Space Week, where more than 300 participants organized roughly 1270 space related activities.[150]

Appendix

See Table 12.14.

---

[147] In December 2018, SGAC elected Tania Robles, a Mexican Mechanical Engineering student, as the next Regional Coordinator in North, Central America and Caribbean, which was recognized with the Space Generation Leadership Award the same year, https://spacegeneration.org/regions/north-central-america (accessed 5 April 2019).

[148] See Young Space Activities in Mexico, Space Generation Advisory Council, https://spacegeneration.org/regions/north-central-america/mexico (accessed 5 April 2019).

[149] A list of Astronomy groups in Mexico is available at Grupos de Astronomía en México, Instituto Nacional de Astrofísica en México, https://www.inaoep.mx/~astrofi/astromex.html (accessed 28 May 2019).

[150] Domínguez, Salma, Continúa Semana Mundial del Espacio con 301 sedes en México, A21, 8 October 2018, https://a21.com.mx/aeroespacial/2018/10/08/continua-semana-mundial-del-espacio-con-301-sedes-en-mexico (accessed 28 May 2019). See also Semana Mundial del Espacio'18 México, Mexican Space Agency, https://haciaelespacio.aem.gob.mx/sme-mexico/2018/index.php#queEs.

Table 12.14 Institutional contact information in Mexico (2019)

Institution	Details
Federal Telecommunications Institute (IFT)	1143 Insurgentes Sur Avenue, Nochebuena, 03720, Mexico City, Mexico Website: http://www.ift.org.mx/, Email: atencion@ift.org.mx
Mexican Space Agency (AEM)	1685 Insurgentes Sur Avenue, Guadalupe Inn, 01020, Mexico City, Mexico Website: https://www.gob.mx/AEM Email: contacto@AEM.gob.mx
Ministry of Foreign Affairs (SRE)	20 Plaza Juárez, Centro, 01020, Mexico City, Mexico Website: https://www.gob.mx/sre Email: atencionciurdadana@sre.gob.mx
Ministry of Communications and Transportation (MCT)	1089 Insurgentes Sur Avenue, Nochebuena, 03720, Mexico City, Mexico Website: https://www.gob.mx/sct Email: buzon_ucg@sct.gob.mx
Telecomunicaciones de México (Telecomm)	Telecomm I and II Centres, Telecommunications Avenue, Leyes de Reforma, 09310, Mexico City, Mexico Website: https://www.gob.mx/telecomm/ Email: muycerca@telecomm.net.mx

Chapter 13
Paraguay

Contents

Abstract Less than a decade ago, Paraguay joined the group of Latin America's countries possessing a national space sector. Although in its infancy, Paraguay's space activities have gradually increased in the last years, especially after the creation of a national space agency and the adoption of an institutional plan in the field. The present chapter describes the current state of Paraguay's space sector, focusing on the following issues: the institutional and legal framework of Paraguay's space activities, the space cooperation programs adopted by Paraguay and other international partners, Paraguay's satellite capacities and space technology development, and the capacity building and outreach of Paraguay's space sector.

13.1 General Country Overview

Located in South America, Paraguay is bordered by Argentina to the south, southeast and southwest, Brazil to the east, and Bolivia to the northwest. The Paraguay River divides the country in two regions: the West Region, which is the most populated one, and the East Region, which is part of the Chaco Boreal. Paraguay is a land

© Springer Nature Switzerland AG 2020   401
A. Froehlich et al., *Space Supporting Latin America*, Studies in Space Policy 25,
https://doi.org/10.1007/978-3-030-38520-0_13

locked country, but it has several ports on the Paraguay and Parana Rivers that give exit to the Atlantic Ocean through the Paraguay-Parana Waterway.[1]

Paraguay has a great variety of flora and fauna thanks to the six eco-regions that converge in its territory. Moreover, Paraguay, which means "water that comes from the river" in the Guaraní language, possess a huge marshland in the northern territory that is considered the most important eco region of the world.[2]

Paraguay is a multi-cultural and bilingual country. Its official languages are Spanish and Guaraní and about 73% of the population speak both languages.[3] Other indigenous languages are also spoken and are considered the cultural heritage of the nation.[4]

13.1.1   Political System

Paraguay is constituted as a unitary, indivisible, decentralized and social state of law.[5] It has a democratic regime and is governed by three powers: the executive power exercised by the President of the Republic (currently Mario Abdo Benítez), the legislative power, exercised by the Congress (Chamber of Deputies and the Senate), and the judicial power, exercised by the Supreme Court of Justice, tribunals and courts.[6]

Paraguay is divided in 17 departments (Fig. 13.1) that are composed of municipalities and districts. Asuncion is Paraguay's capital, which is a municipality but is independent from all departments.[7] Each department is governed by a governor, who represents the executive power, and a departmental board.[8]

---

[1] Datos generales del Paraguay, Embajada de la República del Paraguay en la República del Ecuador, http://www.embajadadeparaguay.ec/index.php/informacion-general/datos-generales-del-paraguay (accessed 4 October 2019).

[2] Ibid.

[3] Ibid.

[4] Art. 140 of the National Constitution of the Republic of Paraguay, Asuncion, 22 June 1992.

[5] Art. 1 of the National Constitution of the Republic of Paraguay.

[6] Art. 182, 226 and 247, respectively, of the National Constitution of the Republic of Paraguay.

[7] Art. 156 of the National Constitution of the Republic of Paraguay.

[8] Art. 161 of the National Constitution of the Republic of Paraguay.

Fig. 13.1 Paraguay's departments (2019)

13.1.2 Economic Situation

Paraguay has had stable economic growth in the last fifteen years thanks to its predictable macroeconomic bases, solid financial system, poverty reduction programs and other policies. Although the economic fluctuations of Paraguay's neighbours and main trading partners, Brazil and Argentina, have decelerated the country's economic growth in the last two years, it is projected that economic expansion will return in 2020 and 2021.[9]

---

[9]The World Bank In Paraguay, The World Bank, 16 April 2019 https://www.worldbank.org/en/country/paraguay/overview (accessed 4 October 2019).

13.2 Space Development Policies and Emerging Programs

Like other emerging space countries, Paraguay has recently decided to establish a governmental body for the promotion and development of its national space activities: the Paraguayan Space Agency (AEP).[10] To achieve these goals, AEP has elaborated an Institutional Strategic Plan (ISP) that will be implemented in the coming years.

13.2.1 The Paraguayan Space Agency

Paraguayan Space Agency (AEP) is a self-governing entity that designs, proposes and implements the national space and aerospace policies and programs. Conceived in 2012 under the General Directorate of the Aerospace Defence of the Ministry of National Defence, the AEP currently answers to the President of the Republic.[11] AEP's mission is to plan and execute space programs by taking advantage of outer space in order to achieve Paraguay's goals.[12] AEP's vision is to become a body of excellence that provides permanent innovation services and contributes to R +D + I (research, development and innovation)[13] in the use of outer space and for the benefit of the country.[14] AEP's main objective is to promote and manage the development of national space activities through technological innovation for the peaceful use of outer space.[15]

Some of AEP's main functions are the following:[16]

- Implement the Paraguayan space policy through the elaboration and application of the National Space Activities Program;
- Conduct research for the formation of groups with the necessary disciplines and techniques to access to space technology and its applications;
- Conduct the development of advanced engineering to have an appropriate national space technology;
- Promote the transfer of space technology to public and private entities;

---

[10] Agencia Espacial del Paraguay (AEP).

[11] Art. 1, Law N° 5151/2014.

[12] Resolución AEP/N° 18 por la que se definen la misión y visión de la Agencia Espacial del Paraguay, Presidencia de la República, Asunción, 15 December 2017, Art 1.

[13] I + D + I: investigación, desarrollo e innovación. The ISP also mentions the execution activities (R + D + I + E).

[14] Resolución AEP/N° 18 por la que se definen la misión y visión de la Agencia Espacial del Paraguay, Presidencia de la República, Asunción, 15 December 2017, Art. 2.

[15] Law N° 5151/2014, Art. 2.

[16] Ibid, Art. 3.

- Provide technical support to the state in the space sector;
- Generate orientations and regulations for the exploration, use and exploitation of outer space for the benefit of the economic, political, social and cultural development of the country, and
- Promote and develop cooperation agreements with public and private entities of other countries.

AEP is composed of a board of directors incorporating a president and twelve members of different public and private entities designated by executive decree. This body proposes policies for the peaceful development and use of outer space, creates specialized committees for the fulfilment of its functions, and approves AEP's budget and internal organisation issues.[17]

13.2.2 The Institutional Strategic Plan

The Institutional Strategic Plan (ISP) for the period 2018-2022 was adopted by the AEP in 2017 as a response to the changes and tendencies in the national and international space sector.[18] The ISP is aligned to the 2030 National Development Plan (NDP) and it determines the mission, vision, actions lines and strategic axes that the Agency should observe in the following years to develop Paraguay's space sector (Table 13.1).[19] The ISP was elaborated after carrying out a FODA analysis consisting of the identification of Paraguay' strengths, weaknesses, opportunities and threats[20] that AEP must take into account when implementing the ISP's strategic axes. Critical

Table 13.1 Strategic axes of the National Development Plan and Institutional Strategic Plan 2018–2022

National Development Plan 2030 Strategic axes	Institutional Strategic Plan Strategic axes
1. Poverty reduction and social development	1. Contribute to the development of the public and private aerospace sectors
2. Inclusive economic growth	2. Human talent training to support emerging challenges
3. Paraguay's incorporation in the international arena	3. Creation and development of AEP's infrastructure according to the national reality
	4. Establishment of a sustainable system for AEP's goals

---

[17] Ibid, Art. 5 and 6.

[18] Resolución AEP/N° 19 por la que se aprueba el Plan Estratégico Institucional (PEI) de la Agencia Espacial del Paraguay, Presidencia de la República, Asunción, 15 December 2017.

[19] Plan Estratégico Institucional 2018–2022, pp. 5–6.

[20] FODA: fortalezas, oportunidades, desafíos y amenazas.

problems and major challenges to AEP's role have also been taken into account by the ISP.

In addition, the ISP details the following AEP's objectives:[21]

- To start and strengthen the country's space technology capacity;
- To increase the use of broadband Internet or equivalent technology;
- To standardize land ownership through a reliable register;
- To promote the reincorporation of prominent Paraguayans living abroad in the public, private or scientific sectors;
- To expand exports of cultural and technological goods and services;
- To restore deteriorated ecosystems to at least 20%;
- To increase national income for the sale of environmental services;
- To reduce recovery costs after climate disasters;
- To exploit aquifers through duly monitoring environmental plans, and
- To increase forest areas and protected biomass coverage.

13.3   Current and Future Perspectives on Legal Framework Related to Space Activities

To foster its space sector, Paraguay has adopted several regulations that are mainly related to the national institutional space framework. In addition, Paraguay has ratified some international instruments related to the space and telecommunications sectors.

13.3.1   National Space Regulations and Instruments

Paraguay's space regulations consist of laws, executive decrees and AEP's internal rules. The most prominent regulation is Law N° 5151/2014[22] adopted on 24 March 2014 which governs the creation of the AEP. The law has only two chapters: Chapter one contains general issues concerning AEP's legal status, objective, functions and budget, and Chapter two details AEP's structure and organisation.

---

[21] Plan Estratégico Institucional 2018–2022, pp. 13–14.

[22] Ley N° 5151 de Agencia Espacial del Paraguay, Poder Legislativo, Paraguay, 24 March 2014, http://www.aep.gov.py/application/files/8115/1852/8075/LEY_5151_agencia_espacial.pdf.

Other space related regulations include Law N° 5740/2016[23] on the ratification of the OST by Paraguay; Decree N° 6466 concerning the appointment of AEP's president ad honorem;[24] Decree N° 7364 concerning the appointment of the members of AEP's Board of Directors,[25] and Decree N° 8543 concerning the validation of the appointment of AEP's president.[26] Finally, AEP adopts different internal rules and resolutions for the implementation of its functions.[27]

13.3.2 Paraguay and International Space Instruments

Paraguay's recent entry into the space sector has been accompanied by the ratification of the OST, the most important UN Space Treaty. The country has also ratified the agreement relating to the International Telecommunications Satellite Organisation (ITSO) and ITU's Constitution and Convention (Table 13.2).[28]

---

[23] Ley N° 5740 del 18 de noviembre de 2016.

[24] Decreto N° 6466 por el cual se nombra al Coronel DEM (R) Liduvino Vielman Díaz, como Presidente Ad Honoren (sic) de la Agencia Espacial del Paraguay, Presidencia de la República del Paraguay, Ministerio de Defensa Nacional, 13 Decembre 2016, http://www.aep.gov.py/application/files/6115/1852/7765/Decreto_6466_13_DIC_2016_Cnel_Liduvino_Vielman_AEP.pdf.

[25] The Decree mentions that the AEP's Board of Directors will be the specialized space body responsible for the control of the satellity sovereignty. See Decreto N° 7364 por el cual se nombra a miembros de la Junta Directiva de la Agencia Espacial del Paraguay, Presidencia de la República del Paraguay, Ministerio de Defensa Nacional, 27 June 2017, http://www.aep.gov.py/application/files/5015/1852/7841/DECRETO7364_JUNTA_DIRECTIVA.pdf.

[26] Decreto N° 8543 por el cual se confirma al Coronel Dem (R) Liduvino Vielman Díaz, como Presidente de la Agencia Espacial del Paraguay, Presidencia de la República del Paraguay, Ministerio de Defensa Nacional, 20 February 2018, https://www.presidencia.gov.py/archivos/documentos/DECRETO8543_uc0hokeg.PDF.

[27] For example, AEP has recently published a draft of a Manual on AEP's organisation and functions. See Manual de Organización y Funciones, Agencia Espacial del Paraguay, available at http://aep.gov.py/application/files/1315/2173/9199/MANUAL_DE_ORGANIZACION_Y_FUNCIONES_AEP_GLOBAL_VX_SP_1.pdf (accessed 12 September 2019).

[28] Status of International Agreements relating to activities in outer space as at 1 January 2019, UNOOSA Doc, A/AC.105/C.2/2019/CRP.3, http://www.unoosa.org/documents/pdf/spacelaw/treatystatus/AC105_C2_2019_CRP03E.pdf (accessed 4 October 2019).

Table 13.2 International space related treaties respected by Paraguay (2019)

International space treaties ratified by Paraguay

Treaty[a]	Adoption	Ratification (R)/signature (S)
OST	1967	R
ARRA	1968	–
LIAB	1972	–
REG	1975	–
MOON	1979	–
NTB	1963	S
BRS	1974	–
ITSO	1971	R
IMSO	1976	–
ITU	1992	R

[a]Treaty on Principles Governing the Activities of States in the Exploration and Use of Outer Space, including the Moon and Other Celestial Bodies (OST); Agreement on the Rescue of Astronauts, the Return of Astronauts and the Return of Objects Launched into Outer Space (ARRA); Convention on International Liability for Damage Caused by Space Objects (LIAB); Convention on Registration of Objects Launched into Outer Space (REG); Agreement Governing the Activities of States on the Moon and Other Celestial Bodies (MOON); Treaty Banning Nuclear Weapon Tests in the Atmosphere, in Outer Space and under Water (NTB); Convention Relating to the Distribution of Programme-Carrying Signals Transmitted by Satellite (BRS); Agreement Relating to the International Telecommunications Satellite Organisation (ITSO); Convention on the International Mobile Satellite Organization (IMSO); International Telecommunication Constitution and Convention (ITU)

13.4   Cooperation Programs

Through different cooperation agreements and other mechanisms, Paraguay has approached several international partners to strengthen its space sector. These include Argentina, Mexico, Japan, Taiwan and Russia.

First, in 2018 the Argentinian-Paraguayan Parana River Joint Commission (COMIP)[29] and the National Commission of Space Activities of Argentina (CONAE)[30] signed a framework agreement to promote the development of capacities for the transformation of space data into high value goods for the benefit of different social sectors. Both parties agreed to prioritize the use and availability of satellite data

[29]Comisión Mixta argentino – paraguaya del Río Paraná (COMIP). This Commission was created in 1971 by Argentina and Paraguay to study and evaluate the technical and economic possibilities of the use of Parana River's resources, see ¿Qué es la COMIP?, COMIP, https://www.comip.org.ar/quienessomos/ (accessed 4 October 2019).

[30]Intervención de la República de Paraguay, Agencia Espacial del Paraguay, Presidencia de la República, UNISPACE + 50 Symposium Presentations, UNOOSA, 19 June 2018, http://www.unoosa.org/documents/pdf/unispace/plus50/Presentations_SYMPOSIUM/Presentation18.pdf.

in the following areas: land management and infrastructure, environment, production sector, and strategic areas.[31]

Second, in January 2018, representatives of the space agencies of Paraguay and Mexico signed a letter of intent for cooperation in the peaceful use of outer space. Specifically, this instrument aims to develop cooperation mechanisms in three fields: technical, technological, academic and scientific cooperation; capacity-building programs; and space education projects.[32]

Third, in June 2018, AEP, the National University of Asuncion (UNA) and the BIRDS Project of the Kyushu Institute of Technology (KYUTECH) of Japan signed a letter of intent to promote a future agreement of cooperation to conduct joint activities on teaching, research and outreach within the field of this project.[33]

Fourth, Paraguay and Taiwan have started talks to adopt potential cooperation actions to strengthen their space sectors, namely in the field of I + D + I.[34] In 2019, the Taiwanese government granted a scholarship to a Paraguayan student to study for a master's degree at the Aeronautic and Astronautic Department of the University of Cheng Kung.[35]

Finally, Paraguay and Russia have discussed the possibility of establishing future bilateral cooperation between AEP and ROSCOSMOS (the Russian Space Agency).[36] Russia currently offers 20 scholarships a year to Paraguayan students in order to pursuit their studies in Russian universities.[37]

---

[31] Acuerdo Marco entre la Comisión Mixta Argentino Paraguaya del Río Paraná (COMIP) y la Comisión Nacional de Actividades Espaciales (CONAE), Buenos Aires, 20 March 2018, available at https://www.comip.org.ar/wp-content/uploads/2019/02/Convenio-Marco-COMIP-CONAE.pdf (accessed 4 October 2019).

[32] La AEP en fase de fortalecimiento, Agencia Espacial del Paraguay, 19 January 2018, http://www.aep.gov.py/index.php/noticias/la-aep-en-fase-de-fortalecimiento (accessed 15 September 2019). In 2019, AEP launched a call to Paraguay's university students to participate in a UNOOSA scholarship for the International Course on Space Engineering in Kyutech, see Posgrado en Ingeniería Espacial UNOOSA, Agencia Espacial del Paraguay, http://www.aep.gov.py/index.php/oportunidades/posgrado-en-ingenieria-espacial-unoosa (accessed 15 September 2019).

[33] Letter of Intent between Paraguay Space Agency, Universidad Nacional de Asunción and BIRDS Project at Kyushu Institute of Technology, 21 June 2018, https://www.una.py/wp-content/uploads/2019/02/Carta-de-intencio%CC%81n-entre-la-UNA-y-KYUSHU-Institute-of-Teccnology-Japon.pdf.

[34] Audiencia con el Sr. Embajador de la República de China, Taiwán, Agencia Espacial del Paraguay, 14 February 2018, http://www.aep.gov.py/index.php/noticias/audiencia-con-el-sr-embajador-de-la-republica-de-china-taiwan (accessed 15 September 2019).

[35] Beca de Maestría de la Embajada de Taiwán, Agencia Espacial del Paraguay, http://www.aep.gov.py/index.php/oportunidades/beca-maestria-MOFA-taiwan (accessed 15 September 2019).

[36] Visita del Embajador de la Federación Rusa, Agencia Espacial del Paraguay, 21 February 2019, http://www.aep.gov.py/index.php/noticias/visita-del-embajador-de-la-federacion-rusa (accessed 15 September 2019).

[37] Becas de Grado y Postgrado del Gobierno de la Federación de Rusia, Agencia Espacial del Paraguay, http://www.aep.gov.py/index.php/oportunidades/becas-de-grado-y-postgrado-del-gobierno-de-la-federacion-de-rusia-para-setiembre-de-2019 (accessed 15 September 2019).

13.5   Satellite Capacities

As noted above, the letter of intent signed between AEP, UNA and the KYUTECH's
BIRDS Project of Japan aims to promote a future agreement of cooperation to conduct
joint activities on teaching, research and outreach within the field of this project.
The Joint Global Multi-Nation Birds Satellite[38] project (or "BIRDS Project") is a
UN cross-border interdisciplinary satellite project for non-space faring countries
supported by Japan.[39] Within this project, Paraguay is developing its first satellite[40]
called GUARANISAT-01. With a total investment of USD 280,000 granted by AEP,
GUARANISAT-01 will carry out ten missions, one of which concerns detection of
the Chagas disease's vectors in the Chaco region.[41] GUARANISAT-01 is planned to
be launched into orbit in July 2020.[42]

13.6   Space Technology Development

Apart from the GUARANISAT-01 satellite mission, there are currently several
Paraguayan university projects on space technology development in the field of basic
space engineering and satellite image processing. In this context, the CABUREI—
4S project (capacity building, research and innovation—4 Space) includes several
subprojects such as a thermal vacuum chamber, a CUBESAT's structural design, the
GUARANISAT ground segment, a CANSAT type picosatellite, an attitude control
system, a frictionless platform, and satellite applications.[43]

---

[38]Joint Global Multi Nation BIRDS 1, BIRDS, https://birds1.birds-project.com/ (accessed 14
September 2019).

[39]BIRDS objectives are: teach engineering graduate students the entire process for putting a satellite
into space—from mission planning, to hardware design, to spacecraft testing, to launching, and to
in-orbit operation; lay down the foundation for a sustainable space program in non-space-faring
countries by building-up human capital in universities—i.e., create a university space research and
education program using graduating students; and using the same students, create an international
human network that can assist members develop their infant space programs through cooperation
and by sharing information and experiences. See Maeda G. et al., BIRDS Project Newsletter, Issue
No. 1, Laboratory of Spacecraft Environment Interaction Engineering (LaSEINE), Kyushu Institute of Technology, Kitakyushu, Japan, January 2016, https://birds1.birds-project.com/JGMNB_
NEWSLETTER_NO1.pdf.

[40]El primer satélite paraguayo, Agencia Espacial del Paraguay, http://www.aep.gov.py/index.php/
proyectos/guaranisat (accessed 14 September 2019). See also Satélite paraguayo está listo y preparan
puesta en órbita, Última Hora, 15 September 2019, https://www.ultimahora.com/satelite-paraguayo-
esta-listo-y-preparan-puesta-orbita-n2843857.html (accessed 4 October 2019).

[41]Primer satélite paraguayo entrará en operación a mediados de 2020, Última Hora, 5 October
2019, https://www.ultimahora.com/primer-satelite-paraguayo-entrara-operacion-mediados-2020-
n2847548.html (accessed 9 October 2019).

[42]Ibid.

[43]Proyecto CABUREI-4S, Agencia Espacial del Paraguay, http://www.aep.gov.py/index.php/
proyectos/proyecto-caburei-4s (accessed 15 September 2019).

13.7 Capacity Building and Outreach

Together with regional and international partners, AEP has organised several events to disseminate space knowledge to students, professionals, companies, members of the government and the general public. These include the First Forum on the Space and Aeronautic Cluster,[44] the First Space Symposium of the East,[45] and the Second Space Conference of Paraguay: "In pursuit of the development of space capacity".[46]

Appendix

See Table 13.3.

Table 13.3 Institutional contact information in Paraguay (2019)

Institution	Details
National University of Asunción (UNA)	Campus Universitario, San Lorenzo, Asunción, Paraguay Website: https://www.una.py/ Email: sgeneral@rec.una.py
Paraguayan Space Agency (AEP)	Avenue Mcal. López and 22 Setiembre, block B, seventh floor, Asunción, Paraguay Website: http://www.aep.gov.py/index.php/contacto

[44] I Foro sobre Cluster de la Industria Espacial y Aeronáutica, Agencia Espacial del Paraguay, 26 January 2018, http://www.aep.gov.py/index.php/noticias/i-foro-sobre-cluster-de-la-industria-espacial-y-aeronautica (accessed 15 September 2019).

[45] I Simposio Espacial del Este, Agencia Espacial del Paraguay, 27 March 2018, http://www.aep.gov.py/index.php/noticias/i-simposio-espacial-del-este (accessed 15 September 2019).

[46] II Conferencia Espacial del Paraguay: "En pos del desarrollo de la capacidad espacial", Agencia Espacial del Paraguay, 12 October 2018, http://www.aep.gov.py/index.php/noticias/ii-conferencia-espacial-del-paraguay (accessed 15 September 2019).

Chapter 14
Peru

Contents

Abstract Although its National Commission for Aerospace Research and Development (Comisión Nacional de Investigación y Desarrollo Espacial—CONIDA) was established in 1974 it was not until the last decade that Peru has actually started to develop its space sector through a series of significant achievements, namely the creation of a national satellite system program and the negotiation of several space cooperation agreements. This chapter presents a brief description of Peru's space activities. Section 14.1 provides a general overview of Peru; Sects. 14.2 and 14.3 focus on the main Peruvian space actors and the legal space framework for Peru's space activities; Sect. 14.4 describes the space cooperation programs of which Peru is part; Sect. 14.5 focuses on Peru's satellite capacities, and Sects. 14.6 and 14.7 describes the development of space technology in the country and Peruvian space capacity-building and outreach.

© Springer Nature Switzerland AG 2020
A. Froehlich et al., *Space Supporting Latin America*, Studies in Space Policy 25,
https://doi.org/10.1007/978-3-030-38520-0_14

14.1   General Country Overview

The Republic of Peru (Peru) is located in western South America. Peru's territory covers 1,285,215 km^2. It borders Colombia and Ecuador to the north, Brazil to the east, Bolivia to the southeast, Chile and the Pacific Ocean to the south, and the Pacific Ocean to the west. Peru has a population of 32,495,500 inhabitants (2019). Its capital is Lima, but Cusco is the historic capital.[1] Peru's official languages are Spanish, Quechua, Aimara and other native languages (about 40 languages). The currency is the Sol.[2]

Geographically, there are three regions in Peru: Costa, Sierra and Selva (or Amazonia). The Costa Region is an arid and sandy territory with some fertile valleys; the Sierra Region is occupied by the Andes that cross the country from North to South, and the Amazonian Region is constituted by western hillsides and plains that are part of the Amazon basin.[3]

14.1.1   Political System

Peru is a democratic, social, sovereign and independent republic. Its government is unitary, representative, decentralized and is based on the principle of separation of powers: the executive power is represented by the president who is also head of state (currently Martín Vizcarra); the legislative power, which resides in the congress (unicameral), and the judicial power is exercised by the Supreme Court of Justice and other courts and tribunals. The electoral power, represented by the National Jury of Elections, is considered the fourth power.[4]

Administratively, Peru is composed of regions, departments (Fig. 14.1), provinces and districts, in which there are national, regional and local governments.[5]

---

[1]Ficha país Perú, Oficina de Información Diplomática, Ministerio de Asuntos Exteriores, Unión Europea y Cooperación, España, http://www.exteriores.gob.es/Documents/FichasPais/PERU_FICHA%20PAIS.pdf (accessed 1 November 2019).

[2]Ibid.

[3]Ibid.

[4]Datos de interés, Embajada del Perú en España, https://www.embajadaperu.es/sobre-el-peru/datos-de-interes.htm (1 November 2019). See also Constitución Política del Perú, 1 January 1994.

[5]Art. 189, Constitución Política del Perú, 1 January 1994.

Fig. 14.1 Peru's departments (2019). Image obtained from http://mapadeperu.blogspot.com/2013/05/mapa-de-peru-en-division-politica.html (accessed 2 November 2019)

14.1.2 Economic Situation

Peru's economic situation between 2004 and 2013 was favourable thanks to a series of macroeconomic and structural policies and a positive external context. This economic growth turned Peru into one of the fastest-growing countries in the region.[6] However, between 2014 and 2018 Peru's GDP growth slowed, which led to temporary reduction

---

[6]The World Bank In Peru, The World Bank, https://www.worldbank.org/en/country/peru/overview (accessed 3 November 2019).

of private investment, less fiscal income and a slowdown of consumption. Good management of fiscal, monetary, and exchange policies, as well as an increase in mineral exports, especially copper, the main Peruvian export good, moderated the impact of the deceleration of Peru's economic growth. Although it is expected that Peru's economy will grow in the medium term, current external conditions demand complementary reforms to raise productivity and promote shared prosperity.[7]

14.2   Space Development Policies and Emerging Programs

The lack of a National Space Plan (NSP) in Peru has not prevented the country from developing its own space sector. Several Peruvian institutions have different space related programs that have contributed to the development and strengthening of specific areas of the national space sector, such as the satellite, rocketry, telecommunications, space technology and applications, research, and education areas.

14.2.1   Space Actors

CONIDA is the main space actor in Peru and seat of the Peruvian Space Agency.[8] Other Peruvian space actors include universities and institutions that have fostered the satellite sector, and an important array of astronomical observatories used for different purposes.

14.2.1.1   The National Commission for Aerospace Research and Development

According to Decree-Law N° 20643 of 11 June 1974, CONIDA is a public institution of the aeronautical sector and an integral part of the National Plan for Scientific and Technical Research.[9] Moreover, Law N° 29075 of 1 August 2007 defines CONIDA as a decentralized public body of the Ministry of Defence and gives CONIDA the status of the Peruvian Space Agency.[10] In addition, the Regulation of CONIDA's

---

[7]Ibid.

[8]CONIDA and the Peruvian Space Agency are frequently used as the same body.

[9]Decreto – Ley de Creación de CONIDA, Decreto Ley 20643, 11 June 1974, available at http://www.conida.gob.pe/transparencia/datos_generales/PDF/Decreto%20Ley%20CONIDA.pdf.

[10]Ley N° 29075 que establece la naturaleza jurídica, función, competencias y estructura orgánica básica del Ministerio de Defensa, Congreso de la República, 1 August 2007, available at http://www.leyes.congreso.gob.pe/Documentos/Leyes/29075.pdf (accessed 2 November 2019).

Organisation and Functions states that CONIDA is an executing public body that possesses functional, administrative and economic autonomy.[11]

CONIDA's mission is to promote, research, develop and disseminate space science and technology by generating goods and services that contribute to the socioeconomic development and security of the country, and that foster its space positioning in the region.[12] CONIDA's vision is to become a regional space institutional leader with an important presence in the international space community.[13]

CONIDA's object and functions are the following[14]:

- Promote and develop space research and projects for peaceful purposes;
- Control the elaboration of space research, studies and theoretical and practical works with national and foreign natural and legal persons, and propose implementation with national or foreign entities;
- Adopt cooperation conventions with national and foreign private and public entities related to the space sector;
- Stimulate the exchange of technology and promote the training of specialists;
- Propose national space legislation;
- Elaborate or promote theoretical and practical studies demanded by the Ministry of Aeronautics[15] and participate in space related activities;
- Study and inform on the different space issues formulated by national and foreign entities;
- Supervise the operative work of the National Centre of Satellite Image Operations (CNOIS),[16] and
- Participate as an executory element in the National Security and Defence System.

Concerning CONIDA's organisation, it seems that the Board of Directors identified by Decree-Law 20643 as CONIDA's higher authority no longer exists.[17] On the contrary, according to the Regulation of CONIDA's Organisation and Functions, the Commission's higher authority is its institutional chief.[18]

Apart from several administrative departments, CONIDA has the CNOIS and three Technical Divisions.

---

[11] Art. 1, Reglamento de Organización y Funciones de la Comisión Nacional de Investigación y Desarrollo Aeroespacial – CONIDA – Agencia Espacial del Perú, available at http://www.conida.gob.pe/images/stories/docpdf/2017/transparencia/plan_y_org/rof_2017-1.pdf (accessed 2 November 2019).

[12] Misión y visión, Agencia Espacial del Perú, CONIDA, http://www.conida.gob.pe/index.php/Informacion-General/mision-y-vision (accessed 2 November 2019).

[13] Ibid.

[14] Art. 5, Decreto – Ley de Creación de CONIDA, Decreto Ley 20643, 11 June 1974. Art. 3 of the Regulation of CONIDA's Organisation and Functions enumerates CONIDA's general functions.

[15] According to Art. 3 (a) of the Regulation of CONIDA's Organisation and Functions, CONIDA must answer the direct questions on space science and technology made by the Ministry of Defense and the President of the Republic.

[16] Centro Nacional de Operaciones de Imágenes Satelitales (CNOIS).

[17] Art. 10, Decreto – Ley de Creación de CONIDA, Decreto Ley 20643, 11 June 1974.

[18] Art. 1, Regulation of CONIDA's Organisation and Functions.

CNOIS is the body responsible for the production of satellite images that are used for national development and security.[19]

The Technical Division on Space Sciences and Applications (DICAE) is responsible for the implementation of activities on space sciences research and development, and applications of space technologies. DICAE has two Divisions: Astrophysics and Geomatics. The Division on Astrophysics concentrates on research and development of scientific and educational projects and programs (basic and advanced) in the areas of solar physics, planetary system and minor bodies, connection Sun-Earth, stellar and galactic Astrophysics, cosmic rays, radio astrophysics and space Geophysics.[20] The Division on Geomatics is a research body responsible for the development of applied methodologies to space geotechnologies. Its objective is to promote the use of space geotechnologies to create EO satellite data applications to better use natural resources and the geographic space.[21]

The Technical Division on Space Technology Development (DTDTE) is the body responsible for the research and development of national rocketry and satellite technology. It has two subsidiary bodies: the Division on Scientific Instruments (DINCI) and the Division of Launch Vehicles (DIVLA). DINCI is in charge of the development of a useful payload for the PAULET rocket that aims to capture the physical parameters of the rocket during the flight and serve as platform for the scientific instruments on board.[22] In turn, DIVLA develops, constructs and launches rockets with scientific instruments to give to the scientific community different tools to carry out studies of the medium and high atmosphere.[23]

Finally, the Technical Division on Space Studies (DTEE) aims to strengthen, promote and disseminate knowledge on space science and technology, and promote, implement and control the capacity-building programs at the postgraduate level and the basic, intermediate and advanced training programs. It also manages and supports space research through the promotion of conventions with universities and other public and private institutions.[24]

---

[19] Art. 36, Regulation of CONIDA's Organisation and Functions.

[20] Presentación Astrofísica, Agencia Espacial del Perú – CONIDA, http://www.conida.gob.pe/index.php/Astrofisica/presentacion-astrofisica.html (accessed 2 November 2019).

[21] Presentación Geomática, Agencia Espacial del Perú – CONIDA, http://www.conida.gob.pe/index.php/Geomatica/presentacion-geomatica.html (accessed 2 November 2019).

[22] Presentación Instrumentación Científica, Agencia Espacial del Perú – CONIDA, http://www.conida.gob.pe/index.php/Instrumentacion-Cientifica/presentacion-instrumentacion-cientifica.html (accessed 2 November 2019).

[23] Presentación Vehículos Lanzadores, Agencia Espacial del Perú – CONIDA, http://www.conida.gob.pe/index.php/Vehiculos-Lanzadores/presentacion-vehiculos-lanzadores.html (accessed 2 November 2019).

[24] Sobre Capacitación, Agencia Espacial del Perú – CONIDA, http://www.conida.gob.pe/index.php/Capacitacion/sobre-capacitacion.html (accessed 2 November 2019).

14.2.1.2   Other Actors

Apart from CONIDA, there are several institutions, universities and observatories that contribute to the development of the Peruvian space sector. As regards universities, the Pontifical Catholic University of Peru (PUCP),[25] the University of Alas Peruanas and the National University of Engineering[26] have all had an important role in the development of the first Peruvian satellites (see Sect. 14.5.1), and the National Institute for Research and Training on Telecommunications (INICTEL) of UNI, and its Group of Processing of Signals and Images (G-PSI)[27] have contributed to research in the field of processing satellite imagery. The Geophysical Institute of Peru,[28] the Raúl Rivera Feijoo Organisation, the National University of Trujillo, and the National University Mayor de San Marcos also participate in the development of Peru's space sector through their astronomical observatories.[29]

As to current Peruvian operational astronomical observatories, there are the Astronomical Observatory Afari (Junín Region), which is used for tourism and university training activities; the Astronomical Observatory of Trujillo (Libertad Region), whose activities are related to teaching and social impact; the Municipal Astronomical Observatory (Cusco Region), which also serves as an educational and tourist facility; the Astronomical Observatory OA-UNI (Junín Region), used for stellar photometry and spectroscopy; and the Astronomical Observatory of Cambrune (Moquegua Region) owned by CONIDA—, which is used for different Astronomical studies.[30] Also worth noting is the creation of ICA's Solar Station, a project of the National University of Ica, which received the Nishimira telescope (Flare Monitoring Telescope—FMT) from the Nishi-Harima Observatory (Japan) in 2014. The FMT is used to introduce university students to the study of solar physics.[31]

14.3   Current and Future Perspectives on Legal Framework Related to Space Activities

The few space regulations adopted by Peru are related to the institutional space framework, namely issues concerning CONIDA's powers. To date there is neither

---

[25] Pontificia Universidad Católica del Perú (PUCP).

[26] Universidad Nacional de Ingeniería.

[27] Instituto Nacional de Investigación y Capacitación de Telecomunicaciones and Grupo de Procesamiento de Señales e Imágenes (G-PSI), respectively.

[28] Instituto Geofísico del Perú.

[29] Observatorios Astronómicos del Perú, Agencia Espacial del Perú, Revista Virtual de Astrofísica N° 001, Noviembre 2018.

[30] Ibid.

[31] Ibid.

an NSP nor other space related national projects or policies in the field.[32] Apart from Peru's space regulations, the country is an active participant in the respect of international legal space and telecommunications instruments.

14.3.1 National Space Regulations and Instruments

The main Peruvian space regulation is Decree-Law 20643 on the creation of CONIDA. Adopted on 11 June 1974, Decree-Law 20643 determines CONIDA's legal nature, address, duration, object, functions, organisation and assets.[33] A related law is Law N° 29075 of 1 August 2007, which confers on CONIDA the status of Space Agency of Peru.[34]

Other space related regulations are CONIDA's internal instruments, in particular the Regulation of CONIDA's Organisation and Functions, which details CONIDA's legal nature, competence, general functions and legal basis.[35] It also establishes CONIDA's institutional structure, inter-institutional and coordination relations, and the labour and financial rules of the Commission.[36]

14.3.2 Peru and International Space Instruments

Peru's engagement in international space law is highly visible because it is one of the few Latin American countries that have ratified all the UN Space Treaties. Furthermore, Peru has ratified other important space and telecommunications treaties, such as the ITU's Convention and Constitution and the Treaty Banning Nuclear Weapon Tests in the Atmosphere, in Outer Space and under Water (NTB) (Table 14.1).

---

[32]There are currently some intentions to develop an NSP in the country. See Cueva Rosmery, En busca del desarrollo de una política espacial en el Perú, Marka Magazine, 21 September 2019, https://markamagazine.com/en-busca-del-desarrollo-de-una-politica-espacial-en-el-peru/ (accessed 4 November 2019).

[33]Decreto – Ley de Creación de CONIDA, Decreto Ley 20643, Lima, 11 June 1974.

[34]Ley N° 29075 que establece la naturaleza jurídica, función, competencias y estructura orgánica básica del Ministerio de Defensa, Congreso de la República, 1 August 2007, available at http://www.leyes.congreso.gob.pe/Documentos/Leyes/29075.pdf (accessed 2 November 2019).

[35]It is worthy to note that the Regulation does not make reference of Decree-Law 20643 as one of CONIDA's legal bases, even when this Decree is still in force and the CONIDA's website mentions it as the main legal framework of the Commission. See Art. 4 of the Regulation of the CONIDA's Organisation and Functions of CONIDA, and Marco Legal, Agencia Espacial del Perú – CONIDA, http://www.conida.gob.pe/index.php/Marco-Legal/Marco-Legal/ (accessed 2 November 2019).

[36]CONIDA's Management Instruments are available at http://www.conida.gob.pe/index.php/Transparencia/planeamiento-y-organizacion (accessed 2 November 2019).

Table 14.1 International space and telecommunications treaties respected by Peru

Treaty[a]	Adoption	Ratification (R)/signature (S)
International space and telecommunications treaties respected by Peru		
OST	1967	R
ARRA	1968	R
LIAB	1972	R
REG	1975	R
MOON	1979	R
NTB	1963	R
BRS	1974	R
ITSO	1971	R
IMSO	1976	R
ITU	1992	R

[a]Treaty on Principles Governing the Activities of States in the Exploration and Use of Outer Space, including the Moon and Other Celestial Bodies (OST); Agreement on the Rescue of Astronauts, the Return of Astronauts and the Return of Objects Launched into Outer Space (ARRA); Convention on International Liability for Damage Caused by Space Objects (LIAB); Convention on Registration of Objects Launched into Outer Space (REG); Agreement Governing the Activities of States on the Moon and Other Celestial Bodies (MOON); Treaty Banning Nuclear Weapon Tests in the Atmosphere, in Outer Space and under Water (NTB); Convention Relating to the Distribution of Programme-Carrying Signals Transmitted by Satellite (BRS); Agreement Relating to the International Telecommunications Satellite Organisation (ITSO); Convention on the International Mobile Satellite Organization (IMSO); International Telecommunication Constitution and Convention (ITU)

14.4 Cooperation Programs

One of the most important bilateral space cooperation program adopted by Peru is the convention between the governments of Peru and France under which the company Airbus Defence and Space developed the first Peruvian EO satellite (see Sect. 14.5.2).[37]

Another remarkable cooperation agreement on space issues is the relationship between Peru and the Asia-Pacific Space Cooperation Organisation (APSCO). Peru is a founding member of APSCO, an intergovernmental organisation created in 2005 to promote and strengthen the development of the space programs of its members

---

[37]Caballero Carlos, PERÚSAT-1: La carrera espacial peruana con tareas pendientes, Gestión, 9 September 2019, https://gestion.pe/opinion/perusat-1-la-carrera-espacial-peruana-con-tareas-pendientes-noticia/ (accessed 3 November 2019).

through cooperation mechanisms.[38] In this regard, Peru has actively participated in APSCO's sessions and both partners have organised several events on space related issues. For instance, in 2016 Peru and APSCO organized an international symposium on space technologies and their applications in the city of Lima, where the topics included the sharing of satellite data, small satellites technology, and the benefits for the use of space in Latin America.[39] Similarly, in 2017, APSCO organised a training course on "Remote Sensing Application and Data Sharing Service Platform in Peru", in CONIDA's headquarters, in which APSCO's personnel trained representatives of Peru's governmental entities in the use of satellite images.[40] The same year, APSCO accepted five Peruvians to study a Master's degree on Space Technology in the University of Aeronautics and Astronomy in Beijing, China.[41] In October 2019, Peruvian students from UNI, the Catholic University San Pablo de Arequipa and PUCP participated in the "Microsatellites" contest organised by APSCO at the Northwestern Polytechnical University in China.[42]

Peru has also approached other countries, such as Ukraine, Bolivia and South Korea, with the aim of developing its space sector. In 2011, CONIDA and the Ukraine Space Agency (SSAU) signed a Framework Cooperation Agreement in the field of space activities to develop joint projects, facilitate technical support and exchange information.[43] Peru and Bolivia have initiated talks with the aim of adopting a bilateral cooperation agreement for the sharing of experience in the use of telecommunication and EO satellites,[44] and South Korea and Peru have adopted a space cooperation agreement to undertake several joint space projects developed by

---

[38]Delegación de APSCO en el Perú, 19 September 2016, Agencia Espacial del Perú – CONIDA, http://www.conida.gob.pe/index.php/noticias/delegacion-de-apsco-en-el-peru (accessed 3 November 2019).

[39]8° Simposio International sobre Tecnología Espacial y sus Aplicaciones, LATAM Satelital, 7 September 2016, http://latamsatelital.com/8o-simposio-internacional-tecnologia-espacial-aplicaciones/ (accessed 3 November 2019).

[40]APSCO y la Agencia Espacial del Perú capacitaron a representantes de Entidades del Estado, Agencia Espacial del Perú – CONIDA, 15 August 2017, http://www.conida.gob.pe/index.php/noticias/apsco-y-la-agencia-espacial-del-peru-capacitaron-a-representantes-de-entidades-del-estado (accessed 3 November 2019).

[41]Maestría en Tecnología Espacial, Agencia Espacial del Perú – CONIDA, 14 September 2017, http://www.conida.gob.pe/index.php/noticias/maestria-en-tecnologia-espacial (accessed 3 November 2019).

[42]Estudiantes peruanos viajan a participar en concurso de Microsatélites, Agencia Espacial del Perú – CONIDA, 23 October 2019, http://200.60.23.229/index.php/noticias/estudiantes-peruanos-viajan-a-participar-en-concurso-de-microsatelites (accessed 4 November 2019).

[43]Perú y Ucrania firman un acuerdo de cooperación en materia espacial, infoespacial.com, 2 June 2011, http://www.infoespacial.com/latam/2011/06/02/noticia-peru-y-ucrania-firman-un-acuerdo-de-cooperacion-en-materia-espacial.html (accessed 3 November 2019).

[44]Perú y Bolivia avanzan en acuerdo de cooperación, LATAM satelital, 12 September 2017, http://latamsatelital.com/peru-bolivia-avanzan-acuerdo-cooperacion/ (accessed 3 November 2019).

the Korea Aerospace Research Institute (KARI) and CONIDA for the benefit of both countries.[45]

Finally, in recent years Peru has initiated negotiations to adopt cooperation agreements with Argentina, Brazil, Kazakhstan, Azerbaijan and the USA.[46]

14.5 Satellite Capacities

Peru's entry to the satellite sector started in 2013 with the launch of a small satellite owned by a Peruvian private university. Three other university satellites were launched in the same decade. After the success of these projects, the Government of Peru launched its own satellite system program: PERÚSAT-1.

14.5.1 The First Peruvian Satellites

PUCP-SAT-1 was the first Peruvian satellite launched into outer space. This 1,280 g nanosatellite launched on 21 November 2013 from the cosmodrome Jasny, Russia, was part of a project of PUCP. It cost of USD 100,000 and was assembled in Italy. PUCP-SAT-1's main purpose was to serve as a platform for the deployment of a smaller satellite called POCKET-PUCP. It currently registers temperatures with 19 sensors and calculates the time it takes to orbit the Earth. PUCP-SAT-1 is still monitored from the base control of the Institute of Radio astronomy (INRAS).[47]

On 6 December 2013, fifteen days after the launch of PUCP-SAT-1, the satellite POCKET-PUCP was launched from the first satellite. With a weight of only 97 g, the femtosatellite calculates its distance and time taken around the Earth's orbit.[48]

The third Peruvian satellite was the nanosatellite CHASQUI 1, which was launched from the International Space Station (ISS) by a Russian cosmonaut on 18 August 2014. CHASQUI 1 weights 1 kg and cost of USD 631,000. The satellite

---

[45]Perú y Corea del Sur estrechan lazos en materia de cooperación espacial, infoespacial.com, 28 January 2016, http://www.infoespacial.com/latam/2016/01/28/noticia-corea-estrechan-lazos-materia-cooperacion-espacial.html (accessed 3 November 2019).

[46]Caballero Carlos, PERÚSAT-1: La carrera espacial peruana con tareas pendientes, Gestión, 9 September 2019.

[47]Cruz Yohel, PERÚSAT-1, Los otros cuatro satélites peruanos que también orbitan la Tierra, RPP Noticias, 23 December 2019, https://rpp.pe/peru/actualidad/peru-sat-1-los-otros-cuatros-satelites-peruanos-que-tambien-orbitan-la-tierra-noticia-1171287?ref = rpp. See also Satélites: PUCP-SAT-1, Instituto de Radioastronomía INRAS – PUCP, http://inras.pucp.edu.pe/proyectos/pucp-sat-1/ (accessed 3 November 2019).

[48]Cruz Yohel, PERÚSAT-1, Los otros cuatro satélites peruanos que también orbitan la Tierra, RPP Noticias, 23 December 2019.

was created by students and graduates of UNI. CHASQUI 1 registers images of the Earth.[49]

The fourth Peruvian satellite was UAP SAT-1, a USD 1.6 million satellite created by students and professors of the University Alas Peruanas (UAP) and launched into Earth's orbit from the Wallops Launch Space Center, USA, on 9 January 2014. UAPSAT-1 collects data on the space environment.[50]

14.5.2 The Satellite System PERÚSAT-1

It has been argued that PERU's space history actually started on 16 September 2016 when the country launched its first EO satellite: PERÚSAT-1.[51] Constructed by Airbus Defence and Space and launched from Kourou, French Guyana, PERÚSAT-1 is a high technology satellite that functions on solar energy and offers free images to all Peruvian public entities.[52] PERÚSAT-1 includes a latest-generation optical satellite with a 70 cm high-resolution instrument called NAOMI.[53] The PERÚSAT Satellite System includes the EO satellite and a ground control segment for image reception and processing referred to as CNOIS, located in Pucusana, near Lima, where the satellite is operated by members of CONIDA.[54]

PERÚSAT-1's images are used for several applications, including security and defence, agriculture, mining, detection of illegal deforestation, risk prevention, mapping, urban planning, land use, ecological economy zoning, and countering drug trafficking. To date, PERÚSAT-1's images have been used by more than 70 public entities registered at CNOIS (Table 14.2).[55]

---

[49]Cruz Yohel, PERÚSAT-1, Los otros cuatro satélites peruanos que también orbitan la Tierra, RPP Noticias, 23 December 2019; see also Satélite peruano CHASQUI 1 ya está en el espacio y entró en órbita, RPP Noticias, 18 August 2014.

[50]Cruz Yohel, PERÚSAT-1, Los otros cuatro satélites peruanos que también orbitan la Tierra, RPP Noticias, 23 December 2019; see also NASA lanza con éxito satélite experimental peruano UAP SAT-1, RPP Noticias, 9 January 2014, https://rpp.pe/lima/actualidad/nasa-lanza-con-exito-satelite-experimental-peruano-uap-sat-1-noticia-660645 (accessed 3 November 2019).

[51]Farje Oscar, Lanzamiento de PERÚSAT-1 es primera página de historia espacial peruana, Andina, Lima, 15 September 2016, https://andina.pe/AGENCIA/noticia-lanzamiento-perusat1-es-primera-pagina-historia-espacial-peruana-631195.aspx (accessed 3 November 2019).

[52]Mendoza Susana, El Perú en la era espacial PERÚSAT-1, Andina, 2018, https://portal.andina.pe/edpespeciales/2018/satelite/index.html (accessed 3 November 2019).

[53]From New AstroSat Optical Modular Instrument.

[54]PERÚSAT-1 Earth Observation Minisatellite, EOPortal Directory, ESA, https://directory.eoportal.org/web/eoportal/satellite-missions/p/perusat-1 (accessed 3 November 2019).

[55]Ibid.

Table 14.2 Peru's satellites launched into outer space

Peru's satellites launched into outer space[a]

Satellite	Date	Orbit	Features	Focus
PUCP-SAT-1	21 November 2013	• Nodal period: 96.6 min • Inclination: 97.8° • Apogee: 607,103 km • Perigee: 607,103 km	• Launch vehicle: KOSMOTRAS DNEPR-1 • Launch site: Yasny launch base, Russia • Operator of the object launched: Pontifical Catholic University of Peru	• Status: active • General functions: design, research and development of a satellite. Attempt to take low-resolution photographs. Testing of the microwheel stabilization system. Launch of a femtosatellite (Pocket-PUCP) from PUCP-SAT-1. Very-low-power inter-satellite and Earth-to-satellite connections (10 mW)
POCKET-PUCP	6 December 2013	• Nodal period: 96.6 min • Inclination: 97.8° • Apogee: 607,137 km • Perigee: 607,137 km	• Launch vehicle: launched by PUCP-SAT-1, which was, in turn, launched by UNISAT-5, which was launched by KOSMOTRAS DNEPR-1 • Launch site: Yasny launch base, Russia • Operator of the object launched: Pontifical Catholic University of Peru	• Status: active • General functions: design, research and development of a satellite. Attempt to take low-resolution photographs. Testing of the microwheel stabilization system. Launch of a femtosatellite (Pocket-PUCP) from PUCP-SAT-1. Very-low-power inter-satellite and Earth-to-satellite connections (10 mW)
PERÚSAT-1	16 September 2016	• Nodal period: 98.786 min • Inclination: 98.1995° • Apogee: 727,252 km • Perigee: 699,889 km	• Launch vehicle: VEGA • Launch site: Centre spatial guyanais, Kourou, French Guiana • Operator of the object launched: Peruvian Space Agency	• Status: active • General functions: EO satellite with a non-geostationary orbit and a very high-resolution optical sensor. Provision of information for all sectors

[a]Information obtained from Note verbale dated 10 February 2017 from the Permanent Mission of Peru to the United Nations (Vienna) addressed to the Secretary-General, Doc. ST/SG/SER.E/792, UNCOPUOS, 8 March 2017, https://www.unoosa.org/oosa/en/osoindex/data/documents/pe/st/stsgser.e792.html (accessed 1 November 2019). Peru has only notified about the launch of these three satellites to the United Nations Office of Outer Space Affairs (UNOOSA)

14.6 Space Technology Development

As noted in Sect. 14.2.1.1, CONIDA's Technical Directions on Space Sciences and Applications and Space Technology Development contribute substantially to the strengthening of the national space sector by enabling the country to possess its own space research and technology. For example, the construction of a useful payload for the PAULET rocket by DINCI, and the development of launch vehicles with scientific instruments by DIVLA, support the improvement of the level of space technology in Peru.[56]

In addition, as part of its infrastructure, CONIDA has a scientific base called "Punta Lobos", which is used to carry out rocket activities and astronomical scientific studies,[57] and the CNOIS (the ground segment, located in Pucusana, Lima), where the reception, storage, processing and distribution of satellite images are conducted.[58]

Apart from CONIDA's projects, there are other space technology programs prepared by Peruvian universities, such as the construction and operation of different radiotelescopes (from 2 to 20 m of diameter) by the PUCP's Radio astronomy Institute,[59] and the construction of a constellation of a microsatellite and two nanosatellites by the Peruvian Technological University (UTP)—with the support of CONIDA—together with the University of Beihang, China.[60]

14.7 Capacity Building and Outreach

CONIDA has had an important role in the development and training of professionals, researchers and civil servants on space issues. In this regard, since the launch of satellite PERÚSAT-1, CONIDA has trained more than 2,980 people in the processing of satellite images.[61] In addition, more than 36,000 people and 460 institutions have benefited from the dissemination of satellite information by CONIDA in all Peruvian territory.[62]

---

[56]Presentación Instrumentación Científica, Agencia Espacial del Perú – CONIDA, http://www.conida.gob.pe/index.php/Instrumentacion-Cientifica/presentacion-instrumentacion-cientifica.html (accessed 3 November 2019).

[57]Infraestrutura de CONIDA, Agencia Espacial del Peru – CONIDA, http://www.conida.gob.pe/index.php/Infraestructura/infraestructura-de-conida.html (accessed 2 November 2019).

[58]Ibid.

[59]Proyectos, Radioastronomía y Astrofísica, Instituto de Radioastronomía INRAS – PUCP, http://inras.pucp.edu.pe/proyectos/radioastronomia/ (accessed 3 November 2019).

[60]UTP y la Agencia Espacial del Perú (CONIDA) firman Acuerdo de Cooperación para promover la Investigación Espacial, Universidad Tecnológica del Perú, 24 January 2019, https://www.utp.edu.pe/noticias/utp-agencia-espacial-peru-conida-firman-acuerdo-cooperacion-para-promover-investigacion (accessed 3 November 2019).

[61]Caballero Carlos, PERÚSAT-1: La carrera espacial peruana con tareas pendientes, Gestión, 9 September 2019.

[62]Ibid.

CONIDA's Technical Division on Space Studies (DTEE) is the main institutional body responsible for the dissemination of space science and technology and the implementation of capacity programs on different space issues. In this regard, the DTEE promotes and execute programs to train CONIDA's personnel, including professionals, technicians and support staff. The DTEE also elaborates, proposes and implements the permanent training and specialization programs addressed to public and private institutions, namely university scientific and technological projects. Furthermore, the DTEE promotes national and international scientific and technological events related to the space sector, fosters the adoption of training conventions with other partners, and coordinates the dissemination of technical information in CONIDA's official website.[63]

Every year, the DTEE organizes a short academic program on space issues at different levels (basic, intermediate and advanced). The DTEE also has the goal of developing postgraduate programs through the adoption of conventions with different universities, in particular a Master's degree on Applied Geomatics.[64]

Appendix

See Table 14.3.

Table 14.3 Institutional contact information in Peru (2019)

Institution	Details
National Commission for Space Research and Development (CONIDA)/Peruvian Space Agency	Street Luis Felipe Villarán 1069, San Isidro, Lima, Perú Website: http://www.conida.gob.pe/ Email: atencionalcliente@conida.gob.pe
National Centre for Satellite Images Operations (CNOIS)	Pucusana 15866, Perú Website: http://www.conida.gob.pe/

---

[63] Sobre Capacitación, Agencia Espacial del Perú – CONIDA, http://www.conida.gob.pe/index.php/Capacitacion/sobre-capacitacion.html (accessed 2 November 2019).
[64] Ibid.

Chapter 15
Venezuela

Contents

Abstract With the aim of strengthening its sovereignty, protecting its territorial integrity and security, and guaranteeing telecommunications services to its people, Venezuela decided to develop its space sector. Despite the difficulties of its internal political and economic crisis of recent years, Venezuela continues to carry out several space activities, in particular the development of its national satellite program. This chapter presents Venezuela's current space activities. Section 15.1 presents the basic facts of Venezuela; Sect. 15.2 concerns Venezuela's space history; Sects. 15.3 and 15.4 focus on the country's space programs, laws and actors; Sect. 15.5 describes the cooperation agreements adopted by Venezuela with other countries; Sect. 15.6 presents the Venezuelan satellite programs, and Sects. 15.7 and 15.8 concentrate on the development of space technology and the capacity-building programs of Venezuela.

© Springer Nature Switzerland AG 2020
A. Froehlich et al., *Space Supporting Latin America*, Studies in Space Policy 25,
https://doi.org/10.1007/978-3-030-38520-0_15

15.1   General Country Overview

The Bolivarian Republic of Venezuela (Venezuela) is located on the north coast of South America. It borders the Caribbean Sea and the Atlantic Ocean to the North, Colombia to the West, Brazil to the South and Guyana to the East. Venezuela's territory is 881,050 km^2; the country has a population of 31.5 million inhabitants, the official language is Spanish, although there are several indigenous languages, and the currency is the Bolívar.[1]

Geographically, Venezuela is partitioned into 9 regions: the Andean Region, Cordillera Central, Cordillera Oriental, Guayana, Insular Region, Los Llanos, Maracaibo Basin Region, Orinoco Delta, and the Lara-Falcón Highlands Region.[2]

15.1.1   Political System

Venezuela is constituted as a democratic and social state of law and justice.[3] It is also a federal and decentralised state.[4] Politically, Venezuela is partitioned into 23 states (Fig. 15.1), a Capital District (Caracas), and several federal departments and territories.

The political organisation of Venezuela is divided in several powers: the executive, is headed by the President of the Republic who is also the Head of State [5]; the legislative power is the National Assembly, which is composed of 167 deputies[6]; the Judicial power is exercised by the Supreme Court of Justice and other minor courts; the "Citizen Power" is exercised by the Republican Moral Council, the Ombudsman, the Attorney General and the General Comptroller; and the electoral power is exercised by the National Electoral Council and other subsidiary bodies.[7]

---

[1] Venezuela country profile, BBC News, 25 February 2019, https://www.bbc.com/news/world-latin-america-19649648 (accessed 29 October 2019).

[2] Venezuelatuya, Geografía, Venezuelatuya.com, https://www.venezuelatuya.com/geografia/index.htm (accessed 29 October 2019).

[3] Art. 1, Constitución de la República Bolivariana de Venezuela, 20 December 1999.

[4] Art. 2, Ibid.

[5] Some States recognize Juan Guaidó as the legitimate president of Venezuela, see Venezuela country profile, BBC News, 25 February 2019, https://www.bbc.com/news/world-latin-america-19649648 (accessed 29 October 2019).

[6] In 2017 there was created a National Constituent Assembly, which has rendered powerless the National Assembly, see Venezuela crisis: How the political situation escalated, BBC News, 8 August 2019, https://www.bbc.com/news/world-latin-america-36319877 (accessed 29 October 2019).

[7] Titule V, Chapters I to V, respectively, Constitución de la República Bolivariana de Venezuela, 20 December 1999.

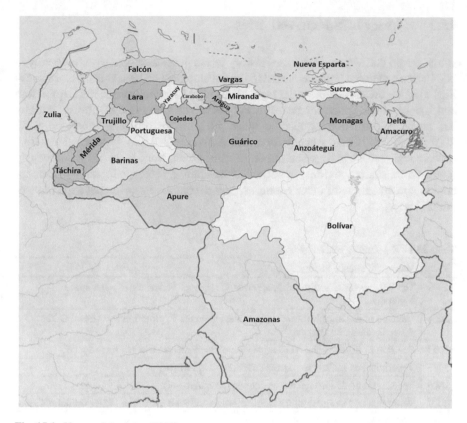

Fig. 15.1 Venezuela's states (2019)

15.1.2 Economic Situation

Venezuela has lived in economic and political crisis in recent years. During the presidential re-election of Nicolás Maduro in May 2018, Venezuela had rampant inflation and lack of basic necessities.[8] The current political situation has escalated the economic crisis, leaving the country with hyperinflation, power cuts and shortages of food and medicine. This situation has prompted more than four million Venezuelans to leave the country.[9]

---

[8] Venezuela country profile, BBC News, 25 February 2019.
[9] Ibid.

15.2   History of Space Activities

It can be said that Venezuela's space history began in the 21st century, specifically in 2004 when a national body dedicated to the development of Venezuela's space activities was first created, namely the Inter-Ministerial Commission on the Exploration and Use of Outer Space for Peaceful Purposes. However, this institution initially had no formal structure or specific space programs or projects. It was not until 2007 that the Bolivarian Agency for Space Activities (ABAE)[10] was created with a specific mandate, functions and structure. Since then, Venezuela has continuously increased its space activities, in particular its satellite sector; adopted space cooperation agreements with other countries, and developed its own space technology (Table 15.1).

Table 15.1   Venezuela's space history

Year	Venezuela's space history[a]
1999	The Venezuelan Constitution recognizes the right of Venezuela to outer space sovereignty
2004	Creation of the Inter-Ministerial Commission on the Exploration and Use of Outer Space for Peaceful Purposes[b]
2005	Creation of the Presidential Commission on the Peaceful Use of Outer Space[c]
2005	Venezuela and India sign a space cooperation agreement
2005	Venezuela and China sign a space cooperation agreement
2006	Venezuela and Uruguay sign an agreement for the joint use of the 78° W orbital position
2006	Creation of the Venezuelan Space Centre[d]
2007	Creation of ABAE
2008	Venezuela and Brazil sign a space cooperation agreement
2008	Launch of satellite VENESAT-1 (SIMON BOLIVAR) from Xichang, China
2008	Inauguration of the Guarico and Bolivar satellite control stations
2009	Venezuela obtains control of the SIMON BOLIVAR satellite
2011	Venezuela and Bolivian sign a space cooperation agreement
2011	Venezuela and China sign a contract for the fabrication of satellite MIRANDA
2012	Launch of satellite VRSS-1 (satellite MIRANDA) from Jiuquan, China
2013	Inauguration of the satellite MIRANDA ground station for the reception of images
2014	First Venezuelan Congress on Space Technology

(continued)

---

[10] Agencia Bolivariana para Actividades Espaciales (ABAE).

Table 15.1 (continued)

Year	Venezuela's space history[a]
2015	Creation of the Centre for Space Research and Development in Borgurata for the creation of new satellites
2017	Launch of satellite VRSS-2 (satellite SUCRE) from Jiuquan, China

[a]Agencia Bolivariana para Actividades Espaciales, Línea de Tiempo ABAE, http://www.abae.gob.ve/web/timeline/index.html (accessed 30 October 2019)
[b]Comisión Interministerial para la Exploración y Utilización del Espacio Ultraterrestre con Fines Pacíficos
[c]Comisión Presidencial Venezolana para el Uso Pacífico del Espacio
[d]Centro Espacial Venezolano

15.3 Space Development Policies and Emerging Programs

Although there is no formal national space policy (NSP), the 2025 Venezuelan Socialist Development Plan or "Homeland Plan" (2025 NDP) contemplates the development of a national satellite policy to put this field of activities into the service of the general development of the nation.[11] Moreover, Venezuela has developed multiple space activities through the work of several specialized institutions and other actors.

15.3.1 The Satellite Policy

Venezuela already possesses a national satellite program that comprises space and ground infrastructure and specialised capacity building (see Sect. 15.6). However, in order to establish a formal satellite policy, the 2025 NDP contains two goals to be attained in the next six years:

(1) Strengthen the peaceful use of satellite technology to give the country sovereign management of its telecommunications and remote sensing systems and associated tools, so as to consolidate national development in different areas such as education, health, security, and food. To this end, the country must maximise the use of available technologies for data transmission and satellite monitoring, and finish the establishment of the small satellite fabric.[12]

(2) Stimulate the development of space technology through training, applications and infrastructure. In this regard, Venezuela must develop the necessary infrastructure for the implementation of telemedicine and tele-education. Similarly,

---

[11]Plan de la Patria 2025: Hacia la prosperidad económica, Venezuela, Objectivo nacional 1.6, meta 1.6.4, p. 84, http://www.mppp.gob.ve/wp-content/uploads/2019/04/Plan-Patria-2019-2025.pdf (accessed 4 November 2019).
[12]Ibid., goal 1.6.4.1.

the country must take advantage of satellite technology for the development of geomatics and the creation of an inventory of natural resources.[13]

15.3.2   Space Actors

Venezuela's space activities are mostly developed by two national bodies: ABAE and the National Telecommunications Commission (CONATEL).[14] Other actors also contribute to a greater or lesser extent to the development of this sector.

15.3.2.1   The Bolivarian Agency for Space Activities

Bolivarian Agency for Space Activities (ABAE) is a specialized, technical and advisory body responsible for the implementation of national space policies and regulations emanating from the governing body on science and technology. It also establishes space programs, projects and policies, and generates space regulations and other related norms.[15] ABAE started operations on 1 January 2008[16] after the dissolution of the Foundation Venezuelan Space Centre, whose functions and powers were moved to ABAE.[17] Currently, ABAE is under the Ministry of Popular Power for Education, Science and Technology.

ABAE's main powers include[18]:

- Design, elaborate and propose an NSP draft to the governing body on science and technology;
- Elaborate, design, advice and execute strategies, plans, projects and programs concerning the exploration, use and exploitation of outer space for peaceful purposes, as well as all space research and development issues;
- Coordinate and implement programs and projects with other public or private space entities;
- Promote and stimulate space scientific research and technology development;
- Coordinate and articulate the development of space activities with operational and research centres;
- Ensure compliance with international space related treaties signed and ratified by Venezuela;

---

[13] Ibid., goal 1.6.4.2.

[14] Comisión Nacional de Telecomunicaciones (CONATEL).

[15] Art. 2 and 3, Ley de la Agencia Bolivariana para Actividades Espaciales, Caracas, 9 August 2007.

[16] Official Gazette N° 38.796 on 25 October 2007.

[17] Final provisions, Chapter V, Ley de la Agencia Bolivariana para Actividades Espaciales, Caracas, 9 August 2007.

[18] Art. 5, Ibid.

- Propose, advise and supply to the National Executive orientations for the formulation of international cooperation policy on space issues;
- Establish coordination and exchange mechanisms with national and international partners specialized in space capacity-building;
- Promote the development, strengthening and expansion of the national industry of space technology, and
- Promote the participation of the private sector and the organized collective in the design and implementation of space development initiatives.

It is also important to note that ABAE is responsible for the operation of the Venezuelan satellites SIMON BOLIVAR and MIRANDA.[19]

Concerning ABAE's organisation, the Agency is composed of a Board of Directors and the administrative and operative bodies required for the fulfilment of its functions. The Board of Directors is the highest ABAE's direction body[20] and is composed of a president and eight directors.[21]

15.3.2.2 The National Telecommunications Commission

CONATEL is the national body responsible for the management, regulation, ordering and control of the radio electric spectrum associated with satellite networks, as well as access and use of the orbit-spectrum resource for space networks assigned by Venezuela and registered in its name.[22] CONATEL is also the competent body to carry out the procedures to make available the orbit-spectrum resource for the establishment of national security networks and provision of telecommunication services for social purposes.[23] Moreover, CONATEL grants licenses for the use of the radio electric spectrum for the provision of direct satellite telecommunications services.[24]

15.3.2.3 Other Actors

There are several actors who participate directly or indirectly in the development of the Venezuelan space sector. Among the most relevant are the Ministry of Popular Power for University Education, Science and Technology; the Foundation Centre on Astronomy Research "Francisco J. Duarte" (CIDA); the Centre of Space Research and Development (CIDE); the Presidential Council for Science, Technology and

---

[19]Bolivarian Agency for Space Activities, http://www.abae.gob.ve/web/Historia.php (accessed 30 October 2019).

[20]Art. 7, Ley de la Agencia Bolivariana para Actividades Espaciales, Caracas, 9 August 2007.

[21]Art. 8, Ibid.

[22]Art. 118, Organic Law of Telecommunications, 7 February 2011.

[23]Art. 119, Ibid.

[24]Art. 124, Ibid.

Innovation; the National Centre for Development and Research on Telecommunications (CENDIT); the National Experimental University of the Armed Forces (UNEFA), and the state telecommunications company CANTV.[25]

15.4 Current and Future Perspectives on Legal Framework Related to Space Activities

Few national legal instruments govern space and telecommunications activities in Venezuela, and these regulations are mainly institutional. At the international level, Venezuela has signed and ratified several space related treaties showing the country's engagement in respecting the international legal framework on the matter.

15.4.1 National Space Regulations and Instruments

The Law on the establishment of the Bolivarian Agency for Space Activities was adopted on 9 August 2007 and entered into force on 1 January 2008. It creates the Agency as an institute responsible for the implementation of the policies and guidelines issued by the science and technology national body for the peaceful exploration and use of outer space and other areas that cannot be considered as common heritage of humankind, as well as everything related to space. The five chapters of the law determine the functions, responsibilities, structure, assets, and other administrative issues of the Agency.[26]

---

[25]Ministerio del Poder Popular para Educación Universitaria, Ciencia y Tecnología; Fundación Centro de Investigaciones de Astronomía "Francisco J. Duarte" (CIDA); Centro de Investigación y Desarrollo Espacial; Consejo Presidencial para la Ciencia, la Tecnología y la Innovación; Centro Nacional de Desarrollo e Investigación en Telecomunicaciones (CENDIT); Universidad Nacional Experimental de las Fuerzas Armadas (UNEFA), and Compañía Anónima Nacional Teléfonos de Venezuela (CANTV), respectively.

[26]Ley de la Agencia Bolivariana para Actividades Espaciales, La Asamblea Nacional de la República Bolivariana de Venezuela, 9 August 2007, http://www.abae.gob.ve/web/leyes/SANC-%20LEY-%20DE-%20LA-%20AGENCIA-%20BOLIVARIANA-PARA-%20ACTIVIDADES-%20ESPACIALES-09-08-07.pdf (accessed 31 October2019).

In June 2018, authorities of ABAE and the Commission on Sovereignty and Territorial Integrity of the National Constituent Assembly recognised the necessity of presenting a proposal for a law on Aerospace Development in the country, as well as a National Aerospace Plan.[27] These initiatives have not yet materialised. However, as noted in Sect. 15.3.1, the 2025 NDP already contemplates the development of a national satellite policy.

Concerning the telecommunications sector, the Organic Law of Telecommunications adopted on 7 February 2011 stipulates that the telecommunications and radio electric spectrum integral regime is the responsibility of the National Public Power, and determines the functions of CONATEL. It also includes the rights and duties of users and operators, the provision of services, and the establishment and exploitation of telecommunications networks, the development of the telecommunication sector, the participation of the government in the field, and several administrative, financial and procedural issues. In particular, Chapter IV regulates all issues related to satellite use and the manner in which the radio electric spectrum and the orbit-spectrum can be exploited.[28]

15.4.2 Venezuela and International Space Instruments

Venezuela's strong commitment to respecting the international legal rules in the fields of space and telecommunications is demonstrated by its large number of ratifications of the UN Space Treaties and other relevant instruments such as the Convention on the International Mobile Satellite Organization and the ITU's Convention and Constitution (Table 15.2).

---

[27] ANC y ABAE activarán la construcción de una propuesta de Ley y Plan Nacional Aeroespacial, Prensa ABAE, 14 June 2018, http://www.abae.gob.ve/web/noticias.php?id=324 (accessed 1 November 2019).

[28] Ley Orgánica de Telecomunicaciones, Official Journal N° 39.610, 7 February 2011, http://www.conatel.gob.ve/ley-organica-de-telecomunicaciones-2/ (accessed 24 May 2019).

Table 15.2 International space and telecommunications treaties respected by Venezuela

International space and telecommunications treaties respected by Venezuela

Treaty[a]	Adoption	Ratification (R)/signature (S)
OST	1967	R
ARRA	1968	S
LIAB	1972	R
REG	1975	R
MOON	1979	R
NTB	1963	R
BRS	1974	–
ITSO	1971	R
IMSO	1976	R
ITU	1992	R

[a]Treaty on Principles Governing the Activities of States in the Exploration and Use of Outer Space, including the Moon and Other Celestial Bodies (OST); Agreement on the Rescue of Astronauts, the Return of Astronauts and the Return of Objects Launched into Outer Space (ARRA); Convention on International Liability for Damage Caused by Space Objects (LIAB); Convention on Registration of Objects Launched into Outer Space (REG); Agreement Governing the Activities of States on the Moon and Other Celestial Bodies (MOON); Treaty Banning Nuclear Weapon Tests in the Atmosphere, in Outer Space and under Water (NTB); Convention Relating to the Distribution of Programme-Carrying Signals Transmitted by Satellite (BRS); Agreement Relating to the International Telecommunications Satellite Organisation (ITSO); Convention on the International Mobile Satellite Organization (IMSO); International Telecommunication Constitution and Convention (ITU)

15.5   Cooperation Programs

To develop its space sector, Venezuela has adopted cooperation agreements with four South American countries: Argentina, Bolivia, Brazil and Uruguay. Moreover, Venezuela has strengthened its international relations in the space field with China and India.

15.5.1   Venezuela and Latin America

In 2011, Venezuela and Argentina adopted a Framework Cooperation Agreement for the peaceful use of outer space, space science, technology and applications. In 2013 both countries also adopted a specific cooperation convention in the satellite field.[29] This bilateral cooperation has resulted in the training of two Venezuelans in the Master's program "Use of space technology in the field of Epidemiology" taught

---

[29]Cooperación Internacional de la Agencia Bolivariana para Actividades Espaciales, ABAE, http://www.abae.gob.ve/web/ConveniosInt.php (accessed 31 October 2019).

by the Mario Gulich Institute for Advanced Space Studies, in Argentina, as well as technical support for the incorporation of the MIRANDA satellite ground segment into the infrastructure of the Argentinian SAOCOM[30] satellite constellation.

Bilateral cooperation between Venezuela and Bolivia in the space field materialised in 2011 with the adoption of an MoU for the development of exchange and training activities on space science and technology. In 2013, both countries signed a cooperation convention for training and scientific and technological application in the peaceful use of outer space, observation, territorial physical modelling and Earth sciences.[31] One of the achievements of this cooperation was the exchange of technical experiences with the personnel of the Bolivian Space Agency in the framework of the TUPAC KATARI Space Program.[32]

Concerning space cooperation between Brazil and Venezuela, in 2008 both countries adopted a Framework Cooperation Agreement on space science and technology. Moreover, in 2011 these countries signed an MoU to facilitate exchange activities and scientific and technological training on geomatics applications, space engineering and geosciences.[33] One of the main results of this cooperation has been the training of Venezuelan professionals in the framework of the "International Course on Remote Sensing and Geographic Information Systems" at the National Institute for Space Research of Brazil.[34]

In 2006 Venezuela and Uruguay adopted an Agreement in the Framework of the VENESAT-1 Program for the joint use of the geo orbital position 78° W. A Complementary Agreement to the 2006 Agreement was adopted in 2008, and an Additional Protocol of this Complementary Agreement was signed in 2011.[35] This cooperation has allowed Venezuela and Uruguay to use the 78° W orbital position granting Venezuela the right to place its satellite VENESAT-1 in this orbital position. Meanwhile, Uruguay acquired the right to use 10% of the useful load of this satellite.[36] Additionally, professionals of both countries were taught to monitor and manage the CSM-B station of the VENESAT-1 Program, located in Manga, Uruguay.[37]

---

[30]SACOM is an Earth observation satellite constellation of Argentina's space agency CONAE.

[31]Cooperación Internacional de la Agencia Bolivariana para Actividades Espaciales, ABAE, http://www.abae.gob.ve/web/ConveniosInt.php (accessed 31 October 2019).

[32]Ibid.

[33]Ibid.

[34]Instituto Nacional de Pesquisas Espaciais (INPE).

[35]Cooperación Internacional de la Agencia Bolivariana para Actividades Espaciales, ABAE, http://www.abae.gob.ve/web/ConveniosInt.php (accessed 31 October 2019).

[36]Ibid.

[37]Ibid.

15.5.2 Venezuela and the Asia-Pacific

Strong cooperation in the space field between Venezuela and China has materialized through the adoption of different bilateral instruments. In 2005, both countries signed an MoU on technical cooperation on the peaceful use and exploration of outer space; in 2013, they signed an Agreement for data exchange and satellite remote sensing applications, and in 2014 they adopted the Agreement for the Establishment of the Asia-Pacific Regional Centre for the Training on Space Science and Technology, located in China.[38] It is also important that in 2005 Venezuela and China signed an international contract for the beginning of the VENESAT-1 Space Program, which included the construction and launch of the satellite SIMON BOLIVAR, construction and equipping of the control ground station, and the training of human capital.[39]

Bilateral cooperation between Venezuela and China has achieved several important results, in particular the Satellite Programs VENESAT-1 and VRSS-1, the training of human talent in the Chinese Academy of Space Technology (CAST), construction of the Centre for Space Research and Development and the Satellite Program VRSS-2 (both under development).[40] Finally, in 2018 the Ministry of Popular Power for University Education, Science and Technology (MPPEUCT)[41] of Venezuela signed a letter of intent with the China Great Wall Industry Corporation (CGWIC) for the commercialization of good and services derived from space applications.[42]

Venezuela and India adopted an MoU on the cooperation in space science and technology in 2005. Capacity building has resulted from this bilateral cooperation. For example, 46 Venezuelans participated in the "Certification Program on Geomatics Applied to Different Areas of Geodata sciences"[43] taught in the Indian Institute on Remote Sensing (IIRS)[44] (Table 15.3).

---

[38]Ibid.

[39]VENESAT-1, ABAE, http://www.abae.gob.ve/web/VENESAT-1.php (accessed 31 October 2019).

[40]Cooperación Internacional de la Agencia Bolivariana para Actividades Espaciales, ABAE, http://www.abae.gob.ve/web/ConveniosInt.php (accessed 31 October 2019).

[41]Ministerio del Poder Popular para la Educación Universitaria, Ciencia y Tecnología (MPPEUCT).

[42]El MPPEUCT firma carta de intención con CGWIC para comercializar productos satelitales, Prensa ABAE, 4 December 2018, http://www.abae.gob.ve/web/noticias.php?id=339 (accessed 1 November 2019).

[43]Diplomado en Geomática Aplicada a las diferentes áreas de las ciencias de la Geoinformación.

[44]Cooperación Internacional de la Agencia Bolivariana para Actividades Espaciales, ABAE, http://www.abae.gob.ve/web/ConveniosInt.php (accessed 31 October 2019).

Table 15.3 Venezuela and space cooperation agreements

Venezuela and space cooperation agreements		
Country	Instrument	Adoption
Argentina	Framework Cooperation Agreement for the peaceful use of outer space, space science, technology and applications between the Government of the Bolivarian Republic of Venezuela and the Government of Argentina	1 December 2011
	Specific Cooperation Convention in the satellite field between the Ministry of Popular Power for Science, Technology and Innovation of the Bolivarian Republic of Venezuela through the Bolivarian Agency for Space Activities (ABAE), and the Ministry of Federal Planning, Public Investment and Services of the Republic of Argentina through the National Commission on Space Activities (CONAE)	8 May 2013
Bolivia	Memorandum of Understanding between the Government of the Bolivarian Republic of Venezuela and the Government of the Plurinational State of Bolivia for the development of exchange and training activities in science and technology for the exploration and use of outer space for peaceful purposes	31 March 2011
	Cooperation Convention for training and scientific and technological application in the peaceful use of outer space, observation, territorial physical modelling and Earth sciences	26 May 2013
Brazil	Framework Cooperation Agreement on space science and technology between the Government of the Bolivarian Republic of Venezuela and the Government of the Federal Republic of Brazil	27 June 2008
	Memorandum of Understanding to make possible exchange and scientific and technological activities in the field of geomatics applications, space engineering and geoscience, between the Ministry of Popular Power for Science and Technology of the Bolivarian Republic of Venezuela, and the Ministry of Science, Technology and Innovation of the Federal Republic of Brazil	1 November 2011
China	Memorandum of Understanding on Technical Cooperation on the Peaceful Use and Exploration of Outer Space between the Ministries of Science and Technology, and Communication and Information of the Bolivarian Republic of Venezuela and the National Space Administration of the Popular Republic of China	28 January 2005
	Agreement between the Ministry of Popular Power for Science, Technology and Innovation of the Bolivarian Republic of Venezuela, and the Chinese Space Administration of the People's Republic of China for Data Exchange and Satellite Remote Sensing Applications	22 September 2013

(continued)

Table 15.3 (continued)

Venezuela and space cooperation agreements

Country	Instrument	Adoption
	Agreement on the Establishment of the Regional Asia-Pacific Centre for Space Science and Technology Training (China)	22 November 2014
India	Memorandum of Understanding between the Ministry of Science and Technology (MCT) of the Bolivarian Republic of Venezuela and the Space Department (DOS) of the Indian Republic for cooperation on space science and technology	5 March 2005
Uruguay	Agreement in the field of development of the VENESAT-1 Program (Satellite System SIMON BOLIVAR) for the joint use of the 78° orbital position required by the Eastern Republic of Uruguay for the URUSAT-3 Program, between the Bolivarian Republic of Venezuela and the Eastern Republic of Uruguay	14 May 2006
	Complementary Agreement to the Agreement in the field of development of the VENESAT-1 Program, between the Bolivarian Republic of Venezuela and the Eastern Republic of Uruguay	24 October 2008
	Additional Protocol to the Complementary Agreement to the Agreement in the field of development of the VENESAT-1 Program, between the Bolivarian Republic of Venezuela and the Eastern Republic of Uruguay	30 March 2011

15.6 Satellite Capacities

In less than a decade, Venezuela's satellite capacities have increased rapidly. There are currently three satellite space programs that enable Venezuela independent access to space: VENESAT-1, VRSS-1 and VRSS-2.

15.6.1 The VENESAT-1 Space Program

The VENESAT-1 Space Program is a geostationary telecommunications satellite that enables Venezuela to be interconnected with all its territory and several parts of South America. One of the most important parts of the VENESAT-1 Space Program is satellite SIMON BOLIVAR, the first Venezuelan satellite, which is the result of bilateral cooperation with China.[45] The satellite was launched on 8 November 2008 and on 10 January 2009 Venezuela assumed command of the satellite. Apart from the space segment (satellite SIMON BOLIVAR), the VENESAT-1 Space Program includes a ground segment (ground satellite control stations and teleport) and technology transfer and human capital formation.[46]

---

[45]VENESAT-1, ABAE, http://www.abae.gob.ve/web/VENESAT-1.php (accessed 31 October 2019).
[46]Ibid.

The main services offered by satellite SIMON BOLIVAR are[47]:

- Digital links capability, uplink/downlink, in the C, Ku and Ka bands;
- Direct home television services;
- Internet access;
- Support to the state mobile telephony network.

Satellite SIMON BOLIVAR is controlled by 30 ABAE specialists from the ground station Baemari in Guarico, Venezuela. Thirty operators from Compañía Anónima Nacional Teléfonos de Venezuela (CANTV) are in charge of the teleport, which functions as the direct link to the users.[48]

The construction of the successor to VENESAT-1, the satellite VENESAT-2 "Guaicaipuro" is currently planned for the 2022–2023 biennial.[49]

15.6.2 The VRSS-1 Space Program

The VRSS-1 Space Program is the result of a cooperation agreement between Venezuela and China and a program created under the Strategic Space Development Plan. The VRSS-1 or satellite MIRANDA is the first EO satellite of Venezuela. It is mainly used for resources research, environmental protection, disaster monitoring and management, estimation of agricultural crops, national territorial defence, and urban planning, among others.[50]

15.6.3 Satellite VRSS-2

Satellite VRSS-2 (or satellite SUCRE) is the second EO satellite of Venezuela. After the signature of a cooperation agreement for the design and construction of the satellite VRSS-2 with China in 2014, the satellite was launched from the Jiuquan centre, China, in 2017. Satellite SUCRE complements the work of satellite MIRANDA and it has applications in the environmental, planning, energy, mining, oil, and agricultural areas.[51] In addition, one of the objectives of this project was the transfer of knowledge to Venezuela, an objective that has materialized in the work carried out in the

---

[47] Ibid.

[48] Ibid.

[49] Orden José Feliz Ribas por el éxito del satélite VRSS-2 Sucre, ABAE, 20 February 2019, http://www.abae.gob.ve/web/noticias.php?id=347 (accessed 31 October 2019), see also Venezuela prevé lanzar un nuevo satélite en 2022 con el apoyo de China, RT, 3 July 2019, https://actualidad.rt.com/actualidad/319980-venezuela-china-lanzamiento-satelite-guaicaipuro (accessed 31 October 2019).

[50] VRSS-1, ABAE, http://www.abae.gob.ve/web/VRSS-1.php (accessed 31 October 2019).

[51] VRSS-2, ABAE, http://www.abae.gob.ve/web/VRSS-2.php (accessed 31 October 2019).

Table 15.4 Venezuela's satellites launched into outer space

Venezuela's satellites launched into outer space[a]

Satellite	Date	Orbit	Features	Focus
SIMON BOLIVAR (VENESAT-1)	29 October 2008	• Nodal period: 1436.4 min • Inclination: 0° • Apogee: 35,786 km • Perigee: 35,786 km • Longitude: 78° W	• Launch vehicle: LONG MARCH 3B • Launch site: Xichang Satellite Launch Centre, China • Launching States: Venezuela and China • Operator of the object launched: ABAE	• Status: active • General functions: Geostationary telecommunications satellite for the transmission of television and radio signals, data and Internet material and the provision of multimedia connectivity to tele-education and tele-medicine systems in Venezuela, Central America, South America and the Caribbean
MIRANDA (Venezuelan Remote Sensing Satellite VRSS-1)	29 September 2012	• Nodal period: 97.32 min • Inclination: 98.02728° • Apogee: 639,540 km • Orbit: Sun-synchronous	• Launch vehicle: LONG MARCH CHANG ZHENG-2D • Launch site: Jiuquan Satellite Launch Centre, China • Launching State: China • Operator of the object launched: ABAE • Owner: Bolivarian Republic of Venezuela	• Status: active • General functions: provision of data and images to support decision-making at the government level in strategic areas such as town planning, food security and agricultural planning, natural resource management, border surveillance and natural disaster management
SUCRE (Venezuelan Remote Sensing Satellite VRSS-2)	9 October 2017	• Orbit: Sun-synchronous	• Launch vehicle: LONG MARCH 2D • Launch site: Jiuquan Satellite Launch Centre, China • Launching State: China • Operator of the object launched: ABAE • Owner: Bolivarian Republic of Venezuela	• Status: active • General functions: provision of data and images in strategic areas such as town planning, energy, agricultural planning, natural resource management, border surveillance and natural disaster management

[a]See Note verbale dated 31 December 2008 from the Permanent Mission of the Bolivarian Republic of Venezuela to the United Nations (Vienna) addressed to the Secretary-General, Doc. A/AC.105/INF.417, 10 March 2009; Note verbale dated 13 December 2012 from the Permanent Mission of the Bolivarian Republic of Venezuela to the United Nations (Vienna) addressed to the Secretary-General, Doc. A/AC.105/INF/423, 27 March 2013

Ground Satellite Control Station of the aerospace base "Capitán Manuel Ríos" (BAE-MARI) in Guárico, and the Ground Applications System of the airbase "Generalísimo Francisco de Miranda" (BAGEM) in Caracas, Venezuela[52] (Table 15.4).

---

[52]Orden José Feliz Ribas por el éxito del satélite VRSS-2 Sucre, ABAE, 20 February 2019, http://www.abae.gob.ve/web/noticias.php?id=347 (accessed 31 October 2019).

15.7 Space Technology Development

With the creation of CIDE, the development of space technology in Venezuela has became a reality. Located in Carabobo, in the north-centre coast of the country, CIDE is the first national technology infrastructure with appropriate equipment and necessary programs for the assembly, integration and verification of satellites.[53] CIDE's main objective is the generation of space technology through the promotion of scientific networks integrated into the space sector and industrial development, and brings together research networks, academic centres and other productive sectors. The construction of small satellites in CIDE's facilities will contribute to the strengthening of the National System of Science, Technology and Innovation.[54] In addition, CIDE's importance for the development of the Venezuelan space sector has led the country to train more that 60 professionals in the design, manufacture and assembly of LEO satellites.[55]

It is also pertinent to highlight that in 2014 the First Venezuelan Congress of Space Technology was organized, in which more than 80 participants from academia, government and industry presented proposals on the development of space plans, projects and programs.[56] Due to its success, a second Congress was held in 2017 which was structured in four thematic axes to better discuss the development of space science and technology: space technology and applications, social benefits and big projects, space education, and space policy and proposals.[57]

Finally, in December 2018 ABAE signed a Convention with the Telecommunications Company of Cuba (ETECSA) for cooperation in space science, technology and innovation. This cooperation aims to promote space research and the use of space technology for the benefit of Cuba and Venezuela.[58]

15.8 Capacity Building and Outreach

ABAE contributes to the training of space professionals and dissemination of space knowledge to the general public throughout the country through its Experimental

---

[53]Centro de Investigación y Desarrollo Espacial, ABAE, http://www.abae.gob.ve/web/CIDE.php (accessed 31 October 2019).

[54]Ibid.

[55]Ibid.

[56]1er Congreso Venezolano de Tecnología Espacial, ABAE, http://www.abae.gob.ve/congreso/ (accessed 1 November 2019).

[57]2do Congreso Venezolano de Tecnología Espacial, ABAE, http://2cvte.abae.gob.ve/ejes.php?idiomas=es (accessed 1 November 2019).

[58]La ABAE y ETECSA inician cooperación en materia de ciencia, tecnología e innovación espacial, Prensa ABAE, 6 December 2018, http://www.abae.gob.ve/web/noticias.php?id=340 (accessed 1 November 2019).

Educational Project "Space Classroom".[59] This project consists of workshops on different areas related to the space sector. They are prepared by specialised personnel of the Agency who use didactic strategies, documentation, multimedia, applications and other activities. Other services offered by ABAE are access to the Centre of Space Documentation and Educational Research (CEDOES),[60] virtual visits to ABAE's headquarters, and permanent expositions on space science and technology.[61]

ABAE has also contributed to capacity building in the field of satellites and space technology. For example, in 2015 the Agency, together with CONATEL, trained 50 professionals in satellite issues in the course "Remote Sensing and Digital Processing of the MIRANDA satellite's images".[62] In 2018, ABAE signed a convention with the Venezuelan Armed Forces to train its members in the use of space technology and EO.[63]

Apart from ABAE's activities, there are several institutions and amateur groups devoted to the dissemination of space sciences in Venezuela, such as the University Group of Astronomical Research (GUIA)[64] and the ARVAL Observatory.[65]

Appendix

See Table 15.5.

Table 15.5 Institutional contact information in Venezuela (2019)

Institution	Details
Bolivarian Agency for Space Activities (ABAE)	Complejo Tecnológico Simón Rodríguez, Caracas, Venezuela Website: http://www.abae.gob.ve
National Telecommunications Commission (CONATEL)	Avenue Veracruz, CONATEL Building, Urbanización Las Mercedes, Caracas 1060, Venezuela Website: http://www.conatel.gob.ve/ Email: conatel@conatel.gob.ve

[59] Programa Educativo Experimental Aula Espacial.

[60] Centro de Documentación e Investigación Educativa Espacial "CEDOES" (in development).

[61] Programa Educativo Experimental Aula Espacial, ABAE, http://www.abae.gob.ve/web/EscuelaEspacial.php (accessed 31 October 2019).

[62] CONATEL y ABAE iniciaron formación satelital a distancia, CONATEL, 1 June 2015, http://www.conatel.gob.ve/conatel-y-abae-iniciaron-formacion-satelital-a-distancia/ (accessed 1 November 2019).

[63] BOLTOMIER Isaac et al., ABAE capacitará a la Fuerza Armada en tecnología espacial y observación de la Tierra, ABAE, 3 May 2018, http://www.abae.gob.ve/web/noticias.php?id=322 (accessed 1 November 2019).

[64] Grupo Universitario de Investigaciones Astronómicas de la Universidad Simón Bolívar, see https://guia-usb.blogspot.com/ (accessed 1 November 2019).

[65] Observatorio ARVAL, see http://www.oarval.org/ (accessed 1 November 2019).

Printed in the United States
by Baker & Taylor Publisher Services